0195603

D1686072

(Bro)

✓

085 199 3540

CABI *Publishing* **is a division of CAB** *International*

CABI Publishing
CAB International
Wallingford
Oxon OX10 8DE
UK

Tel: +44 (0)1491 832111
Fax: +44 (0)1491 833508
Email: cabi@cabi.org
Web site: http://www.cabi.org

CABI Publishing
10 E 40th Street
Suite 3203
New York, NY 10016
USA

Tel: +1 212 481 7018
Fax: +1 212 686 7993
Email: cabi-nao@cabi.org

A catalogue record for this book is available from the British Library,
London, UK.

Library of Congress Cataloging-in-Publication Data
CAP regimes and the European countryside : prospects for integration between
agricultural, regional, and environmental policies / edited by Floor Brouwer and Philip Lowe.
    p. cm.
  Includes bibliographical references.
  ISBN 0-85199-354-0 (alk. paper)
    1. Agriculture--Environmental aspects--European Union countries. 2.
Agriculture--Environmental aspects--Government policy--European Union countries. 3.
Agriculture and state--European Union countries. 4. European Union
countries--Economic conditions--Regional disparities. I. Brouwer, Floor. II. Lowe,
Philip.

S589.76.E85 C36 2000
333.76'16'094--dc21

                                                                    99-057217

ISBN 0 85199 354 0

Printed and bound in the UK by Cromwell Press, Trowbridge
from copy supplied by the author.

Contents

Contributors

Erling Andersen is Researcher at the Danish Forest and Landscape Research Institute, Hørsholm Kongevej 11, 2970 Hørsholm, Denmark.

Stuart Ashworth is Senior Agricultural Economist at the Agricultural Economics Department, SAC, Auchincruive, Ayr, KA6 5HW, UK.

David Baldock is Director of the Institute for European Environmental Policy, Dean Bradley House, 52 Horseferry Road, London SW1P 2AG, UK.

Guy Beaufoy is Director of the Asociación para el Análisis y Reforma de la Política Agro-rural (ARPA), Ibiza 17, 7D, E-28009 Madrid, Spain.

Floor Brouwer is Head of Research Unit Environment and Technology at the Agricultural Economics Research Institute (LEI), 2502 LS The Hague, The Netherlands.

Henry Buller is Lecturer in Geography at the Université of Paris 7 and Member of the CNRS Research Laboratory LADYSS at the Université of Paris 10, Département de géographie, UFR GHSS, Case Courrier 7001, Université de Paris VII, 2. Place Jussieu, 75251 Paris Cedex 05, France.

Helen Caraveli is a Lecturer of Agricultural Economics and Economic Geography, Department of Economics, Athens University of Economics and Business, Patission Street 76, 104 34 Athens, Greece.

Thomas Dax is Deputy Director of Bundesanstalt für Bergbauernfragen (Federal Institute for Less-Favoured and Mountainous Areas), Möllwaldplatz 5, A-1040 Wien, Austria.

Katherine Falconer is Lecturer in Countryside Management in the Department of Agricultural Economics and Food Marketing, University of Newcastle upon Tyne, NE1 7RU, UK.

Brendan Flynn is Lecturer at the Department of Political Science and Sociology, University College Galway, Ireland.

Carolyn Foster is Research Assistant at the Organic Farming Unit, Welsh Institute of Rural Studies, University of Wales, Aberystwyth, SY23 3AL, UK.

Ingo Heinz is Senior Research Scholar, Institute for Environmental Research (INFU), University of Dortmund, 44221 Dortmund, Germany.

Petra Hellegers is Research Scholar at the Agricultural Economics Research Institute (LEI), P.O. Box 29703, 2502 LS The Hague, The Netherlands.

Marga Hoogeveen is Research Assistant at the Agricultural Economics Research Institute (LEI), P.O. Box 29703, 2502 LS The Hague, The Netherlands.

Flemming Just is Professor of Contemporary European History at the University of Southern Denmark, Esbjerg, and former Head of Department of Co-operative and Agricultural Research at South Jutland University Centre, University of Southern Denmark, Niels Bohrs Vej 9, 6700 Esbjerg, Denmark.

Werner Kleinhanss is Senior Researcher and Head of the Working Group of 'Quantitative Policy Assessments', Institute of Farm Economics and Rural Studies, Federal Agricultural Research Centre, Bundesallee 50, 38116 Braunschweig, Germany.

Nicolas Lampkin is Senior Lecturer for Agricultural Economics and Co-ordinator of the Organic Farming Unit at Welsh Institute of Rural Studies, University of Wales, Aberystwyth, SY23 3AL, UK.

Wilfrid Legg is Head of Policies and Environment in the Agriculture Directorate, Organisation for Economic Co-operation and Development (OECD), 2 Rue André Pascal, F-75775 Paris, France.

Philip Lowe is Duke of Northumberland Professor of Rural Economy and Director of the Centre for Rural Economy, Department of Agricultural Economics and Food Marketing, University of Newcastle upon Tyne, NE1 7RU, UK.

Maurizio Merlo is Professor of Forest Economics and Policy and Director of the Centre for Environmental Accounting and Management in Agriculture and Forestry, Centro Veneto di Contabilità e Gestione Agraria, Forestale e Ambientale, University of Padua, Agripolis, Via Romea - Legnaro (PD), I-35020 Padova, Italy.

Erica Milocco is Researcher at the Centro Veneto di Contabilità e Gestione Agraria, Forestale e Ambientale, University of Padua, Agripolis, Via Romea - Legnaro (PD), I- 35020 Padova, Italy.

Arie Oskam is Professor of Agricultural Economics and Policy at the Department of Economics and Management, Agricultural Economics and Policy Group, Wageningen University, Hollandseweg 1, 6706 KN Wageningen, The Netherlands.

Susanne Padel is Research Associate at the Organic Farming Unit, Welsh Institute of Rural Studies, University of Wales, Aberystwyth, SY23 3AL, UK.

Teresa Patrício is Vice-President of the Institute of International Scientific and Technological Cooperation and a Member of the Department of Sociology at Instituto Superior de Ciências do Trabalho e da Empresa, Av. das Forças Armadas, Lisbon, Portugal.

Andrea Povellato is Senior Researcher at the National Institute of Agricultural Economics (INEA), Agripolis, Via Romea, 35020 Legnaro (PD), Italy.

Jordi Rosell is Lecturer in the Department of Applied Economics, Universitat Autònoma de Barcelona, Campus de Bellaterra, E-08193 Bellaterra (Barcelona), Spain.

Alastair Rutherford is Head of Rural Policy at the Council for the Protection of Rural England, Warwick House, 25 Buckingham Palace Road, London SW1W OPP, UK.

Aida Valadas de Lima is Lecturer in Environmental Sociology and Rural Sociology at ISCTE/Lisbon and Invited Researcher at ICS/University of Lisbon, Portugal.

Lourdes Viladomiu is Lecturer in the Department of Applied Economics, Universitat Autònoma de Barcelona, Campus de Bellaterra, E-08193 Bellaterra (Barcelona), Spain.

Paola Virgilietti is Researcher at the Centro Veneto di Contabilità e Gestione Agraria, Forestale e Ambientale, University of Padua, Agripolis, Via Romea - Legnaro (PD), I- 35020 Padova, Italy.

Michael Winter is Professor of Rural Economy at the Countryside and Community Research Unit, Cheltenham & Gloucester College of Higher Education, Francis Close Hall, Cheltenham, GL50 4AZ, UK.

Preface

This book deals with CAP regimes and the European countryside. The overall objective is to review assessments made on the environmental effects of the Common Agricultural Policy (CAP). Emphasis is given to synoptic, comparative or analytical studies, rather than descriptive studies limited to one country. This should improve our understanding of the contribution of agricultural policy to further enhance environmental quality in the European Union (EU). The prospects for integration between agricultural, regional and environmental policies are examined. We will focus on the role of agricultural policy in reducing harmful effects and/or creating beneficial effects to the physical environment, landscape and nature. The book covers the research frontier concerning both dilemmas facing European policy and options available for policy reform.

Part of the book consists of edited and revised versions of papers presented at a workshop on CAP and Environment in the EU, held in The Netherlands (5-8 February 1998). Revisions were made in light of the reforms in 1999 of the CAP with the adoption of Agenda 2000. In addition, the editors identified a series of additional issues of importance and identified key experts.

The workshop was organised by the Agricultural Economics Research Institute (LEI) in the context of a research project entitled 'Thematic network on CAP and environment in the European Union'. The objective is to improve research methods to assess the effects of the CAP on the environment, nature and landscape, and to communicate the present understanding with representatives of national and regional governments, the European Commission and interest groups. The emphasis is on the post-1992 changes to the CAP, their impact on farming systems and consequences for the environment, nature and landscape in the EU. The book has also been prepared as part of that project, with financial support from the Commission of the European Communities, Agriculture and Fisheries (FAIR) specific (RTD) programme, FAIR3-CT96-

1793, Thematic network on CAP and environment in the EU. This support is gratefully acknowledged. It does not necessarily reflect the Commission's views and in no way anticipates its future policy in this area.

The editors would like to thank the following persons for their advice to select and revise first drafts of the papers: Professor Dave Abler, Pennsylvania State University (USA), Dr Clive Potter, Wye College, University of London (UK) and Professor Martin Whitby, Department of Agricultural Economics and Marketing, University of Newcastle upon Tyne, Newcastle (UK). They provided advice in targeting the book towards integrative analyses and exploring the role of the operation of the CAP in different institutional, cultural and environmental settings.

The editors are very grateful for the secretarial assistance provided by Ms Mei Li Tai, Agricultural Economics Research Institute. She took responsibility for guiding the publication process, and prepared the several drafts of the chapters.

Floor Brouwer and Philip Lowe

CAP and the Environment: Policy Development and the State of Research

1

Floor Brouwer and Philip Lowe

INTRODUCTION

Agriculture not only produces food and fibre; it also shapes the rural environment. Of the total territory of the European Union (EU), 50.5% is agricultural land. A managed environment that ensures the productivity of that land is of course essential to the maintenance of primary production. Increasingly, also, modern society values the environmental resources which arise as joint outputs with primary land use, including water supply, semi-natural habitats, wildlife, the historic pattern of land settlement and its associated cultural artefacts, rural landscapes and open spaces. For these resources to be conserved and made available, the land must be managed and this mainly entails the continuity of certain farming practices. However, rapid changes in primary land use and technology have jeopardised the supply of these resources, leading to problems of pollution and the impoverishment of the landscape in some regions and to land abandonment and problems of erosion in other regions.

Agricultural policy is dominated by an EU policy framework - the Common Agricultural Policy (CAP). The Treaty of Rome (1957) laid down the foundations of the CAP, but did not mention the environment at all. The priority at the time was to increase agricultural productivity; protection of the rural environment was not a concern. The extensive modernisation of European agriculture thus took place with little regard for the environmental consequences. Increasingly, though, the CAP attracted criticism for its role in driving changes in agricultural land use and farming practices that were detrimental to the countryside. Recently, European policy makers have begun to respond to such criticisms. This book sets out to describe the changes made to the CAP particularly in the 1992 reform and to assess the effects on the environment.

The CAP was subject to a major reform in 1992 - the so-called MacSharry reform, named after the then European Commissioner for Agriculture, Ray MacSharry - and for the first time environmental protection was acknowledged as an objective of the CAP. Agriculture had come under gathering pressures in the preceding years to improve its environmental performance. The Single European

1

Act, which came into force in 1987, established a legal requirement to integrate environmental protection into other policy areas. The Maastricht Treaty, which came into force in 1993, strengthened this requirement, as well as embodying the principle of sustainability.

A significant political push was given by the Fifth Environmental Action Programme, adopted by the European Commission in 1992, in which environmental protection was recognised as fundamental to the sound development of the EU (CEC, 1992). The Programme covers the period 1993-2000. Agriculture is one of the five target sectors. The Programme lays down the fundamental objectives of maintaining the basic natural processes indispensable for a sustainable agricultural sector, notably through the conservation of water, soil and genetic resources. The Programme also sets out specific objectives, namely, to reduce chemical inputs, to achieve a balance between nutrient inputs and the absorption capacity of the soil and plants, to promote rural environmental management practices, to conserve biodiversity and natural habitats and to minimise natural risks.

The 1992 reform of the CAP reflected some of these concerns. However, they were not the main objective of the reform. It was primarily aimed at restructuring agricultural markets, to restore market balance and to improve the competitiveness of EU agriculture, in the context of the General Agreement on Tariffs and Trade (GATT) negotiations to liberalise trade in agricultural (and other) commodities which were completed in 1994. One of the central elements of the reform was the encouragement of farmers to use less intensive production methods, through price reductions for some major products, for which producers were compensated by direct payments. It was expected that this would reduce the pressures on the environment as well as cut unwanted farm surpluses. An explicit agri-environment measure was also incorporated into the reformed CAP. The 1992 reform of the CAP has been characterised as 'a turning point' in the movement towards 'an economically sounder and environmentally friendlier policy' (CEC, 1997, p. 8).

LINKAGES BETWEEN AGRICULTURE, THE PHYSICAL ENVIRONMENT, LANDSCAPE AND BIODIVERSITY

The relationship between agriculture and the environment is complex, including both beneficial and benign effects as well as harmful ones (see also OECD, 1997, for an investigation of issues and policies related to the environmental effects of agriculture, and CEC, 1999a, to highlight the available statistical information on agriculture and the environment). Interactions between agriculture and the environment can be classified according to the following themes:

- *Soil quality* (in terms of contamination, erosion, desertification, nutrient supply, moisture balance, and compaction of soils due to the use of heavy machinery). Changes in land use practices - such as deforestation, the removal of hedgerows, overgrazing, neglect of soil conservation methods, the farming of uncultivated land - can damage soils. Soil erosion, often linked to changing agricultural practices, is one of the most severe rural environmental problems in Mediterranean countries. Some 115 million ha are estimated to suffer from water erosion, and 42 million from wind erosion. The application of livestock manure in most cases improves soil quality, but excessive levels may saturate and degrade soils as well as cause water pollution.

- *Water quality* and *water quantity* (leaching of nutrients and pesticides, water extraction and drainage). Contamination of both ground and surface waters caused by high levels of production and use of manure and chemical fertilisers is a serious problem in some parts of Europe, particularly where there are concentrations of intensive livestock production or large areas of specialised crop production. Nitrate and phosphate loading and the run-off of livestock wastes can cause significant water pollution problems. The excessive use of pesticides poses a widespread threat to the environment. High levels of pesticides in water are associated mainly with areas of intensive arable crops and horticulture. Water quantity problems arise in regions where water consumption exceeds critical levels in relation to the available water resources. A growing area of farmland in Europe is irrigated, and agriculture is the single most significant user of water in the Mediterranean parts of Europe. Over-abstraction causes the water table to lower, leading to the drying out of wetland areas and, in coastal regions, the ingress of brackish water and the salinisation of groundwaters.

- *Air quality* (emissions of ammonia and greenhouse gases). Emissions of ammonia contribute to acidification of soils and water, and agriculture is responsible for around 95% of these emissions. Volatilisation from livestock excretions is the major source of this type of pollution. Agriculture is responsible for about 8% of total greenhouse gas emissions but is the principal source of methane (from cattle production) and nitrogen oxide (from grazing livestock), contributing around 40% of these two gases.

- *Biodiversity* (i.e. biological diversity including genetic, species and ecosystem diversity). The biodiversity of much of the EU is found on, or adjacent to, farmland and is thereby considerably affected by agricultural management and practices. The intensification of agriculture has led to widespread reduction of species and habitats. However, about two-fifths of the EU's agricultural area remains under low intensity systems - mainly either grazing land under various systems of livestock management or permanent crops (olive trees, vines and fruit and nut trees) under traditional management. They support semi-natural habitats and wildlife species of conservation importance, but may face the threat of abandonment or of intensification.

- *Landscape* (including preservation of landscapes by farming systems with high nature value). Marginalisation of land used agriculturally, by which certain farmland areas cease to be viable, can lead to the abandonment or neglect of traditionally farmed landscapes. On the other hand, intensification of agriculture may lead to the loss of important landscape features such as hedges and ponds, the enlargement of fields and the replacement of traditional farm buildings with industrial structures.

THE RESPONSIBILITY OF THE CAP

Many of these environmental impacts have been laid at the feet of the European Community, specifically its CAP, and it must be admitted that the CAP has been in place during the most widespread and rapid transformation of the rural environment in the whole of European history. To ascribe all responsibility for this transformation and its myriad changes to the CAP would, however, be naïve (Lowe and Whitby, 1997).

There have been associated changes in rural social and economic structures and in technology with which the CAP has interacted but which would have had profound consequences without the CAP. Arguably, one effect of the CAP has been to moderate some of the more detrimental pressures and, given that many of the environmental benefits from rural land management depend upon the continuity of certain practices, it is possible that without the CAP there would have been even greater environmental losses. For example, the intensification and concentration in the pig and poultry sectors, with all of their attendant problems of disposal of waste products, have occurred without the CAP's commodity price supports, although they have been encouraged indirectly by a common and protected market. On a broader view, though, environmental problems associated with structural or technological changes in agriculture may point to gaps in policy, including at the Community level, especially if other Community policies - such as the Single Market - are implicated.

Three broad areas of concern have been identified about the direct effects of the CAP (Brouwer and Lowe, 1998): the level and efficiency of input use and the consequences for agricultural pollution; the rationalisation of farm size and structure and the consequences for rural landscapes and habitats; and the maintenance and encouragement of farming in marginal areas. Below we consider each of these in turn.

- It is generally agreed that high product prices paid under the CAP have encouraged a greater use of bought-in inputs than would otherwise have been the case. This has led to a less efficient use and hence a greater polluting surplus of chemical inputs (inorganic fertilisers and plant protection products); greater use of purchased feed and thus an encouragement to overstocking; and greater reliance on bought-in fertiliser, leading to even bigger surpluses of manure. Brouwer and Van Berkum (1996) confirm the empirical evidence that the CAP has given significant incentives to increase the use of agrochemicals, though there are considerable variations between farms and regions.
- A long-term objective of the CAP has been the improvement of the structure of farming. This has involved grants and technical aid to revitalise the age structure of the farming population, to modernise farms and to consolidate holdings. This has led to considerable change and rationalisation in rural landscapes including, for example, the spread of irrigation and drainage, and field enlargement, and the consequent loss of many traditional features and micro-habitats (e.g. hedges, trees, field margins and wet areas).

- Structural policies and price support have also helped to maintain rural populations across Europe. This has been particularly important in sustaining farming in marginal regions. The Less Favoured Areas (LFAs), which have been a focus of CAP support since the mid-1970s, cover some 55% of the agricultural area of the EU, including most of the land under low intensity systems which is of nature conservation or landscape value.

Efforts to reform the CAP from an environmental perspective have been aimed both to overcome the negative externalities associated with production supports and to incorporate positive environmental aims into the objectives of the CAP. The 1992 reform sought to stimulate less intensive production methods, primarily through the reforms of the arable crops regime, as well as the beef and sheep regimes, and it was anticipated that this would have indirect environmental benefits. In particular the reduction in prices was expected to lead to a less intensive use of pesticides and nutrients in the crop sector and a reduction of emissions (methane, ammonia, nitrate, animal wastes) from livestock farming. The payment of beef premiums was subject to limits on the number of eligible male animals per farm and the stocking density, meant to encourage extensive production methods. Provisions were also made for environmental protection by allowing for conditions to be placed on compensatory payments for price reductions in the beef and sheep sectors. In the arable sector, quasi-compulsory set-aside became a key means of production control; it was seen to be potentially beneficial for the environment through alleviating the pressure exerted by farming activities. In addition, the set-aside scheme included provisions regarding the maintenance and use of the land set-aside to protect the environment. As part of the 1992 reform, specific measures were incorporated centrally in the CAP to promote environmentally beneficial farming under the agri-environmental Regulation 2078/92. Aid was thereby made available to reduce agro-chemical inputs, to assist organic farming, to facilitate shifts to extensive forms of crop production or grassland management and to support production methods that protect the environment and maintain the countryside.

This book seeks to assess the environmental effects of these policy developments, looking forward also to the prospective changes that Agenda 2000 is likely to make to agricultural policy (CEC, 1999b). The CAP is a complicated set of policy instruments whose reform is not straightforward. It operates through a great variability of farming systems across Europe (e.g. farm types, breeds, production systems, farm management). The environmental effects depend upon circumstances and conditions which may be specific to the region, the locality or the site. Uniform responses or single environmental outcomes to a change in policy are not to be expected. Instead, the consequences are likely to be complex, including quite unanticipated results, and to be subject to a range of factors extraneous to agricultural policy. There is an obvious need for sound scientific research.

RESEARCH AND THE POLICY MAKING PROCESS

Policy makers look to researchers to provide pertinent and sound evidence about problems and robust concepts to inform public decisions and choices. Science is expected to provide clarity and certainty. This can be exemplified by the use of research in the context of multilateral trade negotiations. The settlement on agriculture in the last GATT Round, for example, was influenced decisively by research results and discussions between European and American economists concerning the effects of domestic market supports on international trade. This transatlantic consensus between economists paved the way for agreement on proposals to reform agricultural policy. Before new policy is negotiated at EU level, there is a need to assess the impact of current policies. The achievement of consensus is critical if scientific results and ideas are to play their part in the policy debate.

When the focus of CAP reform was essentially economic, achieving a scientific consensus was more straightforward than nowadays, when there is much less agreement on the environmental implications of policy reform. Opinions are more divided and there is a greater range of disciplinary perspectives reflecting the increasing complexity of the issues involved. Environmental linkages remain less integrated into policy formulation than economic linkages. However, the report of the group of experts charged with outlining the principles that might guide the transition of the CAP towards the integration of environmental and rural development objectives provides a possible model to follow (European Economy, 1997). We return to consider this in the final chapter of the book.

IDENTIFICATION OF RESEARCH GAPS

European agri-environmental research is flourishing (Brouwer and Lowe, 1998). The research effort is still in its infancy and inevitably it is not yet systematic in its coverage. Individual studies have been conducted in response to specific issues or particular circumstances. Therefore, we need to be modest on what we know of linkages between the CAP and the environment. Compiling an overview reveals gaps as well as the potential for more synoptic approaches. The partiality of the existing research effort can be characterised in the following terms:

First, there is a northern bias in the research coverage, reflecting the strength of northern European concerns. This bias emerges in:

- the geographical coverage of studies (with many studies for countries such as Germany, the UK and The Netherlands, but many fewer for southern Europe);
- the sectors and systems studied (temperate rather than Mediterranean crops, intensive rather than extensive systems);
- the problems and issues addressed (with water pollution particularly by pesticides and nitrates well covered but not air pollution nor problems of water supply and over-abstraction; with landscape deterioration better covered than

biodiversity; and with little attention to such agri-environmental problems as soil erosion, desertification, flooding, fire hazards or salinisation).

Second, there is a strong interest in agri-environmental measures compared to the effects of other elements of the CAP. This bias towards what is novel in policy development is understandable. The point should not be missed, however, that the Agri-environment Regulation was introduced as an accompanying measure to the 1992 CAP reform. It remains a minor component of the CAP. Not only do the main commodity regimes have much greater consequences for the environment, but also there is evidence that commodity supports actively discourage take up of agri-environment measures. Thus beneficial effects of the Agri-environment Regulation may be swamped by the environmental impact of the rest of the CAP. There is a risk, though, that the novelty and the intricacies of implementing the Regulation will distract attention away from the bigger picture.

Third, little if any work is being done on the environmental effects of:

- certain commodity regimes, e.g. tobacco and sugar;
- the other accompanying measures (the early retirement and afforestation schemes);
- the horizontal socio-structural measures (e.g. LFAs);
- regional and rural policy;
- other measures such as incentives for alternative crops, quality and label policy, biomass production and farm diversification.

Fourth, there are biases also in the style of research, with a tendency for single country studies, for an orientation to specific policy measures and for single disciplinary approaches. These lead to narrow or partial analysis, including:

- A lack of comparative studies (except, that is, for agri-environmental policy). Such studies are needed, for example, to understand variations in the implementation of common policies.
- A lack of integrated studies, focusing on a region or a farming system or a rural community and seeing how the complete set of policy changes interact upon it. Not only does this deficiency detract from a holistic assessment of the CAP but it is also linked to a poor understanding of the agricultural sector in its wider socio-economic context. For example, to what extent do the pressures of the CAP reflect broader social and technological changes and to what extent are these directly impinging on farmers and rural communities?
- A lack of linkage between agricultural economic analysis and farming systems/agro-ecology analysis. There is thus a disjuncture in understanding between the effects of policy on farm decision making and the effects of farmers' actions on the environment.

Comparative, holistic and dynamic studies are required at various levels, from the global, EU and national levels down to the regional and farm levels. Increasingly, the thrust of policy is towards integrated territorial measures. To accommodate variability in social and environmental circumstances calls for subsidiarity to be respected in the implementation of policy. This requires understanding of the articulation between different levels, i.e. the ways in which actions and decisions taken at one level constrain or inform those taken at another level. Generally speaking, there is a need to understand better the diversity of farming across Europe, including the role of different types of farming in their specific rural contexts. That could help inform options for policy makers. There is thus a need for a systematic comparative regional geography of farming systems and their environmental and rural development relations. Agri-environmental indicators should be developed to operationalise policy-relevant concepts and generally to improve our understanding regarding links between causes and effects. They could help guide responses to changes in environmental conditions and the move from a sectoral approach to more of a territorial policy (Brouwer and Crabtree, 1999).

The policy framework has to be sensitive to the variability within agriculture and its multi-functionality. In the future, policy may increasingly distinguish between commercial and non-commercial farming. Commercial farming is likely to be subject to market forces and environmental regulation; while support is given to non-commercial farming to supply beneficial services in the public domain, such as environmental quality, landscape maintenance, resource management and the provision of nature. The implications of such a dual structure and efforts at geographical rebalancing the forces of intensification and concentration call for careful analysis and better information.

Policy makers need to understand what constitutes sustainable agriculture, and how policies could be designed to provide beneficial effects for the environment (CEC, 1999c). Research could certainly make a contribution to this, identifying what farming practices are more or less sustainable, specifying which systems would be sustainable in which areas, and clarifying the relationship between rural and agricultural development. We also need to identify which farming systems are effective, and which problematical, in achieving the goals of environmental Directives.

Above all, there is a requirement for a clear vision of the future role of European agriculture for the provision of food, nature, landscape and a healthy environment. This must be set within a context of an enlarged EU and global markets. Such a vision will be vital in defining the role that policy can and/or should play, and therefore in guiding future CAP reforms.

THE ORGANISATION OF THE BOOK

It is against this backcloth of what research is needed that the present volume has been compiled. It particularly addresses the requirement to assess the implications of policy reform for the physical environment, landscape and nature. The book has four parts: (i) the environmental reform of agricultural policy; (ii) the environmental performance of the CAP regimes; (iii) institutional factors in reorienting agriculture; and (iv) outlook.

The first part of the book examines pertinent developments in agricultural policy and the extent to which these have taken account of the environment. The chapter following this one (Chapter 2) presents an overview by Wilfrid Legg of contemporary agricultural policy reforms in OECD countries and their linkages to the environment. The author shows the general trend towards a more market-oriented agriculture and argues that such policy liberalisation is a necessary, but not always a sufficient condition to improve the environmental performance of agriculture. The diversity and site specificity of agro-ecological conditions are such that uniform responses or single environmental outcomes from a change in agricultural policy are unlikely. Key policy lessons are drawn in this chapter, to guide policymakers in achieving a more efficient and sustainable agriculture. Chapter 3 reviews progress in integrating environmental concerns into agricultural policy making within the EU. Philip Lowe and David Baldock recognise the challenge that environmental issues present for traditional functions of government that are organised along sectoral lines. They look at where the pressures have come from to integrate the environment into the CAP and the response of agricultural policymakers. After assessing the extent to which the reformed CAP has helped achieve the objectives of EU environmental policy, they conclude that so far the integration of the environment into agricultural policy has largely been indirect and defensive.

The second part of the book is on the environmental performance of the CAP regimes. Emphasis is given to synoptic, comparative and analytical studies, focusing on a specific production or commodity regime but covering several countries. Each chapter gives an account of the evolution of the regime in question focusing on the changes made in the 1990s, particularly on the MacSharry reform, and making references where appropriate to the Agenda 2000 proposals. The key questions addressed are:

- What is the impact of the regime on the physical environment, landscape and nature, including both beneficial and harmful effects?
- What are the key mechanisms in the regime producing environmental effects?
- Are there obvious transnational or regional differences in the way in which the regime operates and the environmental consequences?

Chapter 4 examines the beef regime. Erling Andersen and his co-authors contrast the impact of the 1992 reforms in the UK and Denmark, with particular reference to the implications for beef cattle and wildlife habitats. Beef production is highly intensive in Denmark but in the UK there is a far wider range of production systems. In general, the impacts of the stocking density rules have been far more significant in Denmark than in the UK. The authors conclude that attempts to encourage more extensive production and to reward more environmentally sustainable systems have proved to be exceedingly weak and have largely failed to have any significant impact due to a range of difficulties, not least the poor design of the extensification mechanisms themselves. They suggest a number of improvements that could be made, including adjusting EU-wide stocking density

rules to reflect national, regional or local conditions and the targeting of Extensification Premiums particularly on permanent grassland.

Stuart Ashworth and Helen Caraveli, in their contribution (Chapter 5) on the
sheepmeat and goatmeat regime, examine production and structural changes in
sheep and goat farming that have taken place alongside developments in the regime and their environmental consequences. The authors consider the way in
which a common policy on sheepmeat and goatmeat has had a diversity of impacts among Member States. Examples are drawn from the UK, as representative
of the northern Member States, and Greece, as representative of the southern
ones. Responses to policy, and the environmental consequences, have been regionally specific and depend upon the significance of sheep and goat production
in total production. Since the 1992 reform the expansion in sheep numbers has
stopped. However, the environmental consequences are mixed and uncertain. Reduced grazing pressures are not necessarily leading to the revitalization of
degraded moor and grassland. Some of the changes which had previously occurred - such as pasture improvement, land abandonment or soil erosion - are not
readily reversible. It is also not clear that the reform has tackled the twin underlying problems of land abandonment of some areas alongside intensification of
grazing in other areas.

Chapters 6, 7 and 8 consider the reforms of the arable crops regime, assessing
in turn their impacts on pesticide use, their potential benefits in reducing nitrogen
pollution from intensive livestock production and the wider countryside implications. Katherine Falconer and Arie Oskam in their chapter focus on the use of
pesticides. Recent decades have seen increased intensification in the use of
chemical inputs on arable crops, linked to high levels of support under the CAP.
It was anticipated that, after the 1992 reforms, lower product prices would reduce
agro-chemical application rates whilst obligatory set-aside would reduce total usage. However, using an aggregate-level model, the authors estimate that the
reforms will have reduced total pesticide use by only about 3%. Technology and
farming systems are more important determinants than adjustments to the CAP
regime. Of course, radical change to the CAP could induce system changes. A
farm-level model linked to a set of environmental indicators suggests that if
commodity markets were to be liberalised, the environmental effects of changes
in pesticide usage would not necessarily be beneficial. Floor Brouwer and Marga
Hoogeveen (Chapter 7) examine the role of the arable crops regime in the effort
to reduce nitrogen pollution from intensive livestock production units. Lower
prices of cereals stimulate the use of cereals in compound feed, which would reduce the protein content of feed to grow pigs and poultry, and subsequently
reduce nitrogen pollution. The authors conclude that the reform in 1992 did provide an incentive to lower the protein content of compound feed and suggest that
the Agenda 2000 changes should provide an additional boost for a wider application of low protein feed, with environmental benefits. The wider countryside
implications of the response of farmers to the reform of the arable crops regime is
examined in Chapter 8. Michael Winter focuses on farm management responses
to the operation of the Arable Area Payment Scheme. The eligibility area rules
did largely limit the expansion of arable land, but there was little or no environ-

mental gain from reduced inputs or arable reversion. This was partly due to the unexpectedly high level of cereal prices on world markets following the reform.

Chapter 9, by Jordi Rosell and Lourdes Viladomiu, is on the wine regime. The environmental implications of wine production are complex and can be either harmful or beneficial. The type of impact depends on the production techniques applied, the location of production and the nature of the local ecosystem. At one extreme are found the intensive models as represented by certain German and Austrian regions, and at the other the extensive models typical of many parts of southern Europe. Intensive systems can require heavy doses of inorganic inputs which may cause serious deterioration in water and soil quality. Extensive systems, on the other hand, are crucial to the land management and landscape conservation of many Less Favoured Areas but may be threatened by marginalisation, grubbing-up schemes or displacement by other crops. The chapter presents a case study of the effects of such pressures on the traditional wine producing region of La Mancha. It also assesses the European Commission's 1998 proposal to reform the wine region (which was agreed, in 1999, as part of the Agenda 2000 package). The reform is judged as favourable to the environment, given that compulsory grubbing-up has been dropped. However, no positive environmental requirements have been incorporated into the regime.

In Chapter 10, Guy Beaufoy considers the environmental issues concerning olive production and the CAP support regime. Production systems vary considerably. Traditional plantations of old trees, managed with few or no chemical inputs, tend to be found in more remote and upland areas, have the highest landscape and biodiversity value and generally represent a sustainable approach to exploiting marginal agricultural land. Semi-intensive and intensive systems, using large inputs of artificial fertilisers and pesticides and often with irrigation, have much less natural value and are often the cause of soil erosion and pesticide pollution. CAP supports have encouraged intensification in the main producing regions, with generally negative environmental consequences, and, to a lesser extent, they have helped to reduce the abandonment of traditional plantations in marginal regions. The current CAP regime incorporates no environmental elements. The chapter evaluates options for its reform including the European Commission's 1997 proposal to replace production subsidies with a direct payment per tree. However, the olive oil regime was not included in the Agenda 2000 reforms (just as it had not been included in the MacSharry reforms). Some adjustments were made in 1998, largely to balance the market and improve monitoring in an attempt to combat fraud, in advance of a wide-ranging reform scheduled for 2001. However, one of the changes introduced - the abolition of the specific aid for small producers - the author judges to be retrograde from an environmental and a social point of view, given the importance of this aid to traditional plantations and the fact that the incentives to intensify are still in place.

Chapter 11 by Thomas Dax and Petra Hellegers is on policies for Less Favoured Areas and their implications for the environment, with particular regard to support for low intensity farming systems that indirectly conserve nature and cultural landscapes. There is a high incidence of such farming systems in LFAs. They face pressures of marginalisation and intensification. The LFA Directive (75/268), introduced in 1975, provides for the payment of selective incentives to farmers in certain agriculturally disadvantaged areas in order to achieve the continuation of farming, thereby counteracting depopulation or conserving the countryside. The chapter brings out the considerable disparity in implementation of LFA supports across the EU, with much greater levels of take-up and financial support in northern Europe than southern Europe. Though the LFA measure has changed little (up to Agenda 2000), its justification has increasingly referred to its environmental importance. The authors argue in favour of extending LFA payments (to help overcome low agricultural incomes in LFAs) and to integrate these supports within regional strategies combining rural development and nature conservation objectives.

In Chapter 12 Henry Buller examines the implementation of Regulation 2078/92. The chapter demonstrates how agri-environmental policy has expanded since 1992. By 1997, 1.3 million contracts had been signed with farmers covering 17% of total agricultural land in the EU, although with great variation between countries. The roll-out of the Regulation has thus been quite impressive. Agri-environmental measures remain nonetheless a marginal component of the CAP, accounting for about 4% of the total budget. The environmental gains are not always clear, and more durable gains will take time. The biggest schemes in terms of budgets and participants seek merely to maintain existing extensive production methods. Relatively few schemes are directly concerned with reducing farm pollution. The author concludes that much more will need to be done to integrate environmental objectives into agricultural and rural development policy.

Chapter 13 by Nicolas Lampkin and colleagues reviews EU policies supporting organic farming and what effect they have had. It covers direct support to the organic sector (mainly under 2078/92 and the structural funds) as well as the indirect boost to organic farming from the 1992 reforms to CAP commodity regimes. Although there were some negative effects, the shift of aid for crops from output to area payments and the increases in headage support payments were of benefit to organic producers. All in all, there has been a very rapid growth in organic farming in recent years. By 1997 it accounted for 1.5% of the total agricultural area in the EU. The authors argue that a 10% share is achievable by 2005, but this will require integrated support for production, for market and regional development and for organic support networks.

The third part of the book considers key institutional factors in reorienting agriculture. It explores the different means for managing and regulating the agricultural environment, including the role of voluntary measures and of markets. It also considers how cultural and institutional factors shape national positions towards reform of the CAP.

Chapter 14 by Flemming Just and Ingo Heinz looks at the growing use of 'soft' or voluntary measures - including incentives, communicative instruments and self-regulation - in agri-environment policy at the EU and national levels. 'Hard' regulations like prohibitions, orders and taxes are often ineffective, and provoke resistance from farmers. In many circumstances, soft measures are more practical, more effective and more acceptable. They are very dependent, though, on institutional conditions. The strength and openness of institutions at the local level - extension services, research organisations, farmers' networks - can be crucial in facilitating change in farming attitudes and behaviour towards the environment. The authors particularly commend the approach of co-operative agreements between nature conservation or water supply organisations and local farmers, as pursued in Germany and The Netherlands, as a model to be followed.

In Chapter 15 on national cultural and institutional factors, Philip Lowe and colleagues report the results of a comparative investigation of the place of the rural environment in shaping national positions towards reform of the CAP, drawing on a survey of opinions in national policy communities in Denmark, Ireland, Italy and Portugal. The countries were chosen to represent the range of production systems, from intensive to extensive ones; the types of environmental problem associated with different farming systems; and varying degrees of politicisation and institutionalisation of environmental concerns. Not only are there marked national variations in the degree of concern for agri-environmental problems, but the way in which these problems are framed also differs markedly: in Denmark, in terms of a pollution discourse; in Ireland, an agrarian discourse; in Italy, a consumer and regionalist discourse; and in Portugal, a rural development discourse. These separate framings reflect, in part, distinct rural conditions and farming systems. They also reflect differences in the relative institutional positions of agricultural and environmental interests in national polities. Fundamentally they also draw on culturally specific conceptions of the role of farmers and agriculture in contemporary society: ranging from traditions that portray them as guardians of the countryside to those which see them as the primary threat to the rural environment.

In Chapter 16 Maurizio Merlo and his co-authors consider the scope for establishing markets for Environmental and Recreational Goods and Services (ERGS). The chapter analyses a range of examples from Italy where ERGS have been transformed into marketable goods, or where means have been found to remunerate the producers of ERGS through complementary goods and services paid for by the beneficiaries. The authors identify a number of strategies that have been effective. In general, products that are linked to recreation can more easily find market remuneration, while the environment alone is not so easily transformed into a marketable good. The approach developed in the chapter is in keeping with the notion of integrated rural development which is a feature of the Agenda 2000 reforms of the CAP.

The final part of the book is forward looking and prospective in nature. The possible economic and environmental impacts of Agenda 2000 are assessed in Chapter 17. Werner Kleinhanss, using an optimisation model for representative farms in Germany, examines adaptation strategies by changes of intensity, scale and allocation of production. The results indicate that production levels and the intensity of arable and beef production will be affected by Agenda 2000. Beneficial environmental effects could be anticipated with regard to the use of agrochemicals and related pollution problems. On the other hand, decoupled premia favour cereal production and will therefore tend to reduce the diversity of land use. In Chapter 18, Philip Lowe and Floor Brouwer look at the specific changes that Agenda 2000 has made to the environmental aspects of the CAP and seek to provide an assessment of the challenges that remain to be addressed in the light of the findings of this book.

REFERENCES

Brouwer, F. and Crabtree, R. (eds) (1999) *Environmental Indicators and Agricultural Policy.* CAB International, Wallingford.

Brouwer, F. and Van Berkum, S. (1996) *CAP and Environment in the European Union: Analysis of the Effects of the CAP on the Environment and Assessment of Existing Environmental Conditions in Policy.* Wageningen Pers, Wageningen.

Brouwer, F. and Lowe, P. (1998) CAP reform and the environment. In: Brouwer, F. and Lowe, P. (eds) *CAP and the Rural Environment in Transition: A Panorama of National Perspectives.* Wageningen Pers, Wageningen, pp. 13-38.

CEC (1992) *Towards Sustainability: A European Community Programme for Policy and Action in Relation to the Environment and Sustainable Development.* Commission of the European Communities, Brussels.

CEC (1993) *The Agricultural Situation in the Community 1992.* Commission of the European Communities, Brussels.

CEC (1997) *Agriculture and the Environment.* Directorate-General for Agriculture, Working Notes on the Common Agricultural Policy, Brussels.

CEC (1999a) *Agriculture, Environment, Rural Development: Facts and figures - A Challenge for Agriculture.* Commission of the European Communities, Luxembourg.

CEC (1999b) *CAP Reform - A Policy for the Future.* Directorate-General for Agriculture, Brussels.

CEC (1999c) *Communication from the Commission to the Council, the European Parliament, the Economic and Social Committee and the Committee for the Regions: Directions Towards Sustainable Agriculture.* Commission of the European Communities, COM (1999) 22 final, Brussels.

European Economy (1997) *Towards a Common Agricultural and Rural Policy for Europe.* Directorate-General for Economic and Financial Affairs, Reports and Studies, Vol. 5, Brussels.

Lowe, P. and Whitby, M.C. (1997) The CAP and the European environment. In: Ritson, C. and Harvey, D.R. (eds) *The Common Agricultural Policy.* CAB International, Wallingford, pp. 285-304.

OECD (1997) *Environmental Benefits from Agriculture: Issues and Policies.* Organisation for Economic Co-operation and Development, Paris.

Part I

Introduction:
The Environmental Reform of
Agricultural Policy

The Environmental Effects of Reforming Agricultural Policies

2

Wilfrid Legg

INTRODUCTION

The impact of agriculture on the environment has become an important consideration in agricultural policy in the developed economies that are members of the Organisation for Economic Co-operation and Development (OECD). This is partly the result of increasing awareness of the mounting pressures on the environment throughout the economy, and the commitments made by countries within a range of international conventions and agreements (for example, on biodiversity, greenhouse gas emissions and desertification). Given the long period of government intervention in agriculture in most OECD countries, there is a particular interest in the environmental effects of agricultural policy reform. Agricultural policy reform - to increase the market orientation of the sector and liberalise agricultural trade while taking account of the multiple roles played by agriculture - was agreed by the OECD Ministerial Council in 1987, and was reaffirmed in March 1998 by OECD Agriculture Ministers. Reform efforts have already been undertaken, in part to meet the commitments entered into by countries within the Uruguay Round Agreement on Agriculture.

For the past 15 years the OECD has been at the forefront in analysing, monitoring and evaluating agricultural policies in Member countries. This work is widely known, and the annual measurement of support due to agricultural policies, using the 'Subsidy Equivalent' methodology, is central to the analysis. Since the early 1990s, reflecting developments in agriculture and the changing priorities for policies, the OECD has also been examining the linkages between agriculture and the environment from a policy perspective.

Work on 'agri-environment' issues in the OECD seeks to identify and analyse ways in which governments might encourage market solutions, and design and implement appropriate policies to achieve environmentally and economically sustainable agriculture, with least resource cost and trade distortions. The work is both an important element in the overall work on agricultural policy analysis, and the OECD-wide activity on Sustainable Development.

This chapter offers an overview of the OECD country experience on reforming agricultural policies and the environment. It starts with a brief look at some of the important aspects of the relationship between agriculture and the environment, followed by a summary of developments in agricultural policy reform in OECD countries. It then makes some observations concerning the links between agricultural policy reform and the environment. Finally, it identifies the areas of work on agriculture and the environment currently under way in the OECD.

AGRICULTURE AND THE ENVIRONMENT IN OECD COUNTRIES

The basic long-term *environmental issue* is whether agricultural activities can efficiently and profitably produce food to meet growing world demand without degrading natural resources - productive soils, clean and sufficient supplies of water, biodiversity and attractive landscape. Improvements in productivity have often been associated with environmentally damaging cropping and animal husbandry practices, which have led to soil degradation, salinisation, the pollution of groundwater, the farming of environmentally fragile land, greenhouse gas emissions, and loss of genetic diversity. In some of the drier areas of the OECD water for irrigation is being pumped from groundwater aquifers at rates that far exceed its natural replenishment. In other areas, surplus plant nutrients from animal wastes and fertilisers have led to eutrophication in surface waters, and excessive pesticide use has poisoned wildlife. But farmers, as 'stewards of the countryside', have also shaped landscapes and contributed to the maintenance of rural communities (OECD, 1997a). Agricultural land often provides important habitats for wildlife. Agriculture also plays a role in the hydrological cycle, in preventing flooding, and in cleansing the air of noxious gases.

Diversity of environmental problems and perceptions

Because of differences in climate, agro-ecosystems, population density and levels of economic development, the relative importance of particular environmental issues varies widely from one OECD country to another - and within countries. For example, surplus manure production and water pollution is often a major problem in the densely populated areas of northern Europe. Australia, on the other hand, along with New Zealand, suffers from extensive damage caused by rodents and other pests. Soil erosion is a major concern in Canada, Mexico, the US and Mediterranean countries, while many of these latter countries are also concerned with the effects of agricultural practices on water resources.

These differences are also reflected in perceptions across and within OECD countries as to what is meant by the 'environment' in agriculture. For some, the 'environment' covers only biophysical and ecological aspects. For others, in particular in Europe, landscape, cultural features, and the rural development aspects of agriculture are also important. In recent years, some aspects related to the quality and safety of food and the welfare of farm animals have become more prominent policy issues, with links to the environment. More generally, agriculture is increasingly considered as not only a 'food producing' activity, but a joint activity in which food production is an integral part of the production of environmental services.

Markets and the environment

Some of the environmental impacts from agricultural activities occur on the farm where they originate and are confined to the farm itself (for example, the quality of the soil on the farm). In such cases, the impacts should be taken into account by farmers and will influence their behaviour and decisions, to the extent that the present value of farming costs and revenues is affected, and they have sufficient information and financial resources to undertake the necessary remedial action. Some of the environmental services may be provided by farmers which raise the current value of farm receipts through, for example, farm tourism, or from sales of higher valued produce that has been produced in 'environmentally friendly' ways (which can be indicated through labelling, certification and place of origin, for example).

However, many environmental impacts extend beyond the farm. If agricultural activities generate pollution outside the farm generating it, and there is no mechanism to charge the farm (or trade the property rights), then the farm costs of production do not include these external costs. If, on the other hand, other groups in the economy benefit from environmental services provided by farms (through the farmer's provision of landscape, or wildlife habitats, for example), and there is no mechanism to value such services through charging those groups, then this reduces farm revenues. In the former situation, the level of agricultural production and pollution is likely to be higher, and in the latter situation, the level of environmental services is likely to be lower than would otherwise be the case.

The implications of farmers' actions causing harmful environmental effects that are not incorporated in farming costs - for example, agricultural chemicals leaching into groundwater - can be seen in higher costs of treating water for drinking. Similarly, when services provided by farmers are not valued, farmers refrain from taking actions that would add to public goods - such as conserving land as habitat for wildlife, or preserving other landscape features such as hedgerows - because effective ways of making the beneficiaries compensate farmers for the costs of such services are lacking.

Particularly from a policy perspective it is crucial that a distinction can be made between those agricultural activities that benefit and those that harm the environment, and to identify those environmental effects that are not taken into account in farmers' costs and revenues. Whichever 'baseline' is chosen (and this is controversial), the direction of change of an environmental effect will indicate whether there has been an improvement or deterioration in environmental performance. This requires information and indicators, on which the OECD has made significant progress.

The OECD has been exploring the concept and operational use of a *reference level*, as an aid to identify 'appropriate' policy responses. The reference level can be defined as the degree of social responsibility of farmers for the quality of the environment. In other words it defines the distribution of property rights over environmental resources, and thus distinguishes between the circumstances for which the 'polluter-pays' principle (PPP) applies and those for which payments to farmers for environmental services might be made. There are a number of possible ways in which the reference level could be expressed in an operational way, in terms of environmental outcomes or agricultural practices, such as 'codes of good agricultural practice'.

AGRICULTURAL POLICIES AND POLICY REFORM IN OECD COUNTRIES

Agriculture in OECD countries has been a success story - in terms of increasing output, largely achieved through higher yields, rather than an increase in land devoted to agriculture. Agricultural output in the OECD area has, over a long period, tended to increase faster than demand in OECD countries.

However, that increased output has been achieved at a high cost for both consumers and taxpayers: agricultural production has been heavily supported, mainly through commodity programmes and associated trade barriers, although the range of support has varied considerably across OECD countries and commodities. Excess production frequently has been exported or stored with subsidies, creating severe tensions in international trading relations from the end of the 1970s. There has also been growing public concern over the environmental damage caused by agriculture in many countries.

OECD Ministers first committed themselves to the reform of agricultural policies in 1987. The essential elements were to reduce support levels, liberalize agricultural trade, and shift to better targeted measures to achieve those policy objectives for which markets might be inadequate. The commitment to reform was further reinforced by the 1994 Uruguay Round Agreement on Agriculture, which obliged countries to replace non-tariff barriers with bound tariffs, increase access for imports, and reduce subsidies to exports and trade-distorting domestic policy measures. OECD Agriculture Ministers reaffirmed their commitment to further agricultural policy reform in 1998.

Starting with New Zealand, which had already undertaken a comprehensive reform of its agricultural policies in 1984, other OECD countries began to shift support away from output levels, towards more direct means of income support, such as replacing price support by area-based payments, and various forms of income safety-nets. Such reforms have been evident in the US (FAIR Act), in Mexico and Canada, and in the 1992 reform of the European Union's (EUs) Common Agricultural Policy.

The overall cost of support to agriculture in the mid-1980s accounted for around 2.1% of GDP in the OECD area, and support was equivalent to about 41% of the value of agricultural production, as measured by the OECD's Producer Subsidy Equivalent (PSE). Two-thirds of the support was channelled through measures that kept domestic market prices paid by consumers about 60% on average above comparable world prices. The OECD PSE (now called the 'Producer

Support Estimate') has since fallen, and in 1998 it was equivalent to 37% of the value of OECD agricultural production, while the share of overall cost of support in GDP was around 1.4%. Over the last decade there has been a decrease in the numbers of people working in agriculture, and support per 'full-time farmer equivalent', which was on average $13,000 in 1986-1988, is currently around $18,000, which is roughly unchanged in real terms, given inflation. As in the earlier period, the same wide disparities in support remain across OECD countries and commodities. However, support that is to a degree 'decoupled' from specific commodities or inputs now accounts for around 30% of total agricultural support, whereas a decade ago the share was under 20%. This is also reflected in the fact that domestic market prices have on average fallen to 40% above world prices.

However, the OECD average figures disguise the wide range of developments among countries and commodities. In 1998 the PSEs in Australia and New Zealand were 7% and 1% respectively, while they were over 60% in Iceland and Japan, 70% in Norway and over 70% in Switzerland. PSEs were 45% in the EU, 16% in Canada and 22% in the US. The milk, sugar and rice sectors have been heavily supported in many OECD countries for a considerable time, and currently register about 58% for milk, 43% for sugar and 74% for rice.

In many cases environmental problems have been aggravated by agricultural and trade policies that distort price signals through support increasing the prices of agricultural commodities, or reducing the costs of agricultural inputs. The economic distortions created by such policies can lead to environmentally inappropriate farming practices with environmentally harmful use of inputs, and discourage the development and adoption of farming technologies less stressful on the environment.

In New Zealand there is good evidence, given the relatively long period of fundamental policy reform, that illustrates the reduction of pressure on the environment, in particular the shift from inappropriate agricultural production on environmentally fragile land, and decreasing intensity of fertiliser use. It should be noted, however, that New Zealand's agricultural policy reforms were accompanied by economy-wide environmental measures (in the 1991 Resource Management Act).

Policy measures targeted to environmental issues in agriculture

Over the last decade, an important component of the wider package of agricultural policy reforms in most OECD countries has been the introduction of new programmes addressed to environmental issues.

For example, the EU has a wide range of agri-environmental measures, cofinanced and implemented by Member States, which are discussed in detail in other chapters (Buller, Chapter 12; Lampkin *et al.,* Chapter 13).

Under Norway's 'acreage and cultural landscape scheme' farmers receive area-based payments if they grow specified crops and refrain from making major changes to landscape features, such as streams, stone fences, paths and forest edges. In Switzerland, annual payments for various types of 'ecological services' support integrated production methods and organic farming.

Because of its hilly terrain, heavy rainfall, intensive agriculture and high population density, Japan's approach to sustainable agriculture is oriented towards hydrological aspects: preventing flooding and water erosion, minimising nutrient leaching and protecting forests. Its 'Comprehensive Program for the Promotion of Sustainable Agriculture', begun in 1994, seeks to promote environmentally friendly, low input agriculture in ways that minimally affect the productivity of agriculture.

In Canada, the Permanent Cover Program aims at reducing soil deterioration on high-risk land, through payments to shift from annual crops to grassland. The US supports the adoption of more sustainable farming methods through numerous programmes. One of its main programmes, the Conservation Reserve Program (CRP), pays farmers on an annual basis for keeping environmentally sensitive land out of production in order to reduce erosion and sediment loading in streams, and to protect water resources and wildlife. Other programmes award cost-share or incentive payment contracts to farmers who adopt approved land management practices or install structures for controlling animal waste, or encourage farmers to adopt integrated strategies for managing their farm resources, while assisting them in meeting environmental quality standards.

Australia and New Zealand (and Canada, to some extent) place great stress on *community-based approaches* to agricultural resource management which aim to motivate farmers to take on greater responsibility for the local management of land, water and related natural resources. Australia's National Landcare Program is the oldest and largest of its kind, and channels much of its financial assistance through farmer-led community groups. Much stress is placed on improving information flows, upgrading skills and using peer pressure to attain results. In this context, approaches such as Linking Environment and Farming (LEAF) in the UK, and FERTIMIEUX ('Better use of Fertilizers') in France are also ways of promoting the management of farms and the adoption of practices that respect the environment, by using farm chemicals in a responsible way.

Most OECD countries have passed laws to deal with problems relating to pollution of air and water, and threats to ecologically sensitive areas and wildlife. The EUs Nitrate Directive is a good example. Activities that create high levels of pollution - such as dumping slurry, aerial pesticide spraying or burning straw - are frequently proscribed via *regulations,* which have also been used to limit risks of serious damage to human health and the environment. Regulations, for example, are used to constrain the use of new biotechnological products and particularly toxic or persistent pesticides. Regulations also set limits on animal stocking densities and circumscribe periods during which manure spreading is permitted.

Another approach is to set *minimum standards* for farm practices and to tie them to existing agricultural support programmes. Such 'cross-compliance' measures have been used for several years in the US and are becoming increasingly important in European countries.

Research and development (R&D) into technologies and farming methods that are less environmentally stressful (such as crop rotation, intercropping, conservation tillage, agroforestry and silvipasture, precision application of fertilisers, and integrated pest management), and on improving scientific understanding of the physical and biological links between agriculture and the environment, is an essential component of promoting agricultural sustainability. R&D is helping to reduce risks from pesticides: for example, through the development of new strains of crops that are more naturally resistant to attack by pests (and less dependent on pesticides); methods for growing crops that rely on safe yet effective biological processes or agents; and chemical agents that are environmentally safe yet effective and affordable.

Continuing technological innovation in the agricultural sector will be crucial to meeting the goals of sustainable agriculture. But if farming is to become more sustainable, it is certain to become more management-intensive, and possibly more labour-intensive, focused on precision techniques.

Educated and informed farmers (who rely on the market for their incomes) are more likely to be motivated to look after the productivity of their land, to be receptive to policies that constrain their activities in the interest of environmental protection, and to be able to implement any changes required of them. *Extension services* play an important role in this process. Improved information on the proper timing of fertilizer application, for example, has already helped reduce the loss of nutrients into water systems in several countries.

The environmental impact of agricultural policy reform

What conclusions can be drawn at this stage from the OECD work on analysing the effects of agricultural policy reform on the environment? By decoupling support from agricultural commodities and inputs, the reform of agricultural policies should improve the domestic and international allocation of resources, and reduce incentives to over-use polluting chemical inputs and to farm environmentally sensitive land. In other words, through reducing output and input use (through a reduction in output and input price distortions, and changes in relative factor prices), the reforms would tend to reverse the harmful environmental impacts associated with commodity and input-specific policy measures. But in those cases where agricultural policies are associated with maintaining farming activities that provide environmental services, reform of such policies can reduce environmental performance.

However, some caution needs to be exercised in assessing the overall environmental effect. Agricultural policy reforms have been relatively recent in most countries. There are few examples of fundamental or radical policy reforms, and the degree of decoupling of policy measures is limited as, for example, compensation payments are still linked to land use in the EU. The environmental effects of policy measures are complex, often site-specific, take some time to become evident, and the data and evidence to quantify them are extremely patchy. And many other developments have been occurring at the same time, which have effects on the environment, and may or may not be linked to policy change: for example technological advances, farmer information and behaviour, consumer demands and environmental group pressure, environmental regulations, industrial pollution, climate change and, not least, the relative prices of farm inputs (including chemicals) and outputs.

OECD work has outlined some of the trends in the environmental effects of agricultural policy reform. By lowering price support and input subsidies, shifting to policies that are less linked to production, and implementing agri-environmental measures, policy reforms have in many cases generated a double benefit: they have resulted in a more efficient allocation of market resources, and they have reduced negative environmental externalities. They have also increased transparency as to the remaining externalities that have the potential to be addressed through targeted environmental measures. The financial gains resulting from a better resource allocation could be used to support such targeted measures.

In particular, reductions in price support and input subsidies have in many cases lowered the demand for chemical and mechanical inputs, as well as for irrigation water, and have led to a de-intensification of crop production. Yet, some land may have been shifted into the production of fruits and vegetables, which are sometimes produced in input-intensive ways, or into other input-intensive crops. In some cases, the use of farm chemicals has again increased after an initial decline, largely linked to increases in world commodity prices. The effects on the environment are not only driven by policies but depend also on the developments of markets and technical progress. Positive environmental effects may, for instance, result from a switch to environmentally friendly production methods induced by consumers' choices.

Reforms in the livestock sector are likely to have resulted in reduced grazing pressure and manure surpluses and, as a consequence, soil erosion and nutrient leaching. However, where direct payments per head of animal have been provided and the stocking density limits set by governments have exceeded the original densities in the area, increases in stocking densities may in some cases have occurred.

Policy reform has also slowed down, or brought to a halt, the conversion of environmentally fragile or ecologically valuable land to agricultural uses in OECD countries. Significant areas of wetland, forest and natural grassland have thus been preserved. In countries where support had previously favoured cropping over grass-based activities, shifts out of crop production into grazing and forage production have taken place. The grass or tree cover established on erodible land as a result of such shifts has reduced soil erosion rates and, in some cases, has helped restore already degraded soils (OECD, 1998a).

Changes in land use have sometimes been aided by land diversion schemes, which have paid farmers for idling land or replacing arable crops by less intensive forms of production and woodland. Over time, incentives to remove the environmentally most sensitive or ecologically most valuable land from production have been introduced, and farmers have been required to make environmental improvements on the diverted land. As a result, substantial areas of land have been improved, wildlife habitat has been created or restored, and the risk of nutrient leaching has diminished. Some of these improvements appear to have been sustained, while others have disappeared when the land was brought back into production (OECD, 1997b).

However, in certain countries, agricultural support has allowed farmers to maintain farming systems that support a rich variety of flora and fauna as well as scenic landscapes that are valued by the population. Such production systems, which would be unprofitable without support, can extend over large expanses of semi-natural land. Elsewhere, support has maintained agricultural activities that have been associated with land conservation, including landslide and flood prevention. There are concerns that such positive environmental externalities of agriculture could be reduced if reform causes agricultural activity to shrink.

Where reform drives farms out of business and the land is not taken over by other farms in the area for lack of profitability, the land may be abandoned. In some cases, abandoned land will revert back to nature and may actually benefit the environment, as vegetation and wildlife continue to develop along the path of natural succession until a new ecosystem has developed in the absence of agricultural activity. In other cases, environmental degradation, including soil erosion, irreversible damage to wildlife habitat, biodiversity and landscapes, and loss of the flood controlling function of the land, may occur.

Agri-environmental measures appear to have been effective when: the environmental objectives are clearly specified and the actions required of farmers are closely targeted to the objectives; the measures are tailored to the environmental, economic and social situation prevailing in a given area; the lands accepted into the programmes have a high conservation value; the incentives provided to farmers are linked to the size of the benefits or the income foregone by adhering to the restrictions; farmer compliance is closely monitored and the effects on farming practices and the environment are continuously assessed against the stated goals; and training and advice are provided to ensure that farmers are sufficiently informed about the measures and the best ways to implement them.

It is not evident that environmental externalities have been taken into account through policy developments: there is a risk that more attention has been given to 'environmental cost compensation' and payments for environmental services of a 'public good' character, and less attention has been paid to recovering pollution costs originating in agriculture (the application of the 'polluter-pays' principle).

INDICATORS OF ENVIRONMENTAL PERFORMANCE

One of the tools to assess the effects of agricultural activities on the environment is agri-environmental indicators. Work in the OECD is developing a set of environmental indicators for the agricultural sector, which could be of use in the policy domain. A framework has been established, and it is intended that indicators will be calculated by 2000 for a set of 13 'policy issue areas', namely nutrient use, pesticide use, water use and quality, land use and conservation, soil quality, greenhouse gases, biodiversity, wildlife habitats, landscape, farm management practices, farm financial resources, and rural viability. Quantitative work is under way, and there are already some interesting results in the areas of agricultural nutrient (nitrogen) and agricultural pesticide use (OECD, 1997c). Figure 2.1 shows changes in the agricultural nitrogen surplus by country. The surplus, which passes to the environment, is the excess of nitrogen inputs, which are mainly chemical fertilisers, livestock waste, atmospheric deposition and nitrogen fixation, over outputs, which are mainly the nitrogen uptake by crops and pasture. In most OECD countries, the trend in nitrogen surplus over the last decade has been downwards, including for countries with a high surplus (in excess of 100 kg ha^{-1}), such as Belgium, Denmark, Japan and The Netherlands. However, the absolute levels of surplus vary considerably across and within countries. Concerning agri-

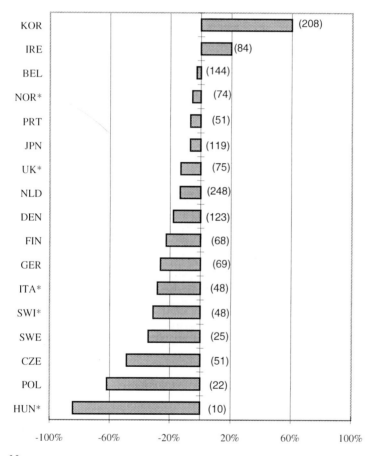

Notes:

(..) kg N surplus per hectare of agricultural land 1994-1996.

Nitrogen (N) use = N inputs (fertiliser manure, etc.) minus N plant uptake, which if > 0 = N surplus; if < 0 = N deficit.

*1986-1988 to 1993-1995.

Figure 2.1   Percentage changes in agricultural nitrogen surpluses, 1986-1988 to 1994-1996

cultural pesticide use (in terms of active ingredients), a similar downward or constant trend is observed (see Figure 2.2), but also with considerable variation across countries and crops. The reasons for these trends relate not only to the effect of agricultural policy reform on input demand, but also to the price of energy, technical changes and farmers' knowledge.

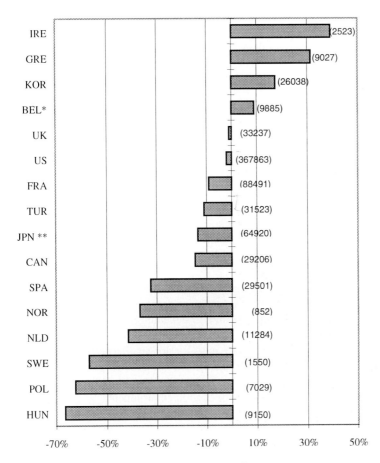

Notes:
*Includes Luxembourg **1986-1988 to 1992
(..) Total use of pesticides in tonnes of active ingredients 1993-1995.

Figure 2.2　Percentage changes in pesticide use in agriculture, 1986-1988 to 1993-1995

PRELIMINARY POLICY CONCLUSIONS

A number of preliminary conclusions emerge from the OECD work on analysing and quantifying the effects of agricultural policy on the environment, which will help policy makers design and implement policies to contribute to achieving an efficient and sustainable agricultural sector (OECD, 1998b):

- The reform of agricultural policies, through reducing the level of support from commodity and input linked policies ('decoupling'), and associated trade measures, will improve resource use efficiency, tend to reduce inappropriate farming practices on environmentally fragile land and the excessive use of farm chemicals, and thereby benefit the environment.
- Switching to agricultural policy measures which provide payments 'coupled' to targeted environmental outcomes, has the potential to reward farmers for those environmental services from agriculture that are not taken into account in their receipts from their commercial activities.
- The current combination of market price support and targeted agri-environmental measures is not the most effective and least-cost way of achieving environmental objectives.
- The 'baseline' to assess environmental performance in agriculture has often been distorted by a long period of significant levels of support in many countries, and it is not clear which environmental services the market would deliver if support were removed.
- The degree to which farmers are held responsible for agricultural pollution, or compensated for improving the environment depends on a 'reference level', which varies across and within countries depending on agri-environmental situations, public perceptions, and the specification of property rights. However, if policies impose charges on farmers for non-marketed externalities that harm the environment, and pay farmers for non-marketed public goods that benefit the environment, then the 'lower' the 'reference level' is set, the higher the cost to taxpayers for producing the desired environment.
- To be consistent, a balanced policy approach is needed: when payments are given for the provision of environmental services over and above those that are remunerated through revenues from agricultural activities, then farmers should also be required to meet the costs of any off-farm environmental damage they cause ('polluter-pays' principle).
- The variety of agri-environmental situations and policy priorities across and within countries suggests that appropriate policy measures and market solutions are likely to differ. This is even more relevant when considering non-OECD countries. But, in general, there remains considerable potential for developing innovative market solutions, localised policy measures, co-operative approaches by farmers, and the dissemination of information from scientific research. Many of these are likely to be among the least costly, and least distorting policy approaches to achieving many environmental objectives.

FURTHER WORK

While the work on the linkages between agriculture and the environment in the OECD has made a significant contribution to the policy debate, there are a number of important areas on the agenda to further advance the work. In brief, future work, as outlined below, should help to find ways of dealing with the fundamental question of knowing how can we obtain the amount of food needed with better environmental performance and less government support:

- completing the methodologies for all of the 13 agri-environmental indicator areas, collecting data and calculating indicators, and using them as a tool in monitoring and evaluating policy developments;
- integrating the modelling work in the OECD, in particular the projections for agricultural markets through the AGLINK model and the work under way in developing the Policy Evaluation Matrix, to undertake scenario analysis of the possible environmental implications (such as greenhouse gas emissions) of market developments, and the agricultural implications of meeting environmental standards;
- establishing criteria and guidelines to determine the appropriate level of remuneration for farmers in cases where environmental services are jointly produced with agricultural products, and to analyse alternative policy options;
- exploring the effects of agricultural policy reform and trade liberalisation on environmental performance in agriculture, and the effects of different domestic environmental standards on trade, and policy measures to minimise possible conflicts between different domestic environmental standards and trade liberalisation;
- analysing the environmental impacts of different agricultural structures (such as large or small, extensive or intensive farms) and farm practices (such as conventional or organic farming), and technological developments (such as genetically modified organisms) from a policy perspective;
- exploring the economic implications for the agricultural sector in implementing sustainable farming practices, and identifying the appropriate mix of market solutions and policy measures to achieve the combination of sufficient (and efficient) agricultural output and good environmental performance.

A final comment

A more market-oriented agriculture through policy reform is a necessary, but not always a sufficient condition to improve the environmental performance of agriculture. Markets may need to be complemented by policies to reach that goal, but many of the policy measures do not emerge as the most cost-effective way. Given the diversity and site specificity of agro-ecological conditions, local, farmer-based approaches, coupled with good research, development, training, information and advice, would appear to be high on the list of 'good policy practices'. These types of approaches focus on the 'public good' aspects of agriculture, reflect the differences across farming, and allow for the development of market-based, innovative approaches.

NOTE

The views expressed in this chapter are those of the author and do not necessarily reflect those of the OECD or its Member countries.

REFERENCES

OECD (1997a) *The Environmental Benefits from Agriculture: Issues and Policies - The Helsinki Seminar.* Organisation for Economic Co-operation and Development, Paris.
OECD (1997b) *The Environmental Effects of Agricultural Land Diversion Schemes.* Organisation for Economic Co-operation and Development, Paris.
OECD (1997c) *Environmental Indicators for Agriculture.* Organisation for Economic Co-operation and Development, Paris.
OECD (1998a) *The Environmental Effects of Reforming Agricultural Policies.* Organisation for Economic Co-operation and Development, Paris.
OECD (1998b) *Agriculture and the Environment: Issues and Policies.* Organisation for Economic Co-operation and Development, Paris.

Integration of Environmental Objectives into Agricultural Policy Making

3

Philip Lowe and David Baldock

INTRODUCTION

Environmental issues and concerns represent a challenge for traditional functions of government that are organised along sectoral lines. This has long been recognised, but the transversal nature of environment policy has proved difficult to express in effective institutional arrangements for policy making. Initially, in the 1970s, the primary effort was directed towards the development of environmental policy in its own right. A new apparatus of environmental ministries and agencies was spawned during this period and the then EEC put in place a substantial body of environmental law subsequently underpinned by changes in the Treaty of Rome. The gathering debate over sustainable development in the 1980s fuelled a more ambitious agenda and underscored some of the limitations of relying primarily on environmental policy instruments. In response, there has emerged a new emphasis on seeking to establish a more synoptic environmental policy with coordinated environmental goals integrated into a number of key sectoral policies. Potentially, this has a number of dimensions: at the strategic level - a global approach to the setting and monitoring of environmental objectives; at the sectoral level - an emphasis on integrating environmental goals into sectoral objectives (e.g. the greening of transport, tourism, etc.); at the level of policy instruments - use of instruments such as cross-compliance, economic mechanisms and environmental assessment procedures.

The commonly identified barriers to integration have both a political and an organisational aspect. In the case of the former, a traditional problem for environmental policy has been its relatively low priority on the political agenda, reflecting the dominant attachment of governments to the imperative of economic growth. Environmental protection has generally been treated as something of a luxury to be afforded in times of economic prosperity, or to be traded-off against material goals. Moreover, environmental problems are often deep-seated and many environmental policies only promise benefits in the long term which means that they are often ignored or sacrificed in political systems geared up for short-term electoral or economic cycles. Sectoral economic policies, on the other hand, tend to be supported by strong producer organisations whose members are usually directly and significantly affected by government action, in contrast to the diffuse public benefits yielded by environmental policy. In consequence, environmental policy may lack the weight to compete for resources or challenge the policy assumptions of other issue arenas and can be marginalised within institutional structures.

In terms of organisational barriers governmental bureaucracies tend to be highly compartmentalised which means they are not well designed to absorb cross-cutting environmental concerns. Although the impetus behind integration may stem from the Parliament, the President or the media and NGOs, the role of traditional sectoral ministries is usually critical. It is not unusual for ministries responsible for agriculture, industry, trade and other sectors to continue to regard the environment as the particular province of the environment ministry, rather than a frame of objectives and values to be incorporated into their own institutional remits and procedures. Consequently the approach to integration adopted often will depend firstly upon the degree of formal power or authority at the disposal of the environment ministry. Among European Union (EU) Member States, most environment ministries have only recently been established: the oldest have been in existence for less than 25 years and some (as in Italy and Spain) for less than ten (Wilkinson, 1997). The approach adopted in environmental policy often differs significantly from those traditions of policy making and styles of regulation within established sectors. The integration of environmental objectives into other sectors is therefore most unlikely to be an overnight process. Some policy sectors will respond more readily than others and the pace of change will vary greatly between different states.

Agriculture might seem to be an ideal candidate for the integration of environmental objectives. There is, after all, an especially close connection between agricultural production, the management of rural resources and the moulding of the landscape, which is in a perpetual state of creation and re-creation. Economists often refer to agricultural goods and the rural environment as being in a relationship of 'joint supply', but even that term implies a separateness whose appropriateness is questionable.

Unfortunately, over a long period, agricultural production has been promoted without due regard for its environmental constraints and consequences. Indeed, because of the intimacy of the links betweeen the two, rather than a force to conserve the rural environment, agriculture has become one of the prime causes of its degradation. The European Community's Fifth Environmental Action Programme (5EAP) presented a detailed indictment of contemporary agriculture, and concluded that 'farming practices in many regions of the Community have led to over-exploitation and degradation of the natural resources on which agriculture itself ultimately depends: soil, water and air' (European Commission, 1992; p. 15, para 25).

During the recent debate over the Agenda 2000 reforms, the Agriculture Commissioner was at pains to emphasise the 'multi-functional' nature of European agriculture. However, many of the specific changes proposed to strengthen the environmental dimension of the Common Agricultural Policy (CAP) have been resisted by agriculture ministers or heads of state, and the thrust towards integration lacks the support of the core agricultural policy community. There is thus a continuing tension between the emerging sustainable development agenda and the traditional preoccupations of agricultural policy.

EUROPEAN DECISION MAKING PROCESSES IN AGRICULTURE AND THE ENVIRONMENT

Decision making in the EU, as in most Member States, is sectoralised, and European agricultural and environmental policies are largely separate. Each has had its own momentum and has evolved within its own network of established policy actors (Table 3.1).

Agriculture is the most mature and well established policy field in the European Community (Marsh, 1991), and agricultural interests are deeply embedded constitutionally and administratively in the functioning of the Community. Agricultural policy has an array of its own decision making structures, including an important group of specialised committees. The policy network includes agricultural ministries and the major farming unions, as well as the Agricultural Directorate (DGVI) of the European Commission, and it has not been easy for non-agricultural interests to penetrate this network (Neville-Rolfe, 1983). Environmental policy is a more recent development and was only explicitly recognised in the basic EC treaties in 1987, although in fact the development of European environmental policy dates back to the early 1970s (Liefferink *et al.*, 1993). This has been a much more fluid and open policy field (Mazey and Richardson, 1992, 1993). Other EC institutions, apart from the Commission and its Environmental Directorate (DGXI), have been actively involved in its development including the European Court and the European Parliament (co-decision making applies for a large proportion of environmental measures, but not to agricultural policy at all).

The two policy regimes have contrasting policy styles and means of intervention. Table 3.2 is a comparison between EU agricultural and environmental policy as presented by DGVI. It depicts agricultural policy as old, big, predominantly interventionist with the EU (as opposed to the Member States) enjoying exclusive legal competence. In contrast, environmental policy is small, recent and subject to the principle of subsidiarity. While not inaccurate, this characterisation does distract attention from the large number of environmental policy instruments and the willingness of Member States to cede substantial powers to the Community over most areas of environmental policy. It also does not bring out the contrast between the use of Regulations as the main legal instrument in European agricultural policy, which leaves little scope for Member States to develop their own variants of the EU model, and Directives, which are the primary form of legislation in European environmental policy and allow Member States much greater discretion in meeting the objectives required. They usually come into operation over a period of years rather than immediately. These different styles and

Table 3.1   Milestones in the development of EU agri-environment policy

Agricultural policy	Environmental policy
	1973 EC's 1st Environmental Action Programme recognised the need to tackle emerging problems of agricultural pollution
1975 Less Favoured Areas Directive, to compensate farmers for working in areas deemed to be disadvantaged	
	1979 Birds Directive
	1980 Drinking Water Directive specifies MACs for pesticides and nitrates
	1983 EC's 3rd Environmental Action Programme stated the need to: 'promote the creation of an overall strategy, making environmental policy a part of economic and social development, (resulting) in a greater awareness of the environmental dimension, notably in the field of agriculture (and)...enhance the positive and reduce the negative effects on the environment of agriculture'. (O.J. C46, 17 February 1983)
	1985 Environmental Assessment Directive
1985 Commission's Green Paper on the future of the CAP proposed that agricultural policy should: 'take account of environmental policy, both as regards the control of harmful practices and the promotion of practices friendly to the environment' (European Commission, 1985)	
1985 Article 19 of Council Regulation 797/85 on Agricultural Structures authorised Member States to introduce 'special national schemes in environmentally sensitive areas' to subsidise farming practices favourable to the environment	
1987 Regulation 1760/87 made ESA payment schemes eligible for a maximum of 25% reimbursement from the EAGGF	**1987** Single European Act required the integration of environmental policy into other EC policy areas
1988 The Commission's *Future of Rural Society* and *Environment and Agriculture* concede the need to adapt agriculture to the requirements of protecting the environment and maintenance of the countryside	
	1991 Nitrates Directive
	1992 Habitats Directive
1992 The Agri-Environment Regulation (2078/92) introduced as part of the MacSharry reforms	**1992** 5th Environmental Action Programme identifies the need for environmental integration as a major theme. Agriculture is singled out as one of the five target sectors

Table 3.2 Contrasting styles of agricultural and environmental policy regimes

	CAP	Environmental policy
Age	*Old:* set up in 1958	*Recent:* commenced in 1970s, codified in the SEA in 1987
Legal basis	art. 38-47 EC Treaty	art. 130r-130t EC Treaty
Nature of the competence	*Exclusive:* this common policy is decided by EU	*Subsidiary:* EU intervenes only when its action is more efficient than the one of Member States
Decision making procedure	*Qualified majority*	Programmes: *co-decision* Implementation measures: *co-operation* (except: fiscal measures, land use, water resources management, energy supply: *unanimity,* unless Council decides to opt for *qualified majority*)
Scope	*Sectoral:* but moving towards rural areas policy	*Horizontal:* the principles of environmental policy must be integrated in all the other EC policies
Objectives	- increase in agricultural productivity; - fair standard of living for the agricultural community; - stabilisation of markets; - availability of supplies; - reasonable consumer prices	- preserving, protecting and improving the quality of environment; - protecting human health; - prudent and rational utilisation of natural resources
Principles	- market unity; - financial solidarity; - Community preference	- precaution; - prevention; - rectification at source; - polluter-pays principle
Nature of the instruments	*(mainly) Interventionist:* economic instruments like price support and direct payments	*(mainly) Regulatory:* normative instruments
EU budget share	*Big:* since EU has the exclusive competence of managing the agricultural sector through economic instruments	*Small:* due to the subsidiary, horizontal regulatory characteristic

Source: European Commission (1997a).

means of intervention in themselves can present (or be used as) obstacles to the effective integration of the two policy fields.

The two policy fields also differ significantly in their openness to outside influences and pressures. Much of the public pressure for change in agriculture on environmental grounds has come from organised environmental interests, mainly based in northern Europe and primarily concerned with the intensification of agri-

culture and the ecological consequences. To a certain extent, such concerns have
also been taken up by some of the northern Member States' governments. However, while the development of European environmental policy has been fairly
accommodating of the environmental movement, European agricultural policy
making has resisted the direct involvement of environmental interests until very
recently (Baldock and Beaufoy, 1993). The closed nature of national and European agricultural policy communities typically excludes from decision-making
those   organisations   not   directly   involved   in   implementing   policy
(Hervieu and Lagrave, 1992).

The environmental pressures upon agricultural policy makers have been indirect ones. In particular, the development of EC environmental policy has
gradually impinged upon agriculture in ways which have increasingly raised
questions concerning the interactions of agricultural and environmental policies.
For example, the Drinking Water Directive (80/778), proposed in 1975 and finally agreed in 1980, was intended to standardise water quality norms across the
Member States in order to protect human health. Sixty-two standards (or
'parameters') for different substances were laid down along with guidelines for
monitoring water quality. The implementation of the Directive revealed previously unpublicised contamination of drinking water supplies in various parts of
Europe, including from agricultural pesticides and nitrates. This widespread evidence of agricultural pollution and the rising costs of remedial action triggered a
second round of environmental legislation to attempt to tackle the problems at
source. The Nitrates Directive (91/676), intended to alleviate major sources of
water contamination by nitrates, introduced a range of measures designed to reduce leaching and run-off from farmland. There has been no parallel EC measure
to curb pollution from pesticides but there are limits on pesticide residues in food
specified in EC Regulations. Over time, responsibility for the authorisation of
pesticides is being transferred from national to the Community level. Under
Directive 91/414 on the registration of plant protection products, the Drinking
Water Directive standard for pesticide concentrations was introduced as an important criterion in the 'uniform principles' at the heart of the EC authorisation
process.

The other EC environmental measures of particular relevance to agriculture
include the Environmental Assessment Directive, the Birds Directive and the
Habitats Directive. While these have not impinged directly on agricultural policy
to any large degree, they represent an increasing body of environmental legislation influencing agricultural management. The biodiversity of much of the
European Community is found on, or adjacent to, farmland, which accounts for
more than 40% of the total land area in the Community, and is thereby considerably affected by agricultural management and practices. Work on birds, which
have been studied in greater detail than other wildlife, suggests that, at a European scale, agricultural habitats have the highest overall species richness of any
category of habitat (Tucker, 1997).

With the advent of agri-environmental policy, and particularly Regulation
2078/92, more direct references to these environmental measures have come into
play within agricultural policy. The relationship between agri-environment

schemes rewarding farmers for appropriate management and environmental leg-islation introducing obligations on Member States and farmers is coming into focus increasingly as 2078/92 becomes a significant policy measure. As DGXI put it (European Commission, 1994, p. 25):

> It is ... important to link agri-environment measures to the implementation of environmental directives, especially to those relating to water, nitrates, birds and habitats. They cover sensitive and vulnerable areas, designated by Member States. When applied to particular areas, they encourage agricultural practices which help protect water, flora and fauna. These directives can only be effectively implemented if supporting measures are perceived as sufficiently attractive.

During the late 1980s and the 1990s, environmental policy makers have thus looked to the agricultural sector, not only as a source of environmental pressures, but also as a potential mine of opportunities for achieving policy change and redirecting resources towards appropriate management of the rural environment. Agricultural policy adjustment has been given increasing priority as a target in several national environmental agencies and ministries. There has thus been a growing recognition of the need to integrate agricultural and environmental policy.

In analysing integration processes it is necessary to be aware of the models that policy makers have of the interaction between the two policy fields. Two models of the relationship between non-market environmental effects and agricultural production have gained currency in both academic and policy circles. They are readily counterposed in policy debate:

- The 'impact' model, where environmental impacts or negative externalities are directly associated with agricultural activities, especially input use (e.g. more fertiliser use leading to more pollution, or headage payments stimulating overstocking). The model portrays an agriculture operating largely in opposition to the environment. Therefore a reduction in the intensity of production will lead to an improvement in environmental quality. It follows that policies to improve the environment should restrict agricultural activity.
- The 'public goods' model, where environmental attributes or positive externalities take the form of jointly produced public goods alongside production (e.g. the pastoral landscapes maintained by grazing systems, or the management of hedgerows that encourage wildlife). The model is premised on agricultural systems or farming practices that have co-evolved with the environment, often over substantial periods of time, to the extent that there is a close interrelationship between the valued characteristics of the environment and certain features of the farming system. The 'public goods' model therefore suggests more complex and indeterminate relationships with production: it assumes that different levels of intensity will lead to different mixes of environmental quality. The implications are less clear-cut, but often imply that policy should support agricultural systems, especially where the major envi-

ronmental threat arises from a decline or abandonment in agricultural uses. Environmental loss may not always follow from abandonment, however: for example, on moorland or heathland the low intensity grazing role played by domestic animals may be replaced by that of wild animals such as deer or rabbits.

While the 'impact' model tends to be used in assessments of more intensive production systems and such non-market effects as pollution and soil erosion, the 'public goods' model tends to focus on more extensive systems and on such non-market effects as management of landscape and wildlife habitat. The two also point to different strategic responses: the 'impact' model emphasises a combination of market mechanisms (including price liberalisation) and environmental regulations to achieve an efficient yet sustainable agriculture; whereas the 'public goods' model emphasises a combination of market mechanisms and incentives to support conservation-oriented practices.

The two are of course not necessarily incompatible; they are complementary in so far as they highlight different facets of the relationship between agriculture and the environment. In most contexts, indeed, there can be elements of both. For example, in intensive arable areas, the management of field boundaries is more appropriately considered from the perspective of the 'public goods' model. On the other hand, in extensive systems overgrazing of fragile environments exemplifies the 'impact' model.

Changes in agricultural technology or broader farming systems, perhaps induced by shifts in policy or prices, may entail alterations in the underlying relationship between agriculture and the environment and affect the relative relevance of the two analytical models. For example, overgrazing of moorland by sheep would accord with the 'impact' model: a reduction in the level of bought-in feed might be assumed to lead to more extensive and less damaging grazing. But if there were also a significant reduction in labour input, such farming systems could go over to a low level of management such as 'ranching', with consequent environmental losses arising from both under- and overgrazing. The lack of incentive to maintain environmentally appropriate practices would thus accord with the 'public goods' model. The occurrence of such system changes (as well as more obvious movements up or down a spectrum of intensity of inputs) underlines the coarse and sometimes unpredictable impacts of such macro interventions as price adjustments, and points to the need specifically to maintain desired farming systems or practices, to achieve environmental objectives.

These different models have influenced policy in different contexts. The pressures for price liberalisation that shaped the 1992 CAP reform were informed by the 'impact' model. This reflected the dominant perspective on the environmental problems of modern agriculture in the leading liberalising countries, such as Australia, the USA and the UK. It also reflected the judgement of economists and many environmentalists that the CAP's high and protected prices encouraged the over-use and inefficient use of inputs leading to problems of pollution and over-intensification (see, e.g. Bowers and Cheshire, 1983). As DGXI argued:

The integration of environmental considerations into agriculture should be relatively easy ... because the necessity to reduce over production corresponds to the environmental objective of reduced intensity of land use (European Commission, 1994, p. 27).

A similar outlook was behind the extension of EU environmental legislation and the polluter-pays principle to cover various agricultural activities seen to pose unacceptable damage and to be in need of restraint. Within each policy field, though, the 'public goods' model has also gathered some currency. The agri-environment Regulation, as part of the 1992 reform of the CAP, embodied this model. There has also been a growing appreciation in environmental policy circles of the positive role that farming plays in the management of the countryside (Beaufoy *et al.*, 1994).

PRESSURES FOR ENVIRONMENTAL INTEGRATION

Concern about the environmental impact of agricultural policy at a European level can be traced back to the 1980s, having arisen from growing debate in several Member States over adverse changes in the countryside and increasing pollution. Environmental NGOs, the media and independent voices were central in constructing an environmental critique of the CAP, while environment ministries in most European countries remained cautious in confronting often larger and more influential agricultural ministries. Some Member States became active in pursuing usually limited environmental objectives within the CAP from the mid-1980s. The British Government, for example, played a prominent role in pressing for the first CAP measure to legitimise agri-environment payments to farmers in 1985 (Baldock and Lowe, 1996).

At a European level, the influence of DGXI was growing during the 1980s but its capacity to engage in a strategic dialogue with DGVI was limited, especially in the absence of any formal commitment to environmental integration in the Treaty. Nonetheless, the importance of agriculture as a force shaping the European environment was recognised by DGXI from a relatively early stage, as can be seen from the successive 'Action Programmes for the Environment' which provided a vision and structure for forward planning in European policy from the early 1970s onwards.

The ECs very first Action Programme for the Environment, published in 1973, recognised the need to tackle emerging problems of agricultural pollution (O.J. C112, 20 December 1973) - a sentiment reiterated 4 years later in the Second Action Programme (O.J. C139, 13 June 1977). More ambitiously, the Third Action Programme for the Environment stated the need to:

... promote the creation of an overall strategy, making environmental policy a part of economic and social development, (resulting) in a greater awareness of the environmental dimension, notably in the field of agriculture (and) ...

enhance the positive and reduce the negative effects on the environment of agriculture (O.J. C46, 17 February 1983).

These intentions were echoed in the Commission's 1985 Green Paper on the future of the CAP which - following interventions by the Environment Directorate in response to drafts from the Agricultural Directorate - departed from previous such documents in including a section which proposed that agricultural policy should 'take account of environmental policy, both as regards the control of harmful practices and the promotion of practices friendly to the environment' (European Commission, 1985).

The legal requirement to integrate environmental protection into other EC policy areas was established in 1987 by the Single European Act. The subject of environmental integration was first given prominence in the Fourth Environmental Action Programme of 1987. Section 2.3 states:

> It will accordingly be a central part of the Commission's efforts during the period of the Fourth Environmental Action Programme to make major progress towards the practical realisation of this objective - initially at the level of the Community's own policies and actions; secondly at the level of the policies implemented by Member States; but as soon as possible in a more generalised way so that all economic and social developments throughout the Community, whether undertaken by public or private bodies or of a mixed character, would have environmental requirements built fully into their planning and execution.

Initially, emphasis was to be placed on the Community's own policies and, to this end, the Commission was to 'develop internal procedures and practices to ensure that this integration of environmental factors takes place routinely in relation to all other policy areas'. Action to give effect to this promise was not forthcoming until after the publication in 1992 of the Fifth Environmental Action Programme (5EAP), and the formal strengthening of the integration requirement in the Maastricht Treaty.

The 5EAP, commencing in 1993, marked an important change of direction for the Community's environmental policy. Previous Action Programmes had taken the form of lists of proposed legislation often selected in response to events, whereas the 5EAP attempted to address the fundamental causes of environmental degradation as a means of creating a more sustainable economy and society.

The 5EAP set the strategy for the EU's environmental policy until the year 2000. It was focused on ten major environmental problems or themes, and five economic sectors which make a significant contribution both to their creation, and by the same token, their solution. These were agriculture, industry, energy, transport and tourism. For each of the themes and target sectors, the programme presented tables setting out policy objectives, the instruments and timetables for achieving them, and the key actors from whom action is required, including the EU, Member States, local authorities and industry.

Fundamental to the Programme was the principle that the environment must be integrated from the beginning into all the policies and actions of industry, government and consumers, especially in the target sectors. Other important features included a recognition that changes in society's patterns of behaviour must be achieved in a spirit of shared responsibility among all key actors, including central and local government, public and private enterprise, and the general public (as both citizens and consumers). It also emphasised that the range of policy instruments to be applied to the solution of environmental problems should be broadened beyond traditional 'command and control' legislation, to include economic instruments, voluntary agreements, and better information and education to enable the public to make more informed choices.

A further step towards integration was achieved in the Amsterdam Treaty, which was ratified in May 1999. For the first time, 'sustainable development' has been made an explicit objective of the EU, now being included in Article 2. The requirement to integrate the environment into other EU policy sectors has been transferred from Article 130r(2) to form a new Article 3d, clearly applying to all policy sectors. A declaration attached to the Treaty, although not binding in itself, commits the Commission to undertake 'environmental impact assessment studies when developing proposals that might have significant environmental implications'. This has been linked to a re-examination of the Commission's own internal procedures for integrating environmental considerations into EC policy initiatives. The original procedures introduced in 1993 were widely regarded as rather ineffective.

The new procedures require each DG to evaluate their own policy proposals for their likely environmental impact. If this is thought likely to be 'significant', a more detailed assessment should be undertaken and published in the proposal. Every year the Commission should review progress towards achieving the integration objective on the basis of an independent evaluation and each DG is to draw up its own policy statement on the environment and sustainable development and how it will achieve its commitment to integrate environmental considerations into its policy proposals. Significantly, there is to be an analysis of the environmental impact of funding from the EU budget, about half of which is devoted to agricultural expenditure.

THE RESPONSE OF THE AGRICULTURAL COMMUNITY: THE MACSHARRY REFORMS AND THE ENVIRONMENT

In February 1991, the European Commission concluded in relation to the CAP that 'the in-built incentive to greater intensity and further production, provided by present mechanisms, puts the environment at increasing risk' (quoted in European Commission, 1991, p. 3). When agreeing the MacSharry reform package in May 1992, the EC Agricultural Council declared its commitment to 'make environmental protection an integral part of the Common Agricultural Policy'.

It would be wrong, however, to see in this and subsequent policy initiatives the triumph of environmental concerns alone. Instead, environmental arguments have coincided with other powerful arguments for agricultural policy reform and together these have induced notable changes. The chronic funding problems of the EC, the strain placed on the already overstretched budgets by the accession of the southern European states, the mounting costs and public scandal of burgeoning agricultural surpluses, and rising international opposition to the dumping of surpluses on world markets, have demanded consideration of means of curbing overproduction and the public costs of farming supports. Thus, some agricultural policy makers have responded to environmental concerns, not necessarily through any deep convictions, but because of a perceived coincidence between the aims of environmental improvement and the need to reduce agricultural output, thereby contributing to the alleviation of surplus and budgetary problems. At the same time in northern Europe, farming leaders, in a context of chronic oversupply of staple products and falling farm incomes, have begun to look to the provision by farmers of environmental 'products', in order to underpin or renew their claims for public support.

In the Commission's proposals for the package of CAP reforms agreed in 1992 the first objective of the CAP was thus reshaped as follows:

Sufficient numbers of farmers must be kept on the land. There is no other way to preserve the natural environment, traditional landscapes and a model of agriculture based on the family farm as favoured by the society generally (European Commission, 1991, pp. 9-10).

The MacSharry reform was clearly motivated by the need to restrain output of certain key commodities, to contain the rising costs of the CAP and to allow the Community to reach an accommodation with the US in the GATT negotiations in which agricultural policy figured prominently. By changing the balance of the CAP so as to reduce the role of price support and provide farmers with a larger proportion of their income through direct subsidies, some of the pressures for continued expansion were curbed. At the same time, a new debate about the justification for compensation payments was inevitable, especially as these were paid quite separately from the new generation of agri-environment payments which became obligatory for all Member States from 1993 onwards. With transfers from the State to the farming community now more explicit and subject to international as well as domestic scrutiny, the relevance of environmental concerns to the underlying legitimacy of the policy became more apparent.

The 1992 CAP reform was mainly aimed at restoring market balance and improving the competitiveness of the main arable and livestock sectors, through controlling output and lowering of prices. Previously, in the pre-MacSharry period, market and price policies were the main instruments to pursue the objectives of the CAP of increasing agricultural productivity and ensuring a fair standard of living for the farming community. Production levels had not been limited; sugar production was originally the only exception, with quotas on supported production as a core element of its common market organisation. In the 1980s the potentially unrestrained increase of agricultural production came to an end with the introduction of dairy quotas and the 'stabilisation scheme' for cereals.

The level of regulation, especially for arable crops, increased substantially as a result of the 1992 reform. This was achieved mainly by setting limitations on rights to support payments in livestock and arable sectors and introducing quasi-compulsory set-aside that obliged most cereal farmers to withdraw a proportion of their land from production. In addition, reductions were introduced in guaranteed output prices. For this, farmers were compensated: in the arable sector by area-based direct payments; and in the livestock sector by increases in headage payment rates. Eligibility for this aid was restricted to the area normally used for arable crops before the reform and limits were placed on the total number of sheep and beef cattle on which claims could be made. There were also new limits on the number of livestock per hectare on which subsidy could be claimed - primarily to limit expenditure.

The main objectives of the 1992 reform were to control output, introduce a measure of decoupling of farm supports on production and improve market balance. Nonetheless, it was expected that there would be indirect environmental benefits and there were also efforts to integrate some environmental requirements into the CAP. Thus it was anticipated that the changes in market and price supports could also have effects on the use of inputs such as inorganic fertilisers and should therefore stimulate less intensive production methods. In addition, an element of environmental conditionality was introduced for set-aside payments and governments had a new power to introduce cross-compliance for beef and sheep headage payments if they wished to do so. Furthermore, measures were adopted to encourage more directly environmentally friendly farm practices by the establishment of an agri-environment regulation as one of the so-called accompanying measures to the reformed CAP.

Expectations that the 1992 market and price reforms would significantly alleviate pressures on the environment have not been fulfilled. Overall the usage of agro-chemicals has not been greatly affected, as the Commission itself has conceded (European Commission, 1997b). In part this is due to the unexpectedly high level of cereal prices following the reform. Obligatory set-aside was another production-control measure that was expected to be environmentally beneficial by diverting land out of production. Undoubtedly, it can have environmental benefits but this depends largely upon how it is managed. Whereas set-aside managed for conservation objectives can deliver a wide range of benefits (such as improved biodiversity, habitat restoration and reduced water pollution), its management for agronomic or economic expendiency can be environmentally damaging (Firbank, 1997). Council Regulation 2293/92 required Member States to 'apply appropriate measures which correspond to the specific situation of the land set aside so as to ensure the protection of the environment' (Article 10). Potential benefits, though, have been reduced by the lowering of the set-aside obligation (from 15% in 1993/94 to 5% in 1998/99), by the failure of some Member States to establish rules for the appropriate care of set-aside, and by revisions to the regulations that have increased the scope for non-food production (e.g. industrial oilseeds) on set-aside land.

Turning to the reforms to the livestock sector, there were hopes that the new stocking limits applying principally to the increased headage payments for beef animals would provide some impetus towards extensification. A maximum stocking density of 2 Livestock Units (LUs) per hectare was applied to the basic beef premia, with an additional extensification premium available for those farms with densities below 1.4 LUs ha^{-1}. However, these limits applied only to the number of animals for which premia were claimed and not to the actual number on the holding. In any case, small producers were exempt from the limits. For many other farms the stocking rates were too high to bite. In practice, therefore, the measures have had little effect in encouraging the extensification of production. The limits of 2 and 1.4 LUs ha^{-1} have also been critisised for not reflecting the wide variation in carrying capacity of grazing land and for being set too high for environmentally beneficial management. Changes to both the beef and sheep regimes and the Less Favoured Areas regulations permitted Member States to withhold headage payments in cases where damage to the environment was occurring, but these measures have received limited implementation in only a couple of countries. There has rightly been criticism of the inadequacy of production control measures such as livestock density maxima to serve an environmental function as well. However, agricultural ministries potentially had at their disposal additional interventionist measures to protect the environment which they chose not to use.

No environmental conditions were formally applied under the MacSharry reforms to the other commodity regimes, such as milk, wine, tobacco and sugar. As the Commission admitted in its interim review of the implementation of the 5EAP: 'the CAP reform in 1992 did little to systematically integrate environmental concerns' (European Commission, 1995, p. 34). Indeed, it judged that of all the 5EAP's target sectors 'integration of environmental considerations ... is least apparent in agriculture' (p. 36).

Undoubtedly, by far the most significant component of the 1992 CAP reforms from an environmental perspective was the new agri-environment Regulation. Regulation 2078/92 is an aid scheme which aims to 'encourage farmers to make undertakings regarding farming methods compatible with the requirements of environmental protection and maintenance of the countryside, and thereby to contribute to balancing the market; whereas the measures must compensate farmers for any income losses caused by reductions in output and/or increases in costs and for the part they play in improving the environment'.

A number of evaluations have been conducted on the implementation of 2078/92 (De Putter, 1995; BirdLife International, 1996; Whitby, 1996; Baldock, 1997; European Commission, 1998; Buller *et al.*, 2000). This is not specifically our purpose. Given our focus on the integration of environmental objectives into the CAP, we are concerned to assess how effective the agri-environment Regulation has been in addressing the objectives of relevant EU environmental legislation.

LINKAGES AND TENSIONS BETWEEN REGULATION 2078/92 AND THE EC BIRDS, HABITATS AND NITRATES DIRECTIVES

In Article 1 of the Agri-Environment Regulation 2078/92, the objectives of the measure are set out. One of these is to 'contribute to the achievement of the Community's policy objectives regarding agriculture and the environment'. However interpreted, these objectives must include those embodied in the primary EC environmental measures of relevance to the farming sector. These include the Nitrates Directive (91/676), the Birds Directive (79/409) and the Habitats Directive (92/43). Each of these measures requires governments to take effective action to meet certain obligations spelled out in the Directives; the legislation is not directly binding on farmers themselves. For example, the Nitrates Directive requires governments to identify 'nitrate vulnerable zones' where water is vulnerable to contamination by nitrates and then to adopt measures designed to prevent pollution arising from agriculture, particularly within these zones, including mandatory restrictions on farmers' application of fertilisers and manure to the land. While governments have a legal obligation to comply with the Directives within a given period of time, which ranges from two to ten years, many have failed to stick to the timetable in practice. Consequently, the process of implementation of these measures has extended over a considerable period of time. As of July 1999, infringement proceedings were open against 12 Member States, 8 of which faced proceedings in the European Court of Justice, for non-compliance with the provisions of the Nitrates Directive. Even the Birds Directive, agreed back in 1979, is still not implemented fully in several Member States and court cases seeking to improve the level of compliance are still commonplace for this Directive, 20 years on.

In principle, agri-environment schemes provide a means whereby governments can provide farmers with an incentive to adopt practices which are more sensitive environmentally and contribute to the implementation of these Directives. Schemes are to have 'positive effects on the environment and the countryside'. By 1997, it was estimated by the European Commission that about 20% of farm holdings in the EU were participating in a voluntary agri-environment scheme and a similar proportion of the total agricultural land had been enrolled. However, Regulation 2078/92 was initially presented mainly as a farm income aid with environmental justifications. Formally it is intended to finance schemes with three objectives. These are to:

- accompany the changes to be introduced under the market organisation rules;
- contribute to the achievement of the Community's policy objectives regarding agriculture and the environment;
- contribute to providing an appropriate income to farmers (Article 1).

These three objectives taken together have shaped the implementation of the Regulation by DGVI and agriculture ministries. Over time, the environmental objectives of the Regulation have become more prominent and the Commission has become increasingly critical of schemes where payments are not commensurate with environmental obligations. In the measure which superseded 2078 in 1999 (Chapter VI of the Rural Development Regulation 1257/1999) there is no mention of income support, and environmental goals are specified for the support offered to farmers.

While agri-environment schemes are generally intended to maintain or improve environmental standards, the stated aims are often vague. Rarely are the Community's environmental policy objectives referred to. Nor is the relationship between the two entirely straightforward. The Commission's view has been that agri-environment schemes should not be used to provide payments to farmers to comply with their legal obligations to protect the environment. It would be inappropriate to provide payment where a clear obligation already exists, for example not to apply more than a certain level of fertiliser per hectare of land. Payments should be reserved for farmers who voluntarily take on commitments which go beyond the baseline requirements set out in legislation, which apply to the entire farming community. Officials in DGVI in the Commission perceive the agri-environment Regulation as a measure with a clear set of objectives of its own. On this argument the Community's agricultural resources should not be diverted into providing funding for Member States to implement measures to which they were committed irrespective of the availability of Community funding. In sanctioning new aid schemes to farmers they are ever conscious of the propriety of such schemes within the legal framework of the CAP and how the schemes' legitimacy will be scrutinised by international trading partners. They seek to emphasise the public goods model, both to those Member States which are inclined to be generous to their own farmers in paying incentives and to other OECD countries.

Environmental interests do not have the same perspective or preoccupations. NGOs and some environment ministries have been wary of agri-environment schemes with rather loose or imprecise aims and have expressed concern about the extent of environmental benefit being obtained. Schemes aimed at specific nature conservation objectives, for example, tend to have relatively modest budgets and often are restricted to small areas, with the consequence that the proportion of funding devoted to nature conservation schemes has been small in most Member States (Baldock, 1997). In the light of the failures to implement the relevant EC environmental measures thoroughly, environmental interests have looked for more agri-environment funding to be diverted into schemes which impose more specific or exacting requirements on farmers. For example, BirdLife International has argued that:

> [the Regulation] provides Member States with a tremendous opportunity to introduce measures which will benefit species and habitats of high conservation value and, more importantly, meet the requirements of European legislation such as the Directive on the Conservation of Wild Birds ... it is through poor design and implementation of agri-environment programmes, rather than any fundamental criticism of the legislation itself, that many Member States are failing to do this (BirdLife International, 1996).

It is possible to perceive some agri-environment schemes, such as the 'Prime à l'herbe' in France, as a mechanism for assisting farmers to continue with present practices largely unchanged, thereby missing an opportunity to raise standards, such as improving habitat management for birds. Similar observations could be made of other basic agri-environment programmes such as the Austrian ÖPUL and the Finnish GAEPs schemes. On the other hand, few schemes under 2078/92 have been targeted towards converting intensive practices to extensive farming which might have alleviated the sort of pressures that the Nitrates Directive is meant to address. Moreover, basic agri-environment schemes have hardly touched intensive agricultural regions in most Member States, largely as a result of competition from other CAP aid schemes.

Environment officials have been concerned that the absence of Community funding for sometimes demanding measures at farm level has inhibited compliance with the key environmental Directives. DGXI has argued that:

> It is ... important to link agri-environment measures to the implementation of environmental directives, especially to those relating to water, nitrates, birds and habitats. ... These directives can only be effectively implemented if supporting measures are perceived as sufficiently attractive (European Commission, 1994, p. 25).

This concern extends also to environment ministries in Member States. For example, agreement on the Habitats Directive (92/43) was held up for some time during 1991 while certain Member States pressed their concerns about the cost implications. The Spanish government in particular pursued the point that, as their country was disproportionately rich in endemic species and threatened or endangered habitats of Community interest, significant costs would fall upon them from the implementation of the Directive. If the rest of the Community wanted them, as one of the poorer Member States, to preserve such a major part of the European natural heritage, then there should be a contribution to the costs. The Spanish therefore held out for an express reference to Community co-financing in the Directive. This was resisted by other governments, including the UK, which argued that Member States should shoulder the domestic consequences of agreed environmental legislation drawing on aid from the Structural Funds if appropriate. The Spanish got their way but no explicit new funding source has transpired in practice and Member States must prevent damage to priority sites, even in the absence of Community funding (Sharp, 1998).

The use of CAP funds to harmonise farming practices with environmental requirements is not problematic for environmental lobbyists and officials who look to agriculture to put its own house in order. Existing production subsidies under the CAP are seen to be behind the pressures on the rural environment. It therefore seems logical and apt to channel CAP resources instead into agri-environmental schemes that would tackle the acknowledged environmental problems of contemporary agriculture and help establish farming systems that would deliver accepted environmental benefits.

The question of paying farmers to comply with legal obligations remains sensitive: the polluter-pays principle has often been ignored in the agriculture sector and Regulation 2078/92 can be misused if payments are available to farmers who fail to meet the basic standards of pollution control specified in legislation. However, this danger should not obscure the fact that it is open to Member States to meet many of their obligations under EC environmental Directives by means of their own choice of policy instruments. These can include incentives as well as restrictions. If a Directive creates an obligation on the Member State it may be entirely appropriate to offer farmers incentives to achieve the required result provided that the targeted farmers choose to participate and adopt the necessary forms of management. Indeed, it is difficult to imagine how certain aspects of implementation can be achieved without incentive payments. If a farmer is required to make substantial changes in management to improve habitat quality, by raising the water table for example, incentives would appear the best way forward.

Most Member States, though, have avoided making direct linkages between the objectives of national, regional agri-environment schemes and EC environmental Directives, despite the difficulties in implementing them. Although a few schemes have been targeted particularly at nature conservation areas, including Birds Directive sites, no Member State has launched major national schemes intended to assist compliance with the three key directives. DGVI may have opposed schemes of this kind if proposed, but nor do they seem to have appealed to agriculture ministries, which usually have lead responsibility for the formulation and implementation of schemes. Typically, the larger schemes have focused on maintaining broadly extensive systems, constraining intensification and abandonment but eschewing tight environmental targets. Within certain regional schemes farmers with environmental obligations in Natura 2000 sites (designated under the Habitats Directive) have actually been refused access to the measures and supports available (Delpeuch, 1997).

CONCLUSIONS

By and large, schemes devised under the agri-environment Regulation (2078/92) of the CAP have not been designed specifically to underpin the implementation of the main EU Directives that relate to the agricultural environment. The chief reason for this is the separation between the two policy fields and their different priorities and constraints. Undoubtedly, EU environmental policy forms part of the broad policy context in which the agri-environment Regulation operates, but the relevant EU environmental Directives do not frame the immediate objectives for the implementation of the Regulation.

From the separation of these two sets of measures and the failure at different levels to co-ordinate them more closely stem some of the shortcomings of each of them. On the one hand, there are the criticisms that environmental objectives have been insufficiently prioritised in the implementation of the agri-environment Regulation, that many schemes have been poorly targeted from an environmental perspective, and that the environmental benefits have not been substantial. On the other hand, several of the main EU Directives affecting the agricultural environment are mired in implementation difficulties. Many of these stem from the opposition of agriculture ministries and of farming groups to possible restrictions on farming practices, and the lack of resources to compensate farmers for the costs involved - where this is justified under the polluter-pays principle.

Of course, there are other quite legitimate goals for the rural environment and management of the countryside, and it is evident that agri-environment schemes do fulfil a variety of regional and national objectives such as accelerating the role of the conversion from orthodox to organic farming in several countries. However, it is right to question whether agricultural officials and ministries in a wide range of Member States should have such a dominant role in laying down the key environmental objectives for agriculture. In detaching agri-environment policy from the delivery of environmental policy, suspicions are aroused of a policy community that is seeking to preserve its autonomy.

The complementarity between the agri-environment Regulation and the relevant environmental Directives is significant not only because of the formal objectives of the Regulation but also because of the limited degree to which an environmental dimension has been integrated into the rest of the CAP. Environmental concerns were not one of the primary reasons for the 1992 reforms. At their heart was a policy of reducing price support and compensating farmers with more direct income support. One aim was to remove the incentive for ever-increasing production levels. It was anticipated that this would alleviate pressures on the environment. Such thinking drew on the 'impact model' of the relationship between agriculture and the environment. The very uncertain environmental consequences of the 1992 changes to commodity supports reveal the limits of this model. The environmental damage sustained under the CAP in recent decades will not necessarily be reversed by cutting farm prices in future. Continuing technological advances, socio-cultural changes in the farming community, the development of rural infrastructure and new farming styles all make an automatic return to 'traditional' landscapes or habitat features or low input farming methods unlikely. Reducing output prices may lead to less intensive agriculture but this may not be of the kind which is required for environmental reasons and it may not occur in the right spatial location.

The environmental significance of recent CAP reforms is often exaggerated. Even in the relatively dispassionate explanation of the CAP and the environment produced by the European Commission in 1997 it is stated that 'One of the central elements of the CAP reform was the encouragement of farmers to use less intensive farming methods, thereby reducing their impact on the environment and cutting the creation of unwanted surpluses' (European Commission, 1997a). As we have seen, direct encouragement to use less intensive production methods was not a hallmark of the reform and the full impact of reductions in institutional prices was rather different to that originally expected. While it is entirely legitimate to point to the potential environmental gains of a decoupling strategy, this needs to be approached with due caution. Furthermore, opportunities were missed to amend the commodity regimes within the CAP in 1992 to provide more direct assistance for low intensity production. Further progress was made in 1999, for example by introducing the concept of environmental cross-compliance in a more prominent way but there remains scope for substantial further amendment of the commodity regimes if the claims for environmental integration are to be justified.

By contrast, both the Commission and Member States have pursued the development of agri-environment programmes on a scale which would have been difficult to forecast prior to 1992. This has given substance to the 'public goods' model within the CAP, reinforced in 1999 when the objectives of the agri-environment element of the new Rural Development Regulation were tightened and focused clearly on the environment rather than farm incomes. As in 1992, the magnitude of change was exaggerated, with Dr Fischler emphasising the importance of rural development as the 'second pillar' of the CAP while the budget for this critical element of European policy was tightly restricted. ˙

In pursuing integration, there is a requirement for a combination of effective basic regulation, more far reaching incorporation of environmental concerns into the CAP commodity regimes and the provision of adequate and well targeted incentives for providing environmental services. All three elements of this strategy have a central part to play and it is insufficient to rely on one of them alone.

Hey (1997) distinguishes three types of approaches to integration that different sectors have adopted: defensive, indirect and active integration. *Defensive integration* attempts to contain and offset possible environmental side-effects arising from the policies already being pursued in the sector. *Indirect integration* arises where existing sectoral policies do give rise to positive environmental benefits but largely as unintended side-effects. *Active integration* occurs where planned environmental targets, objectives and policy instruments are adopted within a sector. Such embedding of environmental integration into sectoral decision making processes is likely to require modifications to organisational structure. So far, it would seem, the integration of environmental objectives into agricultural policy has involved varying degrees of defensive and indirect integration.

REFERENCES

Baldock, D. (1997) Lessons to be learnt from the implementation of Regulation 2078/92.
    In: Bennett, G. (ed.) *EU Expert Seminar on Agriculture and Natura 2000.* Ministry of
    Agriculture, Nature Management and Fisheries, The Hague. Proceedings.
Baldock, D. and Beaufoy, G. (1993) *Plough On! An Environmental Appraisal of the CAP.*
    A report to WWF UK. Institute for European Environmental Policy, London.
Baldock, D. and Lowe, P. (1996) The development of European agri-environment policy.
    In: Whitby, M. (ed.) *The European Environment and CAP Reform.* CAB International, Wallingford, pp. 8-25.
Beaufoy, G., Baldock, D. and Clark, J. (eds) (1994) *The Nature of Farming: Low Intensity
    Farming Systems in Nine European Countries.* Institute for European Environmental
    Policy, London.
BirdLife International (1996) *Nature Conservation Benefits of Plans under the Agri-
    environment Regulation (EEC 2078/92).* BirdLife International, Sandy, UK.

Bowers, J.K. and Cheshire, P.C. (1983) *Agriculture, the Countryside and Land Use.* Methuen, London.

Buller, H., Wilson, G. and Höll, A. (eds) (2000) *European Agri-Environmental Policy.* Ashgate, Basingstoke.

Delpeuch, B. (1997) Natura 2000 and the role of the Habitats and Birds Directive. In: Bennett, G. (ed.) *EU Expert Seminar on Agriculture and Natura 2000.* Ministry of Agriculture, Nature Management and Fisheries, The Hague.

De Putter, J. (1995) *The Greening of Europe's Agricultural Policy: the 'Agri-Environmental Regulation' of the MacSharry Reform.* Ministry of Agriculture, Nature Management and Fisheries and Agricultural Economics Research Institute (LEI-DLO), The Hague.

European Commission (1985) *Perspectives for the Common Agricultural Policy.* Commission of the European Communities, Brussels, COM (85) 333.

European Commission (1988a) *The Future of Rural Society.* Commission of the European Communities, Brussels, COM (88) 501.

European Commission (1988b) *Environment and Agriculture.* Commission of the European Communities, Brussels, COM (88) 338.

European Commission (1991) *The Development and Future of the Common Agricultural Policy: follow-up of the Reflections Paper; (COM (91) 100 of 1 February 1991) – Proposals of the Commission.* Commission of the European Communities, Brussels, COM (91) 258.

European Commission (1992) *Towards Sustainability: A European Community Programme of Policy and Action in Relation to the Environment and Sustainable Development.* Commission of the European Communities, Brussels, COM (92) 23.

European Commission (1994) *Towards Sustainability: Interim Review of Implementation of the European Community Programme of Policy and Action in Relation to the Environment and Sustainable Development.* Commission of the European Communities, Brussels, COM (94) 453.

European Commission (1995) *Progress Report on the European Community Programme of Policy and Action in Relation to the Environment and Sustainable Development.* Commission of the European Communities, Brussels, COM (95) 624.

European Commission (1997a) *CAP Working Notes: Agriculture and the Environment.* Directorate General for Agriculture, Brussels.

European Commission (1997b) *CAP 2000: Situation and Outlook: Cereals, Oilseeds, Protein Crops.* Directorate General for Agriculture, Brussels.

European Commission (1998) *Evaluation of Agri-environment Programmes. State of Application of Regulation (EEC) 2078/92.* DGVI Commission Working Document VII 7655/98 (available on Internet).

Firbank, L. G. (ed) (1997) Agronomic and environmental monitoring of set-aside under the EC Arable Payments Scheme. Unpublished report to MAFF by ITE, ADAS and BTO.

Hervieu, B. and Lagrave, R.-M. (eds) (1992) *Les Syndicats Agricoles en Europe.* L'Harmattan, Paris.

Hey, C. (1997) Integrating the environment into transport policy. In: Liefferink, J.D. and Andersen, M.S. (eds) *The Innovation of EU Environmental Policy.* Scandinavian University Press, Copenhagen.

Liefferink, J.D., Lowe, P.D. and Mol, A.J.P. (eds) (1993) *European Integration and Environmental Policy.* Wiley, Chichester.

Marsh, J. (1991) *The Changing Role of the Common Agricultural Policy.* Belhaven, London.

Mazey, S. and Richardson, J.J. (1992) Environmental groups and the EC: challenges and opportunities. *Environmental Politics* 1, 110-128.

Mazey, S. and Richardson, J.J. (1993) EC policy making: an emerging European policy style. In: Liefferink, J.D., Lowe, P.D. and Mol, A.J.P. (eds) *European Integration and Environmental Policy.* Belhaven, London, pp. 114-125.

Neville-Rolfe, E. (1983) *The Politics of Agriculture in the European Community.* Croom Helm, London.

Sharp, R. (1998) Responding to Europeanisation: a governmental perspective. In: Lowe, P. and Ward, S. (eds) *British Environmental Policy and Europe.* Routledge, London, pp. 33-56.

Tucker, G. (1997) Priorities for bird conservation in Europe: the importance of the farmed landscape. In: Pain, D. and Pienkowski, M. (eds) *Farming and Birds in Europe: the Common Agricultural Policy and its Implications for Bird Conservation.* Academic Press, London, pp. 79-116.

Whitby, M. (ed.) (1996) *The European Environment and CAP Reform: Policies and Prospects for Conservation.* CAB International, Wallingford.

Wilkinson, D. (1997) Towards sustainable development in the European Union? Steps within the European Commission towards integrating the environment into other European policy sectors. *Environmental Politics* Volume 6, Spring 1997, Number 1.

Part II

The Environmental Performance
of the CAP Regimes

The Beef Regime

4

Erling Andersen, Alastair Rutherford and Michael Winter

INTRODUCTION

Beef cattle may have both positive and negative impacts on the natural environment. At a generic level cattle will have an impact on air, soil and water quality, the amenity value of the countryside, landscape quality and diversity, maintenance of the genetic diversity of domestic breeds and resource use. However, this chapter concentrates on the relationship between beef cattle and wildlife habitats in Europe, with particular reference to two recent studies in the UK and Denmark.

Almost all habitats of value to wildlife in both Britain and Denmark are semi-natural and are dependent on continued human intervention to maintain their value and wildlife interest. Extensive farming systems are particularly important in nature conservation terms in the European Union (EU) because they maintain large areas of semi-natural habitat, which are important for many species of wildlife (Bignal and McCracken, 1996). Beef cattle have an important role in these extensive high natural value farming systems, although more detailed data are required on the type of cattle involved and the systems utilised in the management of these habitats.

In Britain, the more extensive beef systems, both breeding and finishing, have strong associations with Less Favoured Areas (LFAs) and regions associated with high natural value farming systems. These areas of low intensity farming, which form a unique farmland biotope, are often highly dependent on grazing regimes involving beef cattle. In Denmark and in lowland Britain, extensive areas of relatively low productivity grassland are found as fragments in the agricultural landscape, often on soils of poor agricultural value, such as Culm grasslands in the west of England or salt-marshes and river valley grassland in Denmark.

A recent study examined the relationships between beef grazing and certain important wildlife sites (Sites of Special Scientific Interest, SSSIs) in England (Winter *et al.,* 1998a). The study experienced difficulty in obtaining good information from local conservation officers regarding the type of beef system and the management of the site, suggesting a poor level of understanding of the relationship between environmental objectives and agricultural management even on designated sites of high ecological value. However, the research did illustrate the importance of beef cattle for maintaining a range of habitat types, most of them also represented in Denmark and elsewhere in northern Europe, and some of the characteristics that make them particularly suitable for this. It was found that beef are preferred over sheep or horse grazing for certain valuable habitats such as coastal grazing marshes, wet grassland/marsh, lowland bog, lowland heath, acidic grassland, neutral grassland, calcareous/neutral and calcareous grassland. Thus, extensively grazed beef cattle are particularly useful tools for the management of sites of high ecological importance. This appears to be due to a number of factors including their grazing habit, trampling activity, hardiness, ability to forage on low quality or rank swards, tolerance to wet conditions and low management requirements. Additionally, one of the main findings drawn from the analysis of the site surveys was the complex nature of management requirements for nature conservation on SSSIs and the farming systems that provide for those requirements.

Some of these findings are similar to findings in a recent study in Denmark gathering information on the knowledge of county officers accumulated in their work on nature conservation. In general terms beef cattle are preferred to sheep for habitat management. Only for management of poor dry grasslands and in special cases were sheep believed to be more suitable than cattle (Buttenschøn and Hansen, 1998).

In the British LFAs, the situation is somewhat more complex. On pure heather moorland, appropriate numbers of sheep are considered better than cattle as cattle tend to damage the heather through trampling (Welch, 1984). Heather moors can be maintained without direct use of cattle, although use of cattle at the heather/grass interface and wherever there is a potential for bracken infestation is generally considered to be helpful (Briggs and Courtney, 1989; Winter *et al.*, 1998b).

THE REFORMED LIVESTOCK SUPPORT SYSTEM POST-1992

With the Common Agricultural Policy (CAP) reform in 1992 the support of beef and sheepmeat production shifted further from price support towards headage payments (Suckler Cow Premium (SCP), Beef Special Premium (BSP) to a maximum of 90 BSPs per holding, and Sheep Annual Premium (SAP)). In 1991, headage payments amounted to 17% of the expenditure on the beef sector under the EAGGF Guarantee section, and this rose to 64% in 1996 (European Commission, 1997b). In addition to reducing the intervention prices and increasing headage payments, support was made dependent on compliance with stocking density rules. These were gradually phased with maximum stocking density successively reduced to a level of 2 Livestock Units (LU) per hectare of forage area by 1996. Furthermore an Extensification Premium was introduced giving farmers stocking below 1.4 LU ha^{-1} a supplementary premium. From 1997 a higher Extensification Premium - the 'Super-extensification Premium' - was made available for farmers with stocking densities of less than 1 LU ha^{-1}. Farmers claiming premiums for less than 15 LU do not have to fulfil the stocking density limits.

It must be stressed that these stocking density limits were 'virtual' rates, calculated for animals receiving premiums and dairy cows. They were a means to limit payment liability for the European exchequer rather than a means for reducing real stocking densities on the ground. Indeed, a farmer might graze non-eligible animals and be well above the stocking density limit and yet still legally receive payments and even Extensification Premiums. Moreover, forage area is a flexible term. It might include crops, which do not have to be fodder crops, but are crops for which the farmer does not receive Arable Area Payments or other EU subsidies. Rough grazing, including common land, and even woodland that receives occasional grazing might be included in the forage area.

The premium level in 1997 was 144.9 ECU for SCPs, 135 ECU for BSPs, 16.9 ECU for SAPs (1996), 36.2 ECU for extensification at level 1.4 LU and 52 ECU for extensification at level 1 LU ha^{-1}. An additional 30.2 ECU could be added nationally to the SCPs in Objective 1 areas, 86% financed by the European Commission. Four Member States - Denmark, Germany, The Netherlands and the UK (except Northern Ireland) - did not grant the national supplement. Besides these premiums additional payments were available for livestock farmers in LFAs. In 1995 the expenditure on head payments in the LFAs amounted to approximately 25% of the headage payments under the Guarantee section. Denmark was the last country to take up this option, implementing the LFA head payments on 27 small islands in 1998.

Apart from the regulation of the husbandry/land ratio two other aspects of the reformed livestock support had a possible environmental effect. First, ceilings, tied to historical references, set an upper limit to the number of available premiums at a regional level, which might influence stocking density, and the intensity of farming. Secondly, there was the possibility within the regulation to attach further environmental conditions to the payment of premiums - this option had actually existed since 1968.

Information on the actual effects of the introduced stocking densities is very limited in most countries. However, several commentators had highlighted the likely limited effect of the stocking densities soon after their implementation (Brouwer and Van Berkum, 1996; Revell and Crabtree, 1996; Winter, 1996). The main arguments were that the stocking densities were set too high, did not reflect actual conditions on farms and were too broad-brush. 'Therefore the consequences for the environment and the landscape are considered to be rather small' (Brouwer and Van Berkum, 1996, p. 88).

THE REFORMED LIVESTOCK SUPPORT REGIME IN DENMARK AND THE UK

In 1997, 103,000 SCPs, 195,000 BSPs and 82,000 SAPs were paid to farmers in Denmark. Apart from the BSPs, these figures had been stable since the reform in 1992. For the BSPs there had been a reduction by almost a third in the number of the premium payments made since 1993. Less than 5% of the BSPs were second premiums, as steer production is almost non-existent in Denmark. Only approximately 19% of premiums (SCPs and BSPs) were supplemented with Extensification Premiums, 40% of these as Super-extensification Premiums, and there were no additional national premiums or, until 1998, LFA head payments in Denmark. The option to use crops eligible for area payments as fodder crops in order to qualify for headage premiums, was exercised over 24,000 ha in 1997 affecting around 10% of the headage payments (Fødevareministeriet, 1997).

Regional ceilings had not had an impact in Denmark. There had been limitations on the possibility to achieve free rights to premiums from the national reserve, but in 1997 just 10% of applicants - 200 farmers - were unsuccessful in their applications to the reserve, although they still had the option to buy rights. Even so, this limitation might in some cases have had a negative environmental effect. For example, if non-livestock farmers in environmentally sensitive areas (ESAs) have difficulty in obtaining free premium rights they may be less inclined to switch to livestock farming. This problem was exacerbated by the freezing of rights in the national reserve due to the BSE crisis. The possibility of preventing the transfer of rights from ESA farms to non-ESA farms through 'ring-fencing' has not been widely discussed in Denmark.

The option of attaching environmental conditions to premiums had not been implemented in Denmark although an intergovernmental report published in 1996 suggested that payments might be linked to environmental regulations such as the rules concerning the handling of animal manure (Landbrugs- og Fiskeriministeriet, 1996). Denmark was also among the countries pushing forward the idea of cross-compliance in the Agenda 2000 reforms.

The introduction of the headage payment system engendered remarkably little debate in the agricultural policy community in Denmark except for a debate over the stocking density rules. The farmers' organisations argued that the integration of beef and dairy production in Denmark, with very intensive production systems for fodder, caused special problems for the Danish farmers (De danske landboforeninger, 1996). Three major proposals to change this had been put forward:

- Dairy farms should be able to claim BSPs for 15 LU, without having to apply the stocking densities.
- Fodder crops in rotation should count twice (1 ha = 2 ha) in the calculation of the stocking densities.
- Fields used for sugar beet should be included partly in the calculations as some of the production is used for fodder.

None of these suggestions had any success, even the first one which was supported by the Ministry of Food, Agriculture and Fisheries but failed to find favour with the European Commission.

Moreover, attempts by some Danish farmers to surmount the stocking density difficulties in an innovative manner were thwarted. In 1997 farmers from the western part of Jutland leased grassland on the island of Saltholm, an unpopulated island with semi-natural vegetation of high natural value just outside Copenhagen.

The farmers had no intention of actually using the land for grazing, but only leased it in order to obtain more forage area and thus entitlements to BSPs. This led to intervention by the Ministry of Agriculture stressing that the entitlements to BSP required active management and use of the land in question, a move which subsequently led to the end of attempts to lease land in Sweden in order to claim BSPs.

Subsequently, the farmers' organisation, Landbrugsrådet, proposed a radical solution to Denmark's difficulties with stocking density rules. It suggested that premiums for dairy cows (and the abolition of dairy quotas) - as proposed in Agenda 2000 (European Commission, 1997a) - should replace BSPs (Landbrugsrådet, 1997). This suggestion was not based on any environmental considerations, but on the assumption that the central problem for Danish agriculture was to ensure fair competition for its specialised, dairy-cow based production.

Only one issue - the export of live calves - had attracted a wider public debate. Denmark chose to implement the so-called 'slaughter model' for the BSP payments, that is premiums could be claimed only for animals weighing at least 200 kg by slaughter. Under the other possibility, the 'farm model', premiums could be claimed when the animals reached an age of 10 months. With many dairy cows in Denmark being light Jerseys, and the stocking density requirements, the implementation of the slaughter model has been a major push factor for the export of live calves to Germany and The Netherlands.

In contrast to Denmark where most beef production is highly intensive and integrated with dairy farming, in the UK there is a wider range of production systems. Although there are environmental concerns over high stocking densities in some lowland grassland areas dominated by dairy and beef production, there is not the same level of concern amongst farmers in the UK over stocking densities as found in Denmark. Thus most of the early discussion within farming circles revolved around relatively modest ways of manipulating stocking density rules in order to maximise premiums entitlements. The fact that much of this discussion revolved around ensuring entitlement to Extensification Premiums suggests just how different were the circumstances of British farmers in this respect. The National Farmers Union (NFU) was quick to point to the possibilities of managing claims so that some farmers might have higher stocking densities than implied by their claims (Winter, 1996). At an early stage too, the Meat and Livestock Commission advised farmers to consider leasing more grassland to reduce stocking density (Meat and Livestock Commission, 1993). Nonetheless, levels of production remained high relative to previous levels. The regional ceiling in England and Wales for BSPs was exceeded each year until 1997. In Scotland and Northern Ireland regional ceilings continued to be exceeded even after this year, but the level of scale back was now greatly attenuated. Another contrast with Denmark is the importance of specialist beef herds in the UK and thus of claims for Suckler Cow Premiums. In contrast to Denmark, therefore, an important aspect of the beef reforms was the introduction of quota, although the allocation of quota was generous with little pressure for extra allocation. Quota trading was generally at a modest level. In 1997, claims for the Suckler Cow Premium were made for around 1,750,000 animals, an increase of 2% on the previous year. Many of the suckler cow herds are located in or around the fringes of upland Britain. Here, environmental concern regarding cattle is focused not so much on intensification but on the crucial role of cattle in providing for balanced grazing regimes. The trend has been to replace cattle with sheep thereby exacerbating the overgrazing problems of the uplands (Winter *et al.*, 1998a, 1998b). A serious consideration raised by the Royal Society for the Protection of Birds (RSPB) is the freezing of agricultural structures caused by the quota system, leading to difficulties in reducing livestock densities for environmental reasons (Egdell and Dixon, 1993). In principle the combination of stocking density rules and the implementation by the UK government of cross-compliance within the beef and sheep sectors in an effort to check overgrazing should have helped. However, in practice the rules whereby overgrazing led to withdrawal of payments (Ministry of Agriculture, Fisheries and Food, 1994) were hard to enforce and legally cumbersome.

It is important also to remember that, in the UK, the beef sector has been hugely affected not only by the 1992 reforms but also by the special measures taken to deal with the BSE crisis. Thus, by the end of 1998, 2.9 million cattle over the age of 30 months had been removed from the human food chain and slaughtered.

THE RESPONSE OF THE FARMERS TO THE REFORMED LIVESTOCK SUPPORT

In this section the impact of the reformed livestock policy is discussed, drawing on findings from two studies. The first, providing data for the UK, was carried out during 1995/96 and included a sample of 389 beef farmers, 125 located in the LFAs (Winter *et al.,* 1997; Winter and Gaskell, 1998). The second study, based on a sample of 101 livestock farmers throughout Denmark with beef production, was carried out as a part of a project on public planning and regulation regarding permanent grassland (Andersen, 1999). Although the two samples have been constructed for different studies and with partly different objectives, the results can be used to compare the effects of the reformed livestock support in the UK and Denmark.

With the introduction of the stocking density regulations to headage payments in 1992, farmers were obliged to provide details of the specific areas of

Table 4.1 Proportion of livestock farmers affected by the stocking density regulation as of 1995

	No. of farmers in sample	Proportion of farmers affected (%)
UK	389	9.6
Denmark	101	37.6

Table 4.2 Proportion of livestock farmers affected by the stocking density regulation (Table 4.1), who had made an overall reduction in stocking density

	Proportion of affected farmers making changes (%)
UK	17.6
Denmark	44.7

land available for fodder for premium animals. Some farmers had stocking densities above the new limits while others were well below. Table 4.1 shows the proportion of farmers affected by the stocking density regulation: just 9.6% in the UK, but 37.6% in Denmark.

The differences are even more striking if we examine the proportion of these affected farmers who had reduced their stocking levels in response to the stocking density rules (Table 4.2). In Denmark almost half of the farmers with stocking densities higher than the rules had made real changes (rather than paper adjustments) to their stocking levels, whereas less than one in five of the affected British farmers had done so.

The proportion of farmers receiving the Extensification Premium (Table 4.3) also varied greatly between the two countries. In Denmark only one in eight of the farmers was able to claim the Extensification Premium, whereas almost two-thirds of the UK farmers were able to do so. Moreover, in the UK fewer than a quarter of the recipients of Extensification Premiums had to adjust their stocking densities in real terms, whereas in Denmark more than one-third of recipients had to make real changes (Table 4.4).

The UK farmers had thus mostly adapted to the stocking density regulation without making significant changes in their management, whereas the Danish farmers had mostly had to make significant farm management adjustments to meet the new requirements. Furthermore, a larger proportion of the farmers affected had made real changes in Denmark than in the UK. Overall the two studies show clearly that the stocking density regulation has struck with very different intensity in the two countries, a finding linked to the more extensive character of the UK beef industry than that of the Danish livestock sector.

Table 4.3 Proportion of livestock farmers receiving Extensification Premium

	No. of farmers in sample	Proportion of farmers receiving Extensification Premium (%)
UK	389	65.2
Denmark	101	12.9

Table 4.4 Proportion of livestock farmers claiming Extensification Premium, who had made an overall reduction in stocking density

	Proportion of farmers claiming Extensification Premium making changes (%)
UK	23.7
Denmark	38.5

The above results are based on data from 1995, where the farmers had to adjust to a stocking density below 2.5 LU ha^{-1} of forage area. Further changes were to have been expected in 1996 when the stocking density had to be adjusted further to 2.0 LU ha^{-1}. In the Danish study 39% of the farmers receiving premiums stocked above 2.0 LU ha^{-1} of forage area. In the UK study 17% of the farmers would have been affected by the changed stocking density in 1996. Also the effects of the introduction of the Super-extensification Premium in 1997 is not included in the analysis.

So what were the actual changes that the stocking density regulation caused? In the Danish study one-third of the farmers making changes had reduced the number of livestock. Half of these had made a general reduction in the number of livestock, whereas the other half, all small dairy farms, had chosen to sell off the male offspring at birth. This livestock was sold to farmers finishing beef indoors at 7-9 months of age and in many cases without receiving headage payments. The majority of the farmers making changes had however chosen to increase their forage area. Taking a closer look at the farmers increasing their forage area, it turns out that more than two-thirds of the farmers in question had chosen to increase their forage area with permanent grassland, while one-third had chosen to increase the forage area with other crops. In both cases the changes included enlargement of the farmed area through tenancies or buying up land.

In contrast, in Britain most farmers did not have to make such major modifications. Nonetheless, a significant proportion of beef farmers (65%) had made subtle 'paper' changes to qualify for Extensification Premiums but very few had made real changes to stocking densities. In general, the impact of the stocking rules has been far more significant in Denmark than in Britain.

THE WIDER EUROPEAN CONTEXT

The density of livestock eligible for support per ha of forage crops on farms with fattening bulls and suckler cows is 1.7 in Denmark, well above the EU-12 average of 0.9, whereas the UK at 0.8 is below the average. However these figures can mask considerable variation within countries. For example, Germany, with an average of 1.2 LUs ha^{-1}, nonetheless has a highly intensive sector so that 42% of male cattle are located on farms with more than 2 LUs ha^{-1} of forage crops. In

France the figure is even higher at 53% of cattle, although national stocking density is only 0.9 LUs ha^{-1}. Denmark has a more homogenous structure with just 8% of male cattle located on farms with more than 2 LUs ha^{-1}. The equivalent figure for the UK is 27%, and the average for EU-12 is 22% (Brouwer and Van Berkum, 1996).

A general picture of the possible effects of the stocking density requirements can also be given by the percentage of cattle premiums that are supplemented by the additional Extensification Premium in the different Member States. As can be seen from Table 4.5, in 1995 62% of the cattle premiums across the EU were supplemented with Extensification Premiums. For nine countries more than half of the premiums received Extensification Premiums. The figures for The Netherlands (10%) and Denmark (15%) clearly indicate the intensive production systems in these countries. It should, however, also be pointed out that a high share of meat stemming from dairy-oriented farms or specialised bull production occurs in Italy, Germany and Belgium as well.

Table 4.5 Extensification Premiums in EU-15 countries, 1995

	Percentage of premiums (SCPs and BSPs) that is supplemented with the Extensification Premium
Netherlands	10
Denmark	15
Belgium	26
Germany	27
Italy	31
Austria	45
Portugal	55
Sweden	61
Ireland	71
UK	72
Finland	76
Greece	76
France	77
Luxembourg	84
Spain	86
EU-15 average	62

Source: European Commission (1997b).

POLICY OPTIONS

As described above, the headage payments and the attached stocking density requirements are believed to have caused relatively limited changes in farming practices. This raises the question of whether the payment system could be improved or changed to enlarge the possibilities for positive environmental effects. Two policy options are discussed below. The first option is to fine-tune the existing system. The second is a more radical option of shifting to area payments in place of headage payments.

Fine-tuning the design of the stocking densities

The limits set under the 1992 reforms - at 2.0, 1.4 and 1.0 LU ha^{-1} - did not reflect the wide variation in carrying capacity of grazing land and were set too high for environmentally beneficial management on many important habitats in both Britain and Denmark. Many beef farms had stocking densities well below 2.0 LU ha^{-1} because of the environmental constraints within which the farms operated. In such cases the limits and Extensification Premiums did not act as a disincentive to further environmentally damaging intensification. Similar problems have been identified on the Spanish dehesas, where traditional stocking densities were around 1.0 LU ha^{-1}. Here farmers were causing overgrazing and environmental damage by stocking up to 1.4 LU ha^{-1} and remaining eligible for the Extensification Premium (BirdLife International, 1997).

The use of environmentally optimal stocking densities has been discussed in some detail by Goss *et al.* (1997), who identify many difficulties in establishing these, but suggest that for northern European lowlands they should be set at about 0.6 to 1.25 LU ha^{-1}, considerably lower than existing thresholds. Goss *et al.* conclude that a unified system of area payments coupled with 'zoning', or dividing the EU into areas of different agricultural and environmental potential, defining the policy objectives for each zone and developing locally appropriate policy measures, provides the most appropriate way to integrate environmental considerations into livestock support systems. However, the difficulty of establishing such a complex land potential assessment for the whole of the Union should not be underestimated, and would be difficult to achieve in practice.

Area payments

The general idea behind area payments is that, in contrast to the headage payments, they would minimise the incentive for the farmer to keep extra animals, as the marginal benefit would be reduced to the market returns rather than the market returns plus the premiums. Three positive effects on the environmental impact might be anticipated:

- the general incentive to intensify production incorporated in the current system would be removed;
- the threat of abandonment would be smaller, as the support is linked to the land and its use;
- low intensity farming systems - few animals/large agricultural area - would benefit if the support were linked to land rather than husbandry.

Furthermore an overall reduction in animal numbers could be anticipated, and the administration of farm supports could be simplified through an amalgamation of the arable area payments and the headage payments.

Goss *et al.* (1997) suggest a three-tiered system with a basic premium based on 'adjusted forage area', a second tier based on payments for certain minimum and maximum stocking densities, and a third tier based on specific environmental management. The calculation of the adjusted forage area is based on a regional redistribution of the current level of expenditure. In order to limit the extent of the reallocation of the support, the adjusted forage area gives higher payments to more productive crops. This approach makes it necessary to freeze the payment for a given piece of land according to the use in a reference year in order to avoid intensification.

One crucial argument for a shift to area payments would be that the current headage payments do not favour extensive production. Table 4.6 illustrates this point for Denmark. The first two rows show, for 1995, the even distribution of support and eligible animals across farms grouped by their intensity of production. The other rows then illustrate different ways of redistributing support on an area basis. A shift to a 'crude' payment per ha of forage area (method A) would not redistribute the support at all. The reason for this is that the farms with higher stocking densities tend to have large forage areas because they also carry a relatively large number of dairy cows. If the proportion of the forage area sup- porting dairy cows were excluded, as in method B, the shift to area payments would favour the most extensive farms. However, the introduction of dairy cow premiums as proposed in Agenda 2000 would reverse this effect, allocating more money to the intensive farms in the calculation.

Table 4.6 Distribution of headage payments (1995) on livestock farms grouped by stocking densities (calculated as LU of cattle and sheep per ha of forage area) - group 1 = lowest stocking density, group 4 = highest stocking density

	Percentage of payments			
	Group 1	Group 2	Group 3	Group 4
Share of eligible LU	16.2	22.9	27.5	33.4
Share of support 1995	16.2	24.9	25.6	33.3
Share of support with area payment based on:				
A. Total forage area	15.7	24.9	25.8	33.5
B. Forage area for eligible husbandry	46.9	29.7	17.6	5.8
C. Permanent grass	38.3	20.1	36.4	11.2
D. 'Natural' grassland	71.5	16.0	5.7	6.8

Data from farm questionnaire, Andersen (1999).

Alternatively, modified versions of area payments could be envisioned. Targeting the environmental resource directly could be an option, but would create an uneven distribution. Whether the area payments were paid on all permanent grass (method C) or on 'natural' grass only - i.e. grassland that is unfertilised and not renewed (method D) - the shift in support would not only increase the support for the most extensive farms, but would also bring about an unjust and abrupt redistribution on the other categories of farms.

The extremes presented in Table 4.6 can, of course, be modified using combinations of headage payments and area payments. Thus the redistribution of the support could be adjusted to a politically more acceptable level. Furthermore the Extensification Premium could be targeted directly to the grassland. This would not change the overall distribution between the different groups of farms radically, but would give the farmers an incentive to increase the grassland area. The Extensification Premium would probably need to be raised for this to have a significant impact.

Based on data from the Farm Accountancy Data Network it has been calculated in a European context that a shift from headage payments to area payments would mean drastic changes in the distribution of the support. Implementation of flat-rate area payments throughout the EU would give a considerable shift in the distribution of the support from Belgium, West Germany, Greece and France to Italy, Portugal and the UK (no data on Denmark is presented). Implementing a national area payment corresponding to the current level of expenditure would also mean change in the distribution of the support between farm types. The more extensively stocked farms would increase the family farm income by 23% in the category 'mainly beef', 11% for 'beef with dairy' and 50% for 'mainly sheep and goats' (Egdell, 1995).

DISCUSSION

Beef farming is a widespread but highly variable activity both within Member States and across the EU. This means that the relationships between production (including other livestock sectors) and the environment are far from uniform and the response of producers to common changes in policy is also highly variable and sometimes difficult to predict.

Grazing beef cattle are very important tools in the maintenance and management of European farmed landscapes and habitats. On important and designated sites, cattle can be managed for specific environmental objectives, but they also have considerable environmental importance in the non-designated 'wider countryside'. However, the relationships between environmental objectives and agricultural management, even on important designated sites, is often poorly understood.

Despite the scale of the 1992 CAP reforms, the legacy of productivist agricultural policy in the beef sector remains strong, particularly with regard to the continued dependence on headage payments as the main support mechanism. Attempts to encourage more extensive production and to reward more environmentally sustainable systems have proved to be exceedingly weak and have largely failed to have any significant impact due to a range of difficulties, not least the poor design of the extensification mechanisms themselves.

These conclusions suggest a number of considerations for the design of more environmentally sustainable policies for the beef sector in the future. As the chapter has shown, the shift to giving the headage payments a more dominant position in livestock support and the introduction of the stocking density rules have had a very limited but generally positive environmental effect on the management of permanent grassland. Being implemented uniformly throughout the Union's territory, this effect does, however, seem to be rather local and it does not challenge the overall impression that the stocking densities have had a limited EU-wide effect.

It cannot be concluded that the stocking density rules were without use as a management instrument for achieving environmental goals, even though the effects so far have been so limited. But the fact that the stocking densities have been set uniformly throughout the EU regardless of the variation in agricultural systems and environmental values has severely limited the extent of positive benefits. Thus stocking densities could be revised to reflect national, regional, local or even farm-level conditions.

In order to attach the stocking densities more directly to the land use, and avoid the calculations of a 'paper-farm', and as an initial attempt to explore the possibilities of area payments, the Extensification Premiums should be targeted more precisely. From an environmental perspective, the key land use issue in regard to livestock farming is permanent grassland. In Denmark, for example, the 1998 level of payments for Extensification Premiums could be transformed to a 'natural grassland payment' of approximately 40 ECU ha^{-1}.

Such a transformation would cross the undefined border between headage payments and agri-environmental measures under Regulation 2078/92. In the specific Danish case, grassland measures have been implemented under Regulation 2078/92. The agreements can be signed at two levels:

- use of fertiliser is allowed up to a level of 80 kg N ha^{-1};
- no use of fertiliser is allowed.

A 'natural grassland payment' attached as an Extensification Premium to the headage payment, as described above, would increase the incentive for farmers to sign agreements reducing fertiliser applications substantially: the difference in the payment between level 1 and level 2 agreements would increase by 60%. On the other hand, the payment would be too low to accomplish land use changes from arable to permanent grassland, which would require a larger part of the budget to be allocated to payments of this type.

For the future, the experience with such a 'natural grassland payment' might prove that some of the objectives of the agri-environmental measures could be achieved by amalgamating them with the headage payments. This would mix the economic power of the headage payments with the environmental intentions of the agri-environmental measures.

Finally, it should be mentioned, although further research is needed, that area payments would probably be received positively by farmers (Goss *et al.*, 1997). The farmers would find such a regime more flexible and more appealing to their agricultural skills. Even though environmental objectives can be achieved the farmers would consider the regime as agricultural more than environmental, which is a crucial element for acceptance.

AGENDA 2000

In the new reforms of the CAP agreed in 1999, the two most important features concerning the beef sector were the redesigning of the Extensification Premiums and the new 'national envelopes' giving the individual Member States control, within certain limits, over the distribution of parts of the support.

The Extensification Premiums have been redesigned giving Member States two options. The first is to operate with one or other of a range of stocking densities qualifying for premiums: either 1.6 to 2.0 LU ha^{-1} (which will be reduced, from 2002, to 1.4 to 1.8) or less than 1.6 (which will be reduced, from 2002, to less than 1.4). In the first range the Extensification Premium will be 33 ECU (from 2002, 40 ECU) and in the second range 66 ECU (from 2002, 80 ECU). Alternatively, the Member States can chose to pay the farmers stocking less than 1.4 LU ha^{-1} 100 ECU. On the one hand, the new ranges look like a step backwards as the incentive to stock less than 1 LU ha^{-1} is being removed. On the other hand, the new ranges might have better extensification effects for farms stocking between 1.4 and 2.0 LU ha^{-1} with the introduction of more stepwise options. The redesign of the Extensification Premiums also reduces significantly the possibility of making 'paper-solutions'. All cattle and sheep on the farm have to be taken into account in calculating stocking densities, not just animals receiving premiums, and the forage area must be grassland or other non-arable fodder areas. More than 50% of the fodder area furthermore has to be 'pasture', a term whose meaning is not quite clear yet. Finally, the redesign of the Extensification Premiums includes a substantial increase in the payments.

The most innovative tool of the reform regarding the beef sector is the national envelopes allowing Member States 'flexibility to compensate for regional differences in production practices and agronomic conditions which might make restructuring difficult, and also aims to encourage extensive production.' (European Commission, 1999). One of the options within the national envelopes is to use them as area payments for permanent grassland as recommended earlier in this chapter. Another option could be to use them as an incentive for the most extensive farms as headage payments for those stocking below 1 LU ha^{-1}, and thus ensuring the continuation of the current payments for these farms.

In general terms, the 1999 beef reforms might have beneficial effects from an environmental point of view. The more flexible design giving the Member States different options does however make it difficult to predict the consequences. Until now it has been the variability of farmers' responses that have made prediction and analysis a complex business. Now there is the additional complication of the discretion opened up to Member States.

REFERENCES

Andersen, E. (1999) Vedvarende græsarealer - landbruget og reguleringen. Unpublished PhD thesis. Royal Agricultural University, Copenhagen.

Bignal, E. and McCracken, D. (1996) The ecological resources of European farmland. In: Whitby, M. (ed.) *The European Environment and CAP Reform: Policies and Prospects for Conservation.* CAB International, Wallingford, pp. 26-42.

BirdLife International (1997) *A Future for European Rural Environment: Reforming the Common Agricultural Policy.* BirdLife International, Brussels.

Briggs, D. and Courtney, F. (1989) *Agriculture and Environment: The Physical Geography of Temperate Agricultural Systems.* Longman, London, 442pp.

Brouwer, F.M. and Van Berkum, S. (1996) *CAP and Environment in the European Union: Analysis of the Effects of the CAP on the Environment and Assessment of Existing Environmental Conditions in Policy.* Wageningen Pers, Wageningen, 171pp.

Buttenschøn, R.M. and Hansen, B. (1998) *Amternes Naturpleje - Rapport fra en Erfaringsopsamling.* Park og Landskabsserien nr. 21. Forskningscentret Skov & Landskab, Hørsholm, 103pp.

De danske landboforeninger (1996) *Beretning 1996.* De danske landboforeninger, Copenhagen, 168pp.

Egdell, J. (1995) *Switching CAP Livestock Support from Headage to Area Payments: The Implications for the Environment and Farm Incomes.* Paper presented at the Annual Conference of the UK Agricultural Economics Society, Cambridge.

Egdell, J. and Dixon, J. (1993) *Proposals for Changes to EC Livestock Policies.* Studies in European Agriculture and Environment Policy No 6. RSPB, Sandy, 46pp.

European Commission (1997a) *Agenda 2000: For a Stronger and Wider Union.* Commission of the European Communities, Brussels.

European Commission (1997b) *CAP 2000. Situation and Outlook - Beef.* Working Document. Commission of the European Communities, DGVI Brussels, 28pp.

European Commission (1999) *Agricultural Council: Political Agreement on CAP Reform.* Newsletter 11 March 1999. Commission of the European Communities, DGVI, Brussels.

Fødevareministeriet (1997) *Jordbrug og fiskeri 1996.* Fødevareministeriet, Copenhagen, 222pp.

Goss, S., Bignal, E., Beaufoy, G. and Bannister, N. (1997) *Possible Options for the Better Integration of Environmental Concerns into the Various Systems of Support for Animal Products.* European Commission, DGXI, Brussels.

Landbrugs- og Fiskeriministeriet (1996) *Betænkning fra Udvalget om Natur, Miljø og EUs Landbrugspolitik.* Betænkning nr. 1309. Landbrugs- og Fiskeriministeriet, Copenhagen, 97pp.

Landbrugsrådet (1997) *Rapport fra ad hoc Gruppen Vedrørende Kvæg.* Landbrugsrådet, Copenhagen, 77pp.

Ministry of Agriculture, Fisheries and Food (1994) *Press release 220/94, 10 June.* MAFF, London.

Meat and Livestock Commission (1993) *CAP Reform - One Year On.* Meat and Livestock Commission, Milton Keynes, 55pp.

Revell, B.J. and Crabtree, J.R. (1996) Policy pressures and responses in European livestock systems. In: Dent, J.B., McGregor, M.J. and Sibbald, A.R. (eds) *Livestock*

Farming Systems: Research, Development Socio-economics and the Land Manager. EAAP publication No. 79. Wageningen, pp. 23-36.

Welch, D. (1984) Studies in the grazing of heather moorland in north-east Scotland II. Response of heather, *Journal of Applied Ecology.* 21(1), 197-207.

Winter, M. (1996) *Rural Politics. Policies for Agriculture, Forestry and the Environment.* Routledge, London, 360pp.

Winter, M. and Gaskell, P. (1998) *The Effects of the 1992 Reform of the Common Agricultural Policy on the Countryside of Great Britain.* Volume 1: Project overview and main findings. Rural Research Monograph Series Number 4. Cheltenham and Gloucester College of Higher Education, Cheltenham.

Winter, M., Rutherford, J.A. and Gaskell, P. (1997) *Beef Farming in the GB LFA - the Response of the Farmers to the 1992 CAP Reform Measures and the Implications for Meeting World Trade Obligations.* Paper presented at LSIRD conference, Nafplio, Greece, 23-25 January, 1997.

Winter, M., Evans, N. and Gaskell, P. (1998a) *The CAP Beef Regime in England and its Impact on Nature Conservation.* English Nature Research Report No. 265. English Nature, Peterborough, 95pp.

Winter, M., Gaskell, P. and Short, C. (1998b) Upland landscapes and the 1992 CAP reforms. *Landscape Research.* 23 (3), 273-288.

The Sheepmeat and Goatmeat Regime

5

Stuart Ashworth and Helen Caraveli

INTRODUCTION

Sheep and goat farming, for both meat and milk production, are found throughout Europe. They are commonly associated with some of the most environmentally sensitive and peripheral areas. In 1992, 75% of all ewes on which ewe premium was claimed were located in the Less Favoured Areas (LFA) (CEC, 1995).

A diversity of sheep and goat production systems can be found across Europe. They cover a wide range of farming situations, from land with poor grazing value to intensively managed land. Furthermore, they occur across an enormous range of climatic conditions. As a consequence, sheep and goat systems continue to be enormously rich in diversity. The breeds used, the product and the management system are all designed to suit local social needs and environmental conditions. This diversity of production systems is in marked contrast to many other sectors, for example the dairy industry where the breed of cow, the management systems and the feeding systems are becoming remarkably uniform across the world.

Nevertheless, although variations in land, climate and cultures in the European Union (EU) lead to many variations in breeds and production systems, three principal sheep and goat systems can be identified according to whether they are primarily aimed at producing meat, or milk and dairy products, or wool and fine fibres. Systems for which wool and fibre production are the primary motivation are rare within the EU. The meat production systems, whose primary output is heavy lamb, are found throughout the EU, but are the dominant form of small ruminant production systems in northern Europe. Production systems whose primary output is milk and dairy products, are predominantly found in Mediterranean Europe. Significant by-products are light lambs sold at weaning at liveweights of less than 25 kg, but a common variant is for the lambs to be weaned and intensively reared to heavier weights, which allows the farmer to be classed as a heavy lamb producer.

It is therefore appropriate to divide sheep and goat production in Europe into two broad geographical categories. On the one hand, northern European systems, that are found in Ireland, the UK, Belgium, The Netherlands, Luxembourg, Denmark, Germany, Sweden, Finland and central and northern France, are characterised by significant use of grassland, by seasonal breeding (with lambing between January and May) and by the production of heavy lamb. Milk production from sheep and meat or milk production from goats are insignificant in northern Europe. On the other hand, southern European systems are found in Portugal, Spain, Italy, Greece and southern France. The range of systems in these countries is more complex than for northern Europe. There are both sheep and goats producing both heavy and light carcasses with a much greater seasonal spread of lambing and kidding than in the north. Dairy systems are also very important.

It is the objective of this chapter to consider how the evolution of the EUs sheepmeat and goatmeat policy has impacted upon this sector and what the consequences have been for the environment. Attention will be drawn to the way in which a common policy has had a diversity of impacts among Member States. Examples will be drawn from the UK, as representative of the northern Member States, and Greece, as representative of the southern ones.

This chapter will, firstly, summarise the significant developments in the sheepmeat and goatmeat regime. Secondly, it will consider the structure of the sheep and goat industry within Europe before finally considering production and structural changes that have taken place in the UK and Greece alongside the developments in the sheepmeat and goatmeat regime, and their environmental consequences.

THE DEVELOPMENT OF THE EU SHEEPMEAT AND GOATMEAT REGIME

The sheepmeat and goatmeat regime was introduced in October 1980 and its evolution can be considered in three stages: 1980-1985, the establishment of the regime and transition to a common base; 1985-1992, widening its scope and cost effectiveness; and after 1992, the implementation of MacSharry reform.

Development of the regime from 1980 to 1985

The regime, as established by Council Regulation 1837/80, set out to create a common organisation of the market in sheepmeat and goatmeat, to stabilise the market and ensure a fair standard of living for sheep and goat farmers. It comprised a common trading system and a common price system covering both sheep and goats. The common trading system operated through the establishment of levies on third-country imports. This effectively prevented imported products reaching the market at prices below those operating in the Community.

The price support system introduced the payment of an annual ewe premium and also the option of using Private Storage Aid and intervention to support the market. The annual ewe premium, commonly called the Sheep Annual Premium (SAP), is the cornerstone of the regime. It was designed to compensate producers for the loss of income resulting from the establishment of a common market organisation. Initially, there was no provision to pay headage premiums to goat keepers. The Private Storage Aid was intended to support the market at 90% of the basic price, while intervention was intended to support the sheepmeat market at 85% of the basic price.

The basic Regulation recognised that there was a need for a transitional period to allow for prices to converge between Member States. To this end, it created five regions for which separate reference prices were established. These prices were based on representative regional market prices in 1979, and would converge over a four year period to the basic price. Thus five different regional annual ewe premiums could be paid. Subsequently, the number of regions was raised to seven with the accession of Greece in 1981 and the demarcation of Northern Ireland from the UK in 1982/83, following problems with policing trade between Northern Ireland and the Irish Republic.

Development of the regime from 1985 to 1992

A review of the regime carried out in 1984 resulted in modifications being introduced in 1986. Council Regulation 3523/85 recognised that in certain parts of the Community sheepmeat and goatmeat production and marketing techniques were very similar. As a result the Regulation required that in these areas the annual ewe premium should be extended to she-goats, but at 80% of the rate paid to breeding ewes to reflect the lower production costs for goats.

In 1988 the Agriculture Council became concerned that the expansion in sheepmeat and goatmeat production which had occurred since 1980, combined with the EUs commitment to Voluntary Restraint Agreements with traditional trading partners, was leading to a situation where the market could become unstable through over supply. It was concluded that it was becoming important not to encourage the further expansion of sheep and goat production (Council Regulation 1115/88). To this end a stabiliser was introduced whereby for every 1% increase in the number of ewes over the December 1987 figure the basic price would be reduced by 1%, thus reducing the level of support.

In 1990 the regime was further consolidated and amended, under a new basic Regulation, Council Regulation 3013/89. Two substantial changes were adopted. Firstly, it was recognised that account must be taken of different specialisations of production systems in the Community. Thus two categories of sheepmeat producer - a heavy lamb producer and a light lamb producer - were recognised. For light lamb producers, defined by the Regulation as 'any sheep farmer marketing sheep's milk products' the premium was reduced to 80% of the full premium rate. Secondly, in order to contain budget costs, provision was made to limit the full rate of the premium to 1000 animals per producer in the Less Favoured and Mountain and Hill Areas and to 500 animals per producer in other areas. Payment of half rate premium was to be made beyond these limits.

Concern that the reduction in support would have unfavourable consequences for the LFA and particularly in those areas where there was no alternative to sheep and goat farming led to the introduction of a further payment to sheep and goat farmers in the LFA in 1991. This payment, made under Regulation 1323/90, has become known as the LFA or rural world supplement, and is paid only in these areas (and is over and above the compensatory allowances paid under the LFA scheme itself, see Dax and Hellegers, Chapter 11).

Development of the regime since 1992

The sheepmeat and goatmeat regime was completely reviewed in 1992, as part of the MacSharry reform of the Common Agricultural Policy (CAP). At the conclusion of the review the Council of Ministers observed that the upward trend in ewe numbers was leading to a substantial drop in prices with serious repercussions on the market balance, and that the increase in production was resulting in a steady increase in support expenditure. They concluded that more severe measures were needed to create a balanced market and control expenditure. As a result Council Regulation 2069/92 was adopted. This Regulation imposed finite limits, or quotas, on the number of animals to be supported. This was implemented at the level of the producer, and based on the number of ewes on which an individual producer claimed premium in a given reference year, chosen from 1989, 1990 or 1991.

Established producers can buy, sell or temporarily lease quota rights, if they so wish. However, to avoid quota being moved from sensitive zones or regions where sheep production is especially important to the local economy, a producer is only allowed to trade entitlement rights within circumscribed areas - a principle which has become generally known as 'ring-fencing'. Most Member States have ring-fenced their LFA as a single area and thus created two trading zones, an LFA zone and a non-LFA zone. The most notable exceptions are France, which designated its whole territory as a single sensitive zone, and the UK, which specified seven separate regions.

Although there is general agreement that the 'ring-fence' principle is valid, there are some concerns that it may result in some environmental damage (CEC, 1996). The concern arises from a view that accumulation of rights within parts of a sensitive zone may lead to localised damage. Equally, the application of ring-fencing prevents the movement of rights to other zones which could potentially benefit from increased sheep numbers, as is the case with some lowland areas of Britain. At the same time, and following from a recognition that difficulties existed in monitoring sheep numbers under the regime, the definition of an eligible ewe was simplified to any animal which has lambed once or is over one year old at the end of the retention period. This made it possible for full premium payments to be received on non-productive ewes.

Agenda 2000 developments

Agenda 2000 proposals make no reference to modification of the sheepmeat and goatmeat regime. Nevertheless, cross-compliance with stocking density calculations applied to extensification payments under the beef regime may influence sheep producers in some Member States. This is likely to occur in those countries, such as the UK and Ireland, where sheep and cattle are co-dominant.

IMPACT ON THE SHEEP AND GOAT INDUSTRY

The most obvious impact of the introduction of the regime has been the rapid growth in breeding sheep numbers within Europe (Table 5.1). However, what Table 5.1 also illustrates is the wide dichotomy in the change in breeding sheep numbers that has occurred between Member States. Thus the impact of the sheepmeat and goatmeat regime has been far from uniform across the EU. Equally, work by Ashworth *et al.* (1997) has shown significant variations in the regional growth in ewe numbers within Member States.

Table 5.1 Index of breeding sheep numbers in the six most significant sheep producing Member States in the EU

	EU-11 [a]	UK	Ireland	France	Spain	Greece	Italy
1996	121	118	220	84	134	86	104
1992	123	127	245	84	152	99	93
1989	123	125	211	92	144	102	106
1986	116	111	135	98	104	103	106
1983	100	100	100	100	100	100	100

Source: Eurostat (1994, 1997).
[a] EU-11 excludes Portugal.

Impact pre-1992

In the UK, Ireland and Spain the increase in breeding ewe numbers between 1983 and 1992 was more rapid than the EU average. However, some caution must be expressed about the accuracy of the Spanish numbers prior to joining the EU in 1986. Ewe numbers in Greece and Italy expanded a little through the mid-1980s but have declined since then. In France the picture is one of steady decline since the early 1980s, except for between 1985 and 1986 when, in common with the other countries, there was a sharp increase. Indeed, the largest increase in breeding ewe numbers in the EU occurred at this time when farmers were coming to terms with the application of milk quotas to dairy cow herds.

The decline in breeding sheep numbers in France may be a reflection of the competitive advantage achieved by the UK and Ireland, and the export opportunities that this created for these two countries. Higher market prices in France for a product of a similar specification and size to that traditionally produced in Britain and Ireland (Table 5.2), and French levels of self-sufficiency falling below 50% (MLC, 1997), make France a very attractive export market for Britain and Ireland. The number of French holdings with sheep has declined by 28% between 1983 and 1993 (Table 5.3) while increases occurred in the UK and Ireland (Eurostat, 1997).

Post-1992

Since 1992 breeding ewe numbers have declined in all Member States, with the exception of Italy. Much of the decline is likely to be due to the modification to the eligibility criteria which has made it possible to receive support on young female sheep intended for breeding, but which are not recorded in the census data as breeding sheep. What the 1992 reform has done, by establishing individual limits on entitlement to support, has been effectively to freeze the size of the sheep flock in the medium term, given the importance of support to the viability of the sheep enterprise. In addition, the ring-fencing of support has reduced the flexibility of the industry to modify its structure. Equally, combined with restrictions to eligibility for support applied to other agricultural sectors, for example Suckler Cow Premium quota, the establishment of sheep quota makes substitution between enterprises, in response to changing market signals and policy initiatives, more difficult.

The regional variations in the growth of breeding ewe numbers (Table 5.1) leads to the consideration of three hypotheses concerning the impact of the

Table 5.2 Average annual market price for lamb (ECU per 100 kg)

	EU	UK	Ireland	France	Spain	Greece	Italy
1996	363	334	308	380	415	392	347
1992	277	224	205	285	329	372	339
1989	342	276	288	310	357	438	427
1986	367	304	297	351	387	434	452
1983	n/a	258	345	426	n/a	497	446
1981	n/a	259	330	367	n/a	494	450

Source: CEC (1996).

Table 5.3 Number of holdings with sheep (x 1,000) and average flocksize per holding

	1983		1987		1993	
	Holdings	Flocksize	Holdings	Flocksize	Holdings	Flocksize
UK	83.7	278	91.4	305	95.5	307
Greece	242.4	41	191.0	57	151.0	67
Spain	160.8	104	161.0	143	120.4	198
France	172.4	76	145.0	83	124.3	84
Italy	170.4	63	176.2	65	150.4	70
Ireland	48.0	59	45.9	94	53.4	112

Source: Eurostat (1994, 1997).

sheepmeat and goatmeat regime on the structure of the sector and its impact on the environment. These are:

- the impact is related to the importance of the sheep and goat industry to the agricultural economy of the region;
- the impact is associated with structural change and is related to the interaction with other agricultural and environmental policies;
- the impact is related to the speed with which commonality of application was achieved between Member States.

Importance of sheepmeat and goatmeat in the agricultural economy

Sheep production tends to be a subsidiary enterprise throughout Europe. Among those producers who produce sheep and goats, almost 60% gain less than one-third of their total agricultural production from this source (Table 5.4). However, variations do occur. In Greece, the holdings with sheep or goats tend to be more specialised, with 25% of these holdings gaining more than 75% of their output from sheep. This is in marked contrast to the UK and Ireland where a considerably greater proportion of holdings have sheep flocks, but less than 10% of these holdings gain more than 75% of their agricultural output from this source. Within countries, too, there are areas where sheep are of greater importance to the total agricultural output, whereas in other areas they add to the enterprise diversity.

Responses to policy, and the environmental consequences, depend upon the significance of the sheep or goat enterprise to the individual farm business and the significance of sheep and goat production to the regional farming economy. Those businesses taking most of their output from a sheep enterprise are likely to react differently to those with a subsidiary sheep enterprise. Equally, farms that are

Table 5.4 Proportion of total agricultural output derived from sheep and goat production on those holdings with sheep and goats in 1991/92 in the six most important sheepmeat and goatmeat producing Member States of the EU

	Share of holdings with sheep (%)	Share of output from the sheep (%)		
		<33	33-75	>75
EU-12	13	58	25	19
UK	46	67	27	6
Greece	24	44	31	25
Spain	13	35	20	45
France	12	60	26	15
Ireland	32	72	24	4
Italy	7	55	35	11

Source: DGVI/A-3/TV OVINS94.WP (8/6/94).

largely devoted to sheep are mostly concentrated in those areas with fewest options, namely the most disadvantaged and environmentally sensitive, and most dependent upon the support payments offered by the sheepmeat and goatmeat policy.

Interactions with other policy measures

For many holdings sheep and goat production is a minority enterprise (Table 5.4). Thus, in many Member States, and particularly the UK and Ireland, the intensity and size of the sheep enterprise are affected by the possibilities available to farmers to switch between enterprises, and thus by the changes made to other commodity regimes than that for sheep and goats. This interaction between commodity regimes was particularly noticeable following the imposition of milk quotas in 1984. With no restriction on sheep or beef cattle support (until 1992), substitution of sheep and beef animals for dairy cattle occurred as the dairy cow policy became increasingly restrictive.

Speed of establishment of a common policy

It was not until 1992 that a truly common policy towards sheepmeat and goatmeat was achieved within the EU, but some intermediate steps had been taken before then. From 1981 to 1984, transition rules applied as Member States moved towards a common basic price and five regional variations in annual ewe premium were paid. Although a common basic price was achieved in 1985, the use of stabilisers and differing regional technical coefficients meant that it was not until 1992 that all Member States received the same level of annual ewe premium. Table 5.2 shows the considerable regional variation in lamb prices in 1981 and thus the growth in the reference price towards a common level based on French prices resulted in different ewe premium payments (Table 5.5). Italy and Greece, for example, had a much slower growth in their reference price than did the UK and Ireland because they had had much higher lamb prices. Thus, producers in the UK and Ireland received greater encouragement to expand, and ewe numbers in these two countries rose the fastest.

However, Italy and Greece were allowed to take the rate of payment applied to France, which had lower lamb prices than they did. This resulted in higher support payments for Italy and Greece than would have been the case if local market prices had been used to determine income loss during the transition period. Thus, while in the early years there was a modest expansion of ewe members in Italy and Greece, France saw a steady decline.

The most significant variation in the implementation of the sheepmeat and goatmeat regime applied in the UK which was allowed to retain its variable premium scheme until the end of 1991. Under this system support was paid on the finished lamb as well as the ewe, so there was considerable incentive to increase the number of lambs produced per ewe. This resulted in considerable growth in the use of more prolific crossbred ewes. These heavier sheep can have a detrimental impact on the grazing resource.

Table 5.5 Annual ewe premium payments (ECU per head)

	1982/83	1984/85	1986	1989	1992	1996
Great Britain	4.42	7.57	10.59	10.33	18.62	16.87
Ireland	10.65	21.30	24.38	21.39	18.62	16.87
France	0.57	5.95	15.43	18.57	18.62	16.87
Spain			7.20	13.00	18.62	16.87

Source: Annual announcements in the *Journal of the European Communities.*

Structural changes in the UK[1]

In the UK, on the one hand, there has been encouragement to introduce a sheep enterprise onto holdings and into areas not previously engaged in sheep production (Table 5.6). On the other hand, those areas where sheep were already important have increased their level of specialisation and hence their degree of dependence upon sheep (Ashworth *et al.*, 1997). Equally, a dichotomy of environmental impacts can result. On the one hand, the introduction of sheep can lead to a greater diversity of grazing patterns while, on the other, increased sheep numbers can lead to overgrazing.

It is clear that EU policy towards sheepmeat and goatmeat has had an impact on the structure of the industry. In particular headage payments have played a significant role in supporting the gross and net profitability of sheep enterprises.

Equally, on those holdings where a significant proportion of output is derived from sheep production, net farm incomes have been sustained because of direct support payments (Ashworth *et al.*, 1997). By sustaining gross profitability and with no restriction between 1981 and 1992 on the number of breeding ewes that would be supported, the sheepmeat and goatmeat policy has encouraged an ex-

Table 5.6 Shifts in livestock keeping in Great Britain

		Proportion (%) of all holdings with		
	Average sheep flocksize	Breeding ewes	Beef cow	Dairy cows
1981	205	33	27	24
1992	239	37	27	17
1996	256	35	26	15

Source: MAFF annual census.

---

[1] Some of the figures given below are for Great Britain rather than the whole UK, i.e. they do not include Northern Ireland.

pansion in both the proportion of holdings with sheep and flocksizes (Table 5.6). In some areas this has led to environmental degradation through overgrazing as the increase in the density of sheep grazing has led to an overall increase in livestock densities (Table 5.7).

Table 5.7 Changes in average breeding livestock grazing pressures in Great Britain

	1981	1985	1995
Grazing Breeding Livestock Units (GBLU) per forage ha	0.55	0.57	0.63
Ewe GBLU per forage ha	0.19	0.22	0.27
Ewe GBLU as per cent of total GBLU per forage ha	34	38	42

1 GBLU = 1 dairy cow or 1 beef cow or 6.66 breeding ewes.

Since 1992 and the imposition of quotas on annual ewe premium, the expansion of breeding ewe numbers in the UK has ended. The number of holdings with sheep has declined although the average flocksize has continued to increase. The policy is now no longer pushing producers towards a general expansion of ewe numbers. Instead it is likely to encourage greater specialisation and intensification of management to improve income from a restricted production base. In many situations the easiest way of increasing income per ewe is to sell more lambs per ewe rather than add value to the existing level of production.

However, intensification may not be achievable on the most extensive and agriculturally disadvantaged hill farms, mainly because of the constraints of land quality and lack of a suitable infrastructure. In such conditions, a reduction in sheep numbers to a lower level of intensity or even abandonment of sheep farming may occur. One auction market group in Scotland has reported the sale, during 1996, of 17,500 hill ewes from the west of Scotland as a result of management decisions to stop hill sheep farming (Stevenson 1997, personal communication). Equally, in lowland areas reduced sheep profitability may result in fewer subsidiary sheep enterprises as holdings specialise in dairy or arable production. Changes of this nature will have important impacts on the grassland mosaic.

Structural changes in Greece

Although Greece is a very significant sheepmeat producer (the sheep and goat sector contributes 43% of the total gross value of animal production and 13% of total agricultural production), the structure of its industry is radically different from that found in the UK. In 1993, 18% of Greek holdings had a sheep flock compared to 39% of holdings in the UK (Eurostat, 1997). However, in total there are more holdings keeping sheep in Greece and these holdings are much more dependent on sheep production (Table 5.4).

In contrast to the UK, both breeding sheep numbers and the number of holdings keeping sheep have declined in Greece since the mid-1980s (Tables 5.1 and 5.3). Thus the number of holdings with sheep fell by 38% between 1983 and 1993. At the same time, flocksizes steadily grew. Thus, in Greece, the sheepmeat regime has been associated with a considerable rationalisation of the sector.

The nature of sheep production systems in Greece, and in particular the high degree of shepherding carried out, mean that it is much less common to find a mixture of small and large ruminant livestock enterprises occurring in the same farm business than in the UK. Thus, the interaction between different livestock policies has been much less significant in the development of the Greek sheep sector.

Historically, transhumance has been a key element, making extensive use of mountain pastures. However, as flocksizes have increased, more intensive management methods have been introduced. Equally, as young people have left the villages the flocks are attended by fewer, older shepherds who are unable or unwilling to supervise the movement of flocks to more distant and higher pastures. Consequently, the traditional nomadic grazing of Greek flocks has declined and been replaced by more intensive use of pastures close to sheep housing and the villages (MAICH, 1997). This is evident from Table 5.8, which shows the evolution in goat population by type of farming from 1956 onwards.

Table 5.8 Evolution in goat population (%) by type of farming

Type of farming	1956	1961	1970	1980	1985
Home-fed	10.8	17.4	20.7	18.6	19.3
Sedentary	60.8	69.5	71.3	73.2	75.5
Transhumant	28.4	13.1	8.0	8.2	5.2

Source: Hatziminaoglou *et al.* (1995); Caraveli (1998b).

These developments have led to the abandonment or undergrazing of some marginal land, leaving large areas of sub-alpine pasture neglected, as well as to the over-exploitation of other areas. The latter has been mainly the result of the increased incentives provided by the Sheep Annual Premiums and LFA payments. The post-1992 changes have led to a stabilisation in livestock numbers in several areas where they were previously falling and thus to a slowdown in the rate of structural adjustment.

Restricting the transfer of production quotas out of 'sensitive' areas (i.e. LFAs), which the modified regime allows, has probably had similar effects. At the same time, the modified support system implies that farms which are presently over-stocked will have a strong incentive to maintain current livestock numbers (Baldock *et al.*, 1996; Caraveli, 1998a, b).

Sheep and goat grazing, properly managed, exerts a considerable ecological benefit on mountain pastures, woodland and scrubland areas. In particular, it inhibits scrub and woodland invasion, maintains biodiversity and reduces the risk of fires. Both abandonment and over-stocking of marginal grazing land prevents the rational management of pastures, may encourage erosion and reduces their feeding capacity. Nationally it is estimated that 25% of the mountainous and semi-mountainous soils have irreversibly lost their productive capacity, while as much as 40% of all extensive grazing land has been degraded to a dangerous level (as a result of deforestation, undergrazing or over-stocking) and is now unsuitable for grazing (Alexandris, 1985; Louloudis *et al.*, 1994; Hatziminaoglou *et al.*, 1995; Baldock *et al.*, 1996; Caraveli, 1998a, b).

DISCUSSION AND CONCLUSIONS

Most of the important landscapes in the UK, Greece and elsewhere in Europe depend upon extensive livestock grazing and management systems. Many are more specifically dependent on sheep and goats. Without such farming, these landscapes would change. In lowland Britain, for example, livestock herds provide a pleasing part of the landscape and encourage the upkeep of features such as hedges, thus contributing to a more diverse landscape mosaic. In Greece, sheep and goats also perform important landscape functions. These include preventing loss of open meadows, maintaining field structures in more intensively managed pasture and, in less intensively managed areas, preventing fires in both scrub and other woodlands that put at risk the whole landscape. In some situations, the risk of erosion appears to be greater without grazing than with it. The conclusion is that sheep and goats are necessary for maintenance of many landscapes.

This chapter has drawn together data and information about the changing structure of the sheep industry. It has demonstrated the complexity of the structure and production systems found throughout the EU. In particular the analysis has identified a number of critical points in relation to change and environmental impact. These are:

- The economic significance and geographic characteristics of sheep and goat production in the EU are highly variable. Consolidated statistics at the EU-level mask significant variation between and within Member States. This means it is very difficult to categorise environmental significance other than by assessment at a sub-national or a regional level.
- The wide variations in production systems between and within different Member States result in distinct responses to changes in policies with very different environmental consequences.
- Regional responses to policy changes are influenced by a highly complex set of interactions at sub-national level involving a number of factors including:
  - economics and markets;

- farm structure and enterprise structure; and
- linkages with other agricultural and environmental policy sectors.

- The sheepmeat and goatmeat regime is itself very complex and a given change within the regime, for example the definition of an eligible ewe, and the flexibility of individual Member States in implementing the policy (e.g. ring-fencing of quotas and choice of quota reference year), can generate a range of producer responses and consequently a wide range of environmental impacts.

- Opportunities for technological change and enterprise substitution lead to further variability in response.

Prior to 1992, in the UK and Ireland where sheep production is less specialised there has been an increase in both flock sizes and the number of holdings keeping sheep. Higher levels of support in the UK and Ireland, as a result of market prices being lower than the European average prices at the introduction of the sheepmeat and goatmeat regime, are implicated in this expansion. The regime created a level of profitability that made it attractive to introduce or expand a sheep enterprise in these countries. Limitations imposed by other livestock sector policies over the period 1980 to 1992 added to the attraction of adopting a sheep enterprise. In contrast, in those countries where sheep production is a more specialised activity and the proportion of holdings with sheep is much lower, for example Greece and Spain, a decline in the number of people keeping sheep has occurred leading to larger flocks and even greater specialisation. In Greece, however, the structural change has also led to a decline in sheep and goat numbers since the mid-1980s. A similar decline in numbers has been seen in France and Italy too, where noticeable falls in sheepmeat prices have occurred as commonality of policy application has been achieved.

In some cases greater specialisation has resulted in the loss of traditional management practices. In Greece, and other Mediterranean Member States, the traditional seasonal movement of livestock is one of the practices to have declined. This has led to land abandonment in some areas and intensification of grazing in other areas with consequent contrasting environmental impacts. In the UK, increasing sheep numbers have led to increased stocking densities that in some areas have caused localised environmental damage. In other areas increasing sheep numbers have resulted in changing sheep/cattle grazing ratios, land improvement or a combination of both with consequent impact on biodiversity and landscape mosaics.

Since 1992, however, the expansion in sheep numbers has stopped because of constraints on individual entitlement to support. In some Member States breeding sheep numbers have stabilised, for example France and Italy, while in others breeding sheep numbers have declined, for example the UK, Ireland and Spain. Production systems, and in particular the practice of not breeding animals until they are more than one year old, may have influenced actions in Great Britain and Ireland. As these non-productive animals have been eligible for support since 1992, it has allowed producers to retain support income and keep fewer sheep. It is too early to clearly identify environmental benefits of these actions, although in principle reduced stock numbers should reduce grazing pressures and help to improve biodiversity.

By restricting the number of animals supported the volume of production will be constrained. This can result in higher market prices, which in turn reduces the level of annual ewe premium, as occurred for example between 1995 and 1996. In this situation the sheepmeat and goatmeat policy is likely to push producers towards increased management intensity so as to either increase the number of lambs or volume of dairy product sold, or to add value to the product currently produced. Intensification processes leading to increased densities of livestock can directly affect vegetation and create pollution. Greater inputs of fertilisers, veterinary products and purchased feeds, more intensive forage conservation methods and more intensive use of sheep housing are likely to result. These changes will further increase pollution risk and modify habitats. Landscapes in lowland situations may not suffer significantly, but in hill and upland areas already at high stocking intensities landscape degradation is likely to continue.

Conversely, where increases in production are not possible producers may seek to recoup income loss by reducing costs by, for example, employing less labour or spreading costs over a greater number of sheep. The latter option, because of the presence of individual quota limits, can only be achieved by amalgamating flocks through quota transfer from producers withdrawing from production. Where the basic resource to improve profitability is not available, for example in the most agriculturally disadvantaged areas of Europe, and where social and demographic changes are under way, leading to an ageing farming population and a decline in employed labour, further pressure to abandon land may result from the current sheepmeat and goatmeat policy.

A move to more extensive methods may lead to preservation and enhancement of some habitats but can in some regions, particularly those which are already extensively managed, lead to habitat loss and landscape degradation. On lower-lying land, currently managed at a moderate to high intensity, reduced input and decreased management may lead to improved nature conservation value. However, it is likely that current landscape features will change and there is the danger of habitat and landscape loss through a lack of maintenance management. Cessation of low input farming will have detrimental effects on habitats, ecosystems, landscapes and cultures by removing all the grazing and land management activities previously carried out. The polarised responses to the current sheepmeat and goatmeat policy of intensification or extensification and land abandonment are likely to occur in different regions of the Community and create different environmental impacts.

It is extremely difficult to ascribe cause and effect to changes in numbers of livestock and other land use measures. It is even more difficult to ascribe environmental impacts to any change in management practices. In many cases these management changes cannot be directly linked to the sheepmeat regime. Nevertheless, such an important set of policy measures will have impacted indirectly on management changes and the environment. Increases in sheep and goat numbers have been linked to the regime in a variety of countries. However, it must be recognised that other factors, particularly market developments, technological changes, and labour and social issues, influence many of the impacts on the environment. Nevertheless, the speed with which technological changes and responses to market developments have been adopted will have been influenced by the reduced risk provided by assured income from the sheepmeat and goatmeat regime.

As a result there are a considerable range of environmental impacts, both detrimental and beneficial, that have, and could, result from a common sheepmeat and goatmeat regime. Thus, attempting to achieve any degree of coherent environmental management, even at a most basic level, across Europe through the sheepmeat and goatmeat regime has proved extremely difficult, if not impossible.

Acknowledgements

The authors acknowledge the financial support of the Land Use Policy Group of the Countryside Agencies of the United Kingdom in carrying out the research on which this chapter is based. However, the views expressed in this chapter are those of the authors.

REFERENCES

Alexandris, S. (1985) *Rangelands and Environment.* Geotechnica, Proceedings of the Congress on Rangelands and the Economy of Mountainous Regions. Thessaloniki, pp. 17-22 (in Greek).

Ashworth, S.W., Waterhouse, A., Treacher, T. and Topp, C.E.F. (1997) *The EU Sheepmeat and Goatmeat Regime and its Impact on the Environment.* A report prepared for the Land Use Policy Group of the Countryside Agencies of the United Kingdom, SAC, Auchincruive.

Baldock, D., Beaufoy, G., Brouwer, F. and Godeschalk, F. (1996) *Farming at the Margins - Abandonment or Redeployment of Agricultural Land in Europe.* Institute for European Environmental Policy (IEEP) and Agricultural Economics Research Institute (LEI-DLO), London/The Hague.

Caraveli, H. (1998a). Environmental implications of various regimes: The case of Greek agriculture. In: Tracy, M. (ed.) *CAP Reform - The Southern Products.* APS Publications, Belgium.

Caraveli, H. (1998b) Greece. In: Brouwer, F. and Lowe, P. (eds) *CAP and the Rural Environment in Transition - A Panorama of National Perspectives.* Wageningen Pers, Wageningen, pp. 267-283.

CEC (1995) *Special Report No. 3/95 on the Implementation of the Intervention Measures Provided for By the Organization of the Market in the Sheepmeat and Goatmeat Sector.* Official Journal of the European Communities C285 Vol. 38, 28 October 1995, p1.

CEC (1996) COM (96) 44 *Report from the Commission to the Council on the Application of the Individual Producer Limits in the Annual Ewe Premium and Suckler Cow Premium Schemes.* Office for Official Publications of the European Communities, Luxembourg.

Council Regulation 1837/80 OJ L 183, 16.7 80, p1. Office for Official Publications of the European Communities, Luxembourg.

Council Regulation 3523/85 OJ L 336, 14.12.85, p2. Office for Official Publications of the European Communities, Luxembourg.

Council Regulation 1115/88 OJ L110 29.4.88, p36. Office for Official Publications of the European Communities, Luxembourg.

Council Regulation 3013/89 OJ L 289, 7.10.1989, p1. Office for Official Publications of the European Communities, Luxembourg.

Council Regulation 1323/90 OJ L 132, 23.5.90, p17. Office for Official Publications of the European Communities, Luxembourg.

Council Regulation 2069/92 OJ L 215, 30.7.92, p59. Office for Official Publications of the European Communities, Luxembourg.

Eurostat (1994) *Agricultural Statistical Yearbook 1994*. Office for Official Publications of the European Communities, Luxembourg.

Eurostat (1997) *Agricultural Statistical Yearbook 1997*. Office for Official Publications of the European Communities, Luxembourg.

Hatziminaoglou, N., Zervas, N. and Bogiatzoglou, J. (1995) Goat production systems in the Mediterranean area: the case of Greece. In: El Aich, A., Landau, S., Bourbouse, A., Rubino, R. and Morand-Fehr, P. (eds) *Goat Production Systems in the Mediterranean*. EAAP Publication No. 71, Wageningen Pers, Wageningen.

Louloudis, L., Beopoulos, N., Koumas, D.T. and Theoharapoulos, J. (1994) *Report on Extensive Farming Systems in Greece*. Institute for European Environmental Policy, London.

MAICH (Mediterranean Agronomic Institute of Chania) (1997) *Effects of Human Activities on Desertification of Psilorites Mountain*. Department of Environmental and Renewable Resources, Chania, Crete.

MLC (1997) *European Handbook*, Volume 2, section 5. MLC, Milton Keynes.

The Arable Crops Regime and the Use of Pesticides

6

Katherine Falconer and Arie Oskam

INTRODUCTION

This chapter assesses the impacts of the arable crop regime on pesticide[1] use, and discusses the potential implications of policy change for environmental quality, whether beneficial or harmful. The market organisation for cereals, oilseeds and protein crops (COP crops) influences pesticide use along three lines: (i) the price level of cereals; (ii) set-aside for COP crops; and (iii) agri-environmental policy. This chapter concentrates on the first two. The third is mainly driven by the measures under the Agri-environmental Regulation 2078/92, whose scope is far broader than arable production and for which clear results remain limited. Agri-environmental policy is covered in Chapters 12 and 13 (see also Whitby, 1996). Others have considered the impacts in relation to nitrate contamination (e.g. Van Zeijts, 1999).

Several authors have stressed the negative environmental effects of relatively high agricultural prices in the European Union (EU) (e.g. Anderson and Black-hurst, 1992; Heerink et al., 1993; Nutzinger, 1994). Higher prices stimulate more intensive agriculture, which at the very least generates greater emissions per ha and places low input, extensive agriculture or biological farming methods in a less favourable position. Implicitly the reverse is assumed for lower prices. Brouwer and Van Berkum (1996, pp. 43-51 and pp. 71-84) provide a balanced overview of the literature. They conclude that 'lower prices have complex implications for the agricultural system and environmental benefits for one aspect may be offset by others'. It is always difficult to distinguish between technological development and long-term price effects (Oskam and Stefanou, 1997). An overview of research in relation to pesticides, however, makes clear that the impacts of technological changes are more important than price effects (Huffman and Evenson, 1992). Often, system changes contribute most to the reduction of the environmental impact of agro-chemicals (Oskam, 1994; Oskam et al., 1998; Wossink et al., 1998). The environmental effects of set-aside have been investi-

---

[1] Pesticides are taken to encompass all crop protection products and especially herbicides, insecticides and fungicides.

gated less intensively. It is a relatively new instrument within the EU, but has been in use for longer in the US and it is worth reviewing the experience gained there.

The focus of the chapter is the integration of assessments of environmental quality change resulting from changes in arable (COP) production caused by commodity price reductions and set-aside policy. We will assess first the links between arable production and pesticide-related environmental concerns. This is followed by a review of specific developments in agricultural policy. An overall estimate is presented of the effects on pesticide use at the EU level of the MacSharry and Agenda 2000 reforms to the arable regime. This is followed by a farm-level evaluation of the effects of price liberalisation on production, pesticide usage, and the hazards to the environment from pesticide usage.

AGRICULTURAL SUPPORT AND PESTICIDE USAGE

Recent decades have shown continued increases in both land and labour productivity, and pesticide use has increased steadily following technological developments after the introduction of synthetic chemicals in the 1940s. Ecological and health risks may be posed through acute or chronic exposures to pesticides, as well as through a multitude of indirect effects. Concern about current levels of chemical use can be characterised along a number of dimensions, relating, for example, to the contamination of groundwater, surface water, soils and food, and their impacts on wildlife and human health (Reus *et al.*, 1994; Skinner *et al.*, 1997). Widespread and growing concerns about over-use have led to suggestions that reductions in pesticides should be made (Carson, 1962; Dinham, 1993; WWF, 1995), balanced by concerns to maintain world food supplies (Fawcett, 1991; Johnen, 1997).

Oppenheimer *et al.* (1996) have investigated environmental concerns relating to pesticides in six different Member States of the EU. The top three concerns were (1) contamination of water resources used for human consumption; (2) possible adverse effects on the ecology, e.g. on non-target species; and (3) risk to consumers of food with residues. At present, EU policy development is focused on the approval of pesticides in Member States according to uniform principles.

Pesticide policy does not develop in a vacuum, but is influenced by developments in other policy frameworks, at both EU and Member State levels (Oskam *et al.*, 1998). The Fifth European Environmental Action Plan, launched by the EU, states a reduction in chemical use as a major objective, although no actual goals or limits are defined and currently Member States are largely free to address their own priorities. The only EU-wide action taken so far that could reduce pesticide use is agricultural policy reform.

The 1992 MacSharry reform of the Common Agricultural Policy (CAP) recognised the need to encourage more environmentally sustainable forms of agricultural production (Brouwer and Van Berkum, 1996). There were three main elements. The first was a reduction in price support, which may have caused pesticide usage to fall compared to the pre-1992 situation (although European market prices were still above world levels). The second element related to the introduction of arable area payments to compensate for reduced levels of price support. Entry into the area payment scheme was voluntary but a set-aside requirement was introduced as a condition for eligibility for support payments, and this also is likely to have affected pesticide usage.

At the national level, the introduction of taxes on pesticides has been proposed to alleviate some of the adverse environmental effects linked to agricultural production. Only Denmark, and to a lesser extent Sweden and Finland, apply them. But even without taxes there is a clear downward trend in the application of pesticides measured in kilograms of active ingredients. This trend is much stronger for countries operating a pesticide use reduction plan (Oskam *et al.*, 1998). Within these developments it is difficult to determine the specific effects of agricultural policy changes.

There is a growing trend by European farmers to reduce agro-chemical application rates (Rayment, 1995). Various factors may have contributed to this. It is necessary to take into account the effects of the weather, technological developments and changes in farmers' attitudes, as well as the incorporation of environmental concerns into farming practice by the adoption of agri-environmental schemes; such factors will have both short-term and longer-term impacts. Therefore, the impacts in practice of the 1992 reforms on the environment and landscapes may never be clear. Moreover, changes due to support rate adjustments cannot be separated easily from those triggered by exogenous factors such as exchange rate movements. In any case, reductions in usage as measured by weight have occurred in recent years through the introduction of new, low-dose chemicals and more efficient use of older chemicals. Given that these 'modern' pesticides are also generally less persistent but are often more potent, these factors together imply that environmental impacts might now be lower; however, a reduction in environmental effects is by no means guaranteed.

The relation between agricultural policy measures and the environment must be assessed in two stages. The first stage deals with the effect of the policy instruments (i.e. agricultural support prices and set-aside in this study) on agricultural production processes, in terms of the outputs produced and inputs used. The second stage links the use of pesticides with pressures on the environment (groundwater and surface water pollution, run-off of soils, biodiversity, landscape quality, extraction of water resources, etc.). The first stage has been investigated in much more depth than the second stage, and usually concentrates on identifying changes in potentially harmful inputs. It seems almost impossible to prevent chemicals that are deliberately introduced into the environment from reaching waters, whatever the method of application and the agricultural practices used: the only solution is thus to reduce the quantities applied and the toxicity of the chemicals used.

Some studies focus on changes in agricultural production but provide no information with respect to the real environmental indicators; others focus on groundwater, surface water, etc. but make no links to agricultural policy elements or changes in agricultural technology (see Wu and Segerson, 1995). Few studies bridge this gap (Bergman and Pugh, 1994; Vatn *et al.*, 1997), and of those that do, models or approaches are often restricted to one farm, a group of farms, a particular region, or a special product. Moreover, such models are very labour intensive to prepare and highly intricate. It is essential to acknowledge, though, that aggregate input variables like the use of pesticides have a very complex relationship to environmental quality variables. Pesticides are very heterogeneous; some of them are very dangerous in small amounts and others can be used in large quantities without much effect. Moreover, adverse environmental effects might depend on the prevailing climate (and weather) conditions under which they are used, on what soil and by what application method, and how near important environmental objects are to the point of application. However, usually it is assumed that there will be a proportional reduction in environmental harm with input usage.

APPLIED THEORY: OVERVIEW OF THE LITERATURE

In theory, diminishing marginal returns to inputs would imply that a reduction in output prices should induce reductions in production and in the usage of potentially contaminating inputs. At issue, therefore, is whether there is a direct relation between the agricultural changes triggered by the reforms and environmental quality levels; the answer depends very much on how agricultural production changes. For example, lower prices may encourage an increase in relatively chemically-intense crops, depending on the shifts in their relative marginal returns. The problem for analysis is that there is no direct correlation between product price (or net returns per ha) and the amount of potentially problematic inputs, such as pesticides, used in production, and the resulting environmental damage.

Antle and Just (1991) demonstrated theoretically that the environmental effects of commodity policy could be either positive or negative, depending on the joint distribution of productive and polluting characteristics. This aspect is also supported by the conclusions of Just and Bockstael (1991), who compared studies at a high level of aggregation (national or pan-national) (for which the environmental effects of policy reforms seem to be rather small) and more disaggregated approaches (farm-level) (in which type of study, estimations of the effects of policy change can be substantial).

Oude Lansink and Peerlings (1995) suggested that although the 1992 reforms could result in pesticide usage reductions of only 2.8% in The Netherlands, they could instead induce an increase, if producers were to shift to crops that use pesticides more intensively (for example, fruit, vegetables, and potatoes). Jones *et al.* (1995) used a country-level model to predict a decrease in the UK in the intensity of production of both cereals and oilseed rape as a result of the 1992 reforms. They noted that current evidence on the effects of the reforms on the intensity of production is mixed, but that the general view is that, in the longer term, production intensity must fall given lower cereals prices. However, these authors also suggested that production might continue to intensify in spite of the reforms: in the short term through the use of fixed factors of production over a smaller land area, and in the long term through technological development, which may boost yields even while prices fall. Moxey *et al.* (1995) predicted that the CAP reforms would result in a decline in the area of arable land in northern England and a reduction in production intensity. Schou (1998) predicted a substantial reduction in pesticides use in Denmark, following the 1992 reform. Similar indications are provided for Agenda 2000 (see e.g. Agrarwirtschaft, December 1998). However, no systematic empirical analysis has yet been published on the actual environmental consequences of the 1992 reform.

It is important to disaggregate the agricultural sector, to capture better the complexities of agricultural commodity programmes and changes in the composition of agricultural production as a result of modifications to those programmes. Debate really centres on the nature and extent of adjustments at farm level, in terms of the prevailing farming systems, cropping plans and biophysical conditions. The environmental costs or benefits of change depend crucially on the particular production and management regimes initiated by individual farmers. The shifts in commodity prices and direct income supports are highly likely to have a widespread impact on the production decisions of farmers across the Community, but different farming systems are hypothesised to respond differently, in terms of adjustments in the mix of enterprises and production practices. Limited empirical research is available regarding responses at farm level and their consequences in terms of pesticide-related environmental change. The response is likely to vary according to location and farming system (De Haen, 1982; Oskam *et al.*, 1992).

A key feature of the MacSharry reforms was the set-aside requirement for eligibility for arable aid. Such programmes have been operating for longer in the US and it is worth reviewing the American experience. Even there, direct empirical information on the environmental effect of set-aside is scarce. However, it would seem that, in the US context, a reduction of long-term set-aside on highly erodible soils may increase the use of chemicals in agriculture, while a reduction in short-term set-aside may slightly reduce chemical use (Gardner, 1991; Miranowski *et al.,* 1991). The different aspects of commodity policies are linked to each other, making policy evaluation rather complex, and making it difficult to extrapolate to the different cropping context of the EU. Given the size of US production in the international grain market, a set-aside change in the US influences international grain prices; therefore set-aside and prices cannot be analysed separately (Wu and Segerson, 1995). The EU has a less prominent position in international grain and oilseed markets; thus, a reduction in the land set-aside in the EU would induce a smaller effect on product prices.

The next section provides a rough estimate at the aggregate level of the implications for pesticide usage of the MacSharry and Agenda 2000 reforms. It is followed by a detailed farm-level evaluation of the potential effects of further liberalising commodity markets.

EFFECTS OF POLICY REFORM ON PESTICIDE USE

The 1992 MacSharry reforms (which were implemented over the period 1993/94-1995/96) reduced intervention prices for cereals (-32%), compensating producers with income support per unit of base acreage, with a partly compensated set-aside requirement for larger producers (farms with a base area producing more than 92 tonnes) who wish to receive the income-support area payments (see Table 6.1).[2] Oilseeds and some protein crops, which were originally supported through deficiency payments, were brought under a similar regime.

A further step in the reform of EU cereals policy was announced in the Agenda 2000 discussion paper (CEC, 1997a). The Commission proposed a 20% reduction in intervention prices and a 0% set-aside obligation. In the event, the European Council decided on a 15% price reduction and a (maximum) of a 10% set-aside obligation (Agra Focus, April 1999). This is still not a complete liberalisation of cereal prices, because the intervention price provides a 'minimum price guarantee' for the producers, but it brings EU prices much more in line with world market prices. Farmers are partly compensated by means of direct income support.

Table 6.1 Official set-aside obligations for large farms in the European Union in percentage of the base area of COP crops

Year	Short term (1 year)	Long term (5 years)
1993/94	15	20
1994/95	15	20
1995/96	12	17
1996/97	10	10
1997/98	5	5

Source: CEC (1996, 1997b).

The calculations for set-aside reduction and a price decrease (including set-aside reduction) have been separated for clarity. Moreover, a 12% set-aside percentage has been taken as the reference level, i.e. a focus on 1995/96 both as the result of the MacSharry reform and as the base period for Agenda 2000. The difference between the 'official' rate and actual area set-aside is explained by large diffe-rences in farm size of COP crops across the EU Member States. The area share of COP crops of small farmers amounts to circa 24%, but varies between Greece (89%) and the UK (6%). In addition, large farms were able to choose to enter the support scheme as 'small' farms, even if their actual COP area was larger than the equivalent 92 tonnes of cereals. While their compensation payments were reduced, they were not obliged to set-aside any of their base area. Furthermore, there were 'penalty set-aside' requirements for farms in areas above the target regional production level in the previous year. Finally, some land was set-aside under the voluntary five-year scheme prior to the 1992 reforms.

Although the official set-aside rate in 1995/96 was 3% lower than in 1994/95, the base area under the set-aside regime amounted to 13.8% in 1995/96 (at the

---

[2] Previously, set-aside existed in the EU on the basis of a voluntary premium system, sometimes stimulated by additional national or regional subsidies. At the start of the MacSharry reform circa 1.5 million ha of arable land was under long-term set-aside (CEC, 1995).

EU level). Table 6.2 presents an extensive overview of set-aside in the EU. For environmental assessment, it is important to know what proportion of the short-term set-aside (16% in 1995/96; see Table 6.2) was used for industrial crop production (mainly oilseeds). There is also some 'slippage', which implies that the reduced area is larger than the reduction of production. A slippage of 20% has been assumed (see Blom, 1995). This results in an estimate of the production effect for a 12% set-aside obligation of 12x(1-0.24)(1-0.16)(1-0.20), to give 6.1%, taking into account small farmers, industrial crops and slippage.

Evaluation of the effects of set-aside on pesticide use is based on the results of Oude Lansink and Peerlings (1996) for The Netherlands. They calculated that an actual set-aside rate of 2.06% reduced pesticide use by 0.73%. Extrapolating this result to the entire EU with a formal set-aside rate of 12% for large farms, but accounting for about 6.1% of the production capacity of COP crops, suggests that pesticide use under the MacSharry reform decreases by 2.2% due to the set-aside component. Produce Studies (1996, p. 17) suggested that a 32% price reduction would decrease pesticide use additionally by circa 7.5%. Taken together with set-aside the reduction in pesticide use for arable crops is estimated to be 9.7%. The share of arable crops in total pesticide use is around one-third (Brouwer *et al.*, 1994, p. 84), implying a reduction of 3.2% in total pesticide use due to the MacSharry reforms to the arable crops regime.

Table 6.2 An overview of set-aside and the base area in the EU (1995/96)

Country	Set-aside area in 1,000 ha				Base area (1,000 ha)
	5 year	Annual	Of which industrial	Total	
Austria	0	123	18	123	1,203
Belgium	0	22	7	22	479
Denmark	5	247	48	252	2,018
Finland	0	171	1	171	1,591
France	134	1,877	373	2,011	13,526
Germany	151	1,457	359	1,608	10,156
Greece	0	18	0	18	1,492
Ireland	0	32	1	32	346
Italy	471	248	67	719	5,801
Luxembourg	0	2	1	2	43
Netherlands	8	12	1	20	437
Portugal	0	67	0	67	1,054
Spain	41	1,426	39	1,467	9,220
Sweden	0	329	24	329	1,737
UK	37	525	85	562	4,461
EU-15	847	6,556	1,024	7,403	53,564

Source: CEC (1997b, tables 3.5.7.1 and 3.5.7.2).

The estimated environmental effects of a further reform of EU cereal/arable policy (Agenda 2000) are very modest, assuming little change in current crop production practices. With a 15% price reduction and - on average - 5% set-aside, implementing Agenda 2000 will reduce total use of pesticides by an estimated 0.8%. Such a change would be very difficult to observe; particularly since the fluctuations in pesticide use caused by weather variations are larger (Oskam *et al.*, 1998, Appendix III). One reason for the limited reduction of pesticide use is that price decreases and set-aside reduction work in opposite directions. The effects of either component by itself is rather small and taken together they tend to cancel each other out.

Real prices of COP crops may change by less than the reduction made in intervention prices, due to adjustment in world market prices. This was the case in the period after the MacSharry reform (see CEC, 1997b) and it will be even more relevant if intervention prices are lower (in which case actual market prices will be influenced more by world market prices). Compared to EU prices in 1996 prices may decrease by more than 15%. However, the anticipation by farmers of price reductions in the period prior to actual implementation of the reforms should be considered too.

Given the uncertainties in methodology and information, only rough indications of the pesticide usage and environmental effects of a set-aside reduction, a removal of area support payments and a price decrease for cereals can be calculated. This chapter considers only the possible effects of removing support for COP crops; further work should examine the effects of the sugar beet regime, for example.

FARM-LEVEL MODELLING OF THE EFFECTS OF AGRICULTURAL LIBERALISATION ON PESTICIDE USE

As noted above, various models have been developed to assess the effects of agricultural support policy change. Normally, these models operate at a high level of aggregation. However, the environmental impacts of production practices such as pesticide application generally occur at a very local level. This has often been the main reason for farm-level analysis with detailed analyses of pesticides applications (Oskam *et al.*, 1992; Wossink, 1993; Michalek and Hanf, 1994; Wossink and Renkema, 1994).

The case study focuses on the complete elimination of the CAP compared to the reference harvest-year of 1993, as an example permitting some insights into the possible environmental dimension to agricultural liberalisation. We present some results of a farm-level linear programming (LP) model developed to assess the potential effects of policy change in detail for an arable farm in East Anglia in the UK, a region important for cereal production in the EU context. LP is a useful method where the policy requiring analysis has no historical precedent, or if for any other reason projections from past data are not possible, for example, due to rapid technological change (see, e.g. Oskam *et al.*, 1992, p. 114). The case study concerns a mainly-cereals farm of 250 ha with combinable break crops, and fuller details are given in Falconer (1997).

In evaluating how optimal farm plans and pesticide usage might change following a removal of agricultural supports, it was necessary to estimate the levels of world market prices for arable commodities in the 1993/94 marketing year. World output prices were considered a good proxy for prices under a support-policy-free scenario. UK ex-farm prices would be around £2 tonne^{-1} higher than the cost, insurance and freight at Rotterdam prices. Pulses and oilseeds were considered slightly differently, as there are no import tariffs for these crops (support is basically a deficiency payment) so current 1993 prices per tonne were considered to be reasonable proxies for prices following CAP removal. The set-aside constraint, set-aside payments and area payments were all set at zero levels under the NO CAP scenario.

The model was calibrated for two farm production situations, one representing current commercial crop production (CONV) and one representing a more 'progressive' farm, with a wider range of cropping activities encompassing cross-combinations of current commercial practice and reduced input rates for pesticide and nitrogen regimes (PROG). It is very important to the predictive ability of the model to include production functions representing different cropping variants for each crop, for example ranging from 'intensive' to 'ecological', to reflect the different options available to farmers with as detailed a specification of resource use as possible. However, crop protection is very complex, given the number of chemicals, tank mixes, timings of application, cultivations, and other crop protection and weeding practices. The data available were too limited to take all of these aspects into account so the number of activities considered was restricted to five pesticides and two nitrogen regimes per crop.

Understanding the input substitution possibilities is central to the analysis of the environmental problems of pesticides. Policy impacts should differ depending on the compliance options available to producers. Sufficient alternatives need to be included to represent adequately the shifts along the production possibility frontier for any given crop. The low-pesticide-input activities in PROG were calibrated on the basis of trials data collected in East Anglia between 1991 and 1996. The low input crop practice coefficients are only provisional estimates of the potential differences in yields and variable costs compared to 'current commercial practice', given the very limited trial data available. Yields would be expected to vary over seasons, soil types, rotations and field operations. Ideally, a much more detailed simulation model would be used, although one is not yet available. The model allowed output data to be generated with regard to the crop land allocation; the total farm management and investment income; and the total farm spray usage in terms of 'spray units' (where one unit equates to one application of a pesticide at the recommended dose per hectare to the crop).

A limitation of the LP model was the constraint on the number of different activities that could be included (due to low data availability); unless a large number of activities are included in the analysis, input substitutions may be masked by product substitutions. The CONV model was restricted to changes in terms of crop extent; the PROG model was designed to have a number of options for changes in production intensity too. However, the relative gross margins, leading to the inclusion of low-input crops in the baseline solution, meant that in actual fact, changes in this model also had to be in terms of the crop extensive-

ness. Farmers would be expected to adjust their crop plans in terms of both intensity (e.g. of spray usage) and extent. The model was validated using data from the
official farm business survey for the region (Murphy, 1995).

Environmental indicators for pesticides

Establishing the link between environmental impacts and pesticide use in arable
production is of great importance to the analysis. There are a vast number of factors to consider when attempting to predict the likely outcomes for such a
complex area, given, for example, factors such as the stochasticity of natural systems, and changes in input and output prices linked to shifts elsewhere in the
wider economic system. In addition to the quantity of pesticide applied in terms
of weight, treated-area or treatment frequency is the 'chemical type' factor: product characteristics such as persistence, eco-toxicity and water solubility will affect the type and magnitude of the eventual environmental impact of pesticide
use. Consequently, there will be a non-linear relationship between environmental
impacts and the quantity of pesticide used.

The development of meaningful environmental indicators on the basis of currently available data for pesticides is far from straightforward, largely because of
the extent of knowledge gaps with regard to ecological and health impacts (see,
for example, Oskam and Vijftigschild, 1999). There are substantial data requirements, relating to three key relationships: between polluting emissions and
concentrations in environmental media; between pesticide concentrations and
physical damage; and finally between physical damage and social welfare. The
harm from pesticides is generally determined by the combination of a large number of interacting factors, especially the quantity of pesticide applied, its toxicity
and other characteristics, and ecological factors such as species presence and
population levels. It is important to draw the distinction between hazard and impact: impact depends on both hazard and exposure to it. The latter, however, is
very difficult to measure or even estimate for most species. As a result, the focus
so far in many studies has tended to be on hazard, as indicated by parameters such
as the toxicity or lethal dose rates for non-target organisms, or physico-chemical
properties such as water solubility or the soil adsorption coefficient.

The chosen approach to pesticide indicator development used a simple sco-
ring and aggregating approach, as presented, for example, by Kovach *et al.*
(1992) and Penrose *et al.* (1994); full details of the methodology are given in Falconer (1997, 1998). The environmental and health precautions listed on statutory
product labels (summarised in Ivens, 1994) take into account all the available information on pesticide risks and impacts which is used in the approval process.
An environmental indicator based on labelled information should be, therefore, a
fair summary of current knowledge about pesticide hazards.[3] Nine different ecological and human-health dimensions were identified and a tenth 'general'
dimension was included on the basis that all pesticides are thought to present
some risk. The ability to disaggregate hazard along different dimensions allows
the nature and extent of trade-offs between changes in hazards to be identified.

---

[3] Other approaches based on physical and chemical properties (see, for example, Reus and Pak, 1993)
were rejected here on the basis of insufficient available data for the products of interest in the model.

Scores were given for each hazard dimension, which were then aggregated so all pesticides received a score from 1 to 10, higher scores indicating that the pesticide poses a greater hazard. Equal weightings for all dimensions were assumed in their aggregation, in the absence of any basis on which to impose different weightings.

Environmental indicators of the type applied here have advantages in terms of disaggregation, transparency and adjustability in the light of the use of available data. However, there are several general caveats to the use of such indicators. They are linear, and take no account of indirect effects caused by interactions between different dimensions. No account is taken of environmental or other management efforts (such as the use of pesticide-exclusion strips along watercourses), which may reduce pesticide risks and impacts. Ideally the yardstick should be tailored for specific soil and climatic circumstances. Multi-dimensional systems analysis is really needed; it is artificial to isolate the effects of pesticide use from the overall effect of the farming system; different components are interdependent. Spring crops, for example, use less pesticide but give rise to more nitrate leaching and soil erosion, but may maintain higher soil fertility and provide valuable winter habitats and food for wildlife.

The results showed that arable pesticide usage and the degree of hazard it posed at the farm level might either increase or decrease, if the MacSharry CAP regime was removed. The results depended on the assumptions about the range of crop production activities available and their relative pesticide use intensities and gross margin per ha. For both farm models, removal of the MacSharry CAP regime resulted in a significant increase in the area of wheat grown; the switch into cereals was more extreme for the CONV farm. The increase in the cereals area has important implications for pesticide and fertiliser usage: increases appear likely, as cereals are more intensive agro-chemical users than other crops (see Table 6.3). Total pesticide hazard increased for the CONV farm, although it falls significantly for the PROG farm, due to the replacement of break crops (particularly winter oilseed rape) with low-pesticide-intensity wheat. Thus, on farms already using (or able to convert easily to) low input cropping activities, pesticide reduction goals could be achieved to a significant extent by removal of agricultural support policies such as the CAP, although complementary policy measures might be required to address potential problems of greater fertiliser usage and nitrate leaching. Farms unwilling or unable to switch to alternative agricultural practices and systems could actually contribute to a worsening of the situation following agricultural policy reforms (further liberalisation) as their cereals area (with relatively high pesticide usage) increases.

However, only the very short term is reflected; factor prices and output prices would also be expected to change, potentially contributing to a new optimal plan. Changes in price relations will take some time to have effects on production methods. Fixed costs would be expected to fall as labour and machinery are shed; dynamic adjustments might also be expected if the lower prices stimulate new technological developments, causing additional shifts in cropping plans. The direct effects of price changes may also be reinforced by shifts in the production function, following technological development and reductions in inefficient resource use, challenging accurate estimation of the farm-level impacts of removing agricultural support; estimation of changes is far from straightforward. A caveat to interpreting the results is that this was the first year following the 1992 reforms, and, as such, was part of a period of on-going transition. LP assumes profit maximisation but some farmers may still have been in the process of adjustment to a new set of anticipated prices, affecting the fit of the model predictions to actual production. Interestingly, the baseline higher management and investment income achieved by the PROG model compared to the CONV model suggested that farm incomes could be improved by adopting less-intensive farming practices, although higher skill levels would inevitably be needed (with extra costs, only some of which could be included in the model) (see also Oskam *et al.*, 1992; Wossink, 1993).

Table 6.3 Farm-level pesticide usage and environmental indicators: the post-1992 regime and removal of the CAP compared

	CONV		PROG	
	Baseline	No CAP	Baseline	No CAP
Total spray expenditure (£)	23,845.0	26,746.0	13,854.7	13,749.8
Total usage (per-ha spray doses):	1,187.5	1,339.0	543.8	487.5
- Herbicide units	587.5	672.0	393.8	337.5
- Fungicide units	425.0	461.0	100.0	150.0
- Insecticide units	175.0	205.7	50.0	0.0
Fertiliser expenditure (£)	13,986.3	14,190.0	11,055.5	12,557.5
Total hazard score (hazard points)	4,414.1	4,724.8	1,965.5	1,630.5
Environmental hazard dimensions (points):				
- To bees	225.0	166.7	100.0	0.0
- From absorption through skin	525.0	519.5	218.8	243.8
- From ingestion	350.0	286.2	256.3	212.5
- To eyes	387.5	436.2	237.5	318.8
- To watercourses	400.0	433.3	162.5	25.0
- From OP exposure	175.0	255.7	50.0	0.0
- To fish	976.6	979.7	390.5	330.5
- To the respiratory tract	62.5	108.3	6.25	12.5
- To game and wildlife	125.0	200.0	0.0	0.0
- General	1,187.5	1,339.0	543.8	487.0

DISCUSSION AND CONCLUSIONS

The MacSharry reforms in the EU are perceived as a major change of the CAP. Their environmental effects still need to be derived from models. No clear and comprehensive empirical evaluation has been found in the literature. Using calculations at a highly aggregated level, we estimated a reduction of total pesticide use by 3.2% due to the MacSharry reforms to the arable regime. The first indications for the environmental effects of a further reform of the EU support regimes (Agenda 2000) for COP crops are also very modest. If we take the more aggregate results as representative, the policy change would reduce the total use of pesticides by less than 1%. Such a change would be very difficult to observe; fluctuations in pesticide use caused by variation in the weather may be larger (Oskam *et al.*, 1998, appendix III). One reason is that price decrease and set-aside reduction work in opposite directions. Moreover, the degree of change for both parts of the policy change is quite low.

The results of the farm-level modelling here suggest that a removal of the MacSharry agricultural support regime will not necessarily lead to an improvement of the environment, depending on the assumptions about the range of crop production activities available; pesticide usage and the total environmental hazards posed by arable production might increase or decrease. In the case of an arable sector composed largely of 'conventional'-type farms, pesticide reduction policy would fill an important role. However, if the sector is dominated by 'progressive' farms willing and able to switch to low pesticide input production practices, pesticide reduction goals could be achieved to a greater extent by further liberalisation of agricultural markets. Complementary environmental policy measures might still be required to address potential problems of greater fertiliser usage and nitrate leaching. As mentioned earlier, our results are indicative for only a small part of the 'real' environmental effects for the EU. Differences between pesticides, the effects of fertiliser, erosion, water resources, energy use, wildlife and landscape, etc. have not been considered, but play a role with regard to environmental quality.

Other instruments would be more effective to reduce environmental pressures from agriculture (see Oskam *et al.*, 1998). A more open EU market policy for cereals may still be a useful target, but the reasoning should not be derived from the application levels of pesticides. Furthermore, it might be argued that removal of CAP-related market distortions give rise to net welfare gains anyway through leading to reduced output levels. Thus, no environmental gain would be needed to justify the policy change.

Of course results also depend very much on the assumptions which have been used. Given the large number of uncertainties, as indicated in this chapter, the predicted changes are likely to be site-specific. Moreover, long-term reactions could be larger if adjustment patterns in capital and labour and a change of technology are considered. In conclusion, the potential improvements from removal of the CAP appear to depend critically on the range of agricultural systems, and the potential for change within them. Technology and farming systems seem to be far more important than the specific CAP regime for the environmental effects of pesticides in arable production.

REFERENCES

Anderson, K. and Blackhurst, R. (eds) (1992) *The Greening of World Trade Issues.* Harvester Wheatsheaf, New York.

Antle, J. and Just, R.E. (1991) Effects of commodity programme structure on resource use and the environment. In: Just, R.E. and Bockstael, N. (eds) *Commodity and Resource Policies in Agricultural Systems.* Springer-Verlag, Berlin, pp. 97-128.

Bergman, L. and Pugh, D.M. (eds) (1994) *Environmental Toxicology, Economics and Institutions: the Atrazine Case Study.* Economy and Environment Volume 8, Kluwer Academic Publishers, Dordrecht.

Blom, J. (1995) *Een geregionaliseerd graan- en mengvoedergrondstoffenmarktmodel voor de EU-12: Model en toepassingen.* Proefschrift, LU-Wageningen.

Boyd, R. (1987) *An Economic Model for Direct and Indirect Effects of Tax Reform on Agriculture.* USDA-ERS Technical Bulletin 1743, Washington.

Brouwer, F.M. and Van Berkum, S. (1996) *CAP and Environment in the European Union: Analysis of the Effects on the Environment and Assessment of Existing Environmental Conditions in Policy.* Wageningen Pers, Wageningen.

Brouwer, F.M., Terluin, I.J. and Godeschalk, F.E. (1994) *Pesticides in the EC.* Onderzoekverslag 121. Agricultural Economics Research Institute, The Hague.

Carson, R. (1962) *Silent Spring.* Fawcett Crest, New York.

CEC (1995) *The Situation of Agriculture in the Community,* Report 1994. Brussels.

CEC (1996) *The Situation of Agriculture in the Community,* Report 1995. Brussels.

CEC (1997a) *Agenda 2000. Volume I Communication: For a stronger and wider Union.* DOC/97/6, Strasbourg.

CEC (1997b) *The Situation of Agriculture in the Community,* Report 1996. Brussels.

Dinham, B. (1993) *The Pesticide Hazard: A Global Health and Environmental Audit.* Zed Books, London.

Falconer, K.E. (1997) Environmental policy and the use of agricultural pesticides. Unpublished PhD thesis, Department of Land Economy, University of Cambridge.

Falconer, K.E. (1998) Classication of pesticides according to environmental impact. Royal Society for the Protection of Birds monograph. Sandy, UK.

Fawcett, R. (1991) Methods of monitoring soil regimes and the interpretation of data relevant to pesticide fate and behaviour. In: Walker, A. (ed.) *Pesticides in Soil and Water,* Monograph no. 47. British Crop Protection Council, Farnham.

Gardner, B.L. (1991) Redistribution of income through commodity and resource policy. In: Just, R.E. and Bockstael, N. (eds) *Commodity and Resource Policies in Agricultural Systems.* Springer-Verlag, Berlin, pp. 129-142.

Haen, H. de (1982) Economic aspects of policies to control nitrate contamination resulting from agricultural production. *European Review of Agricultural Economics* 9, 443-465.

Heerink, N.B.M., Helming, J.F.M., Kuik, O.J., Kuyvenhoven, A. and Verbruggen, H. (1993) *International Trade and the Environment.* Wageningen Economic Studies 30, Wageningen.

Huffman, W.E. and Evenson, R.E. (1992) Contributions of public and private science and technology to US agricultural productivity. *American Journal of Agricultural Economics* 74, 751-756.

Ivens, G. (1994) *UK Pesticide Guide 1994.* CABI/BCPC, Farnham.

Johnen, B.G. (1997) Risk assessment and crop protection products use reduction. In: Oskam, A.J. and Vijftigschild, R.A.N. (eds) *Proceedings and Discussions of the Workshop on Pesticides,* 24 - 27 August 1995, Wageningen, pp. 459-470.

Jones, P.J., Rehman, T., Harvey, D.R., Tranter, R.B., Marsh, J.S., Bunce, R.G. and Howard, D.C. (1995) *Developing LUAM (Land Use Allocation Model) and Modelling CAP Reforms.* Paper 32, Centre for Agricultural Strategy, University of Reading.

Just, R.E. and Bockstael, N. (eds) (1991) *Commodity and Resource Policies in Agricultural Systems.* Springer-Verlag, Berlin.

Kovach, J., Petzoldt, C., Degni, J. and Tette, J. (1992) A method to measure the environmental impact of pesticides. *New York's Food and Life Sciences Bulletin* 139, 1-8.

Michalek, J. and Hanf, C.-H. (eds) (1994) *The Economic Consequences of a Drastic Reduction in Pesticide use in the EU*. Vauk, Kiel.

Miranowski, J.A., Hrubovcak, J. and Sutton, J. (1991) The effects of commodity programmes on resource use. In: Just, R.E. and Bockstael, N. (eds) *Commodity and Resource Policies in Agricultural Systems*. Springer-Verlag, Berlin, pp. 275-292.

Moxey, A., White, B. and O'Callaghan, J.R. (1995) The economic component of NELUP. *Journal of Environmental Planning and Management* 40, 21-34.

Murphy, M. (1995) *Report of Farming in the Eastern Counties of England*. Agricultural Economics Unit, Department of Land Economy, University of Cambridge.

Nutzinger, H.G. (1994) Economic instruments for environmental protection in agriculture: Some basic problems of implementation. In: Opschoor, J.B. and Turner, R.K. (eds) *Economic Incentives and Environmental Policies: Principles and Practice*. Kluwer, Dordrecht, pp. 175-193.

Oppenheimer, Wolff and Donnelly (1996) *Elaboration on Possible Arguments and Objectives of an Additional EC Policy on Plant Protection Products* (PES-A 2/3), Study for Phase 2 of the project on 'Possibilities for future E.C. environmental policy on plant protection products'.

Oskam, A.J. (1994) Pesticides; issues at stake for agricultural economists. In: Michalek, J. and Hanf, C-H. (eds) *The Economic Consequences of a Drastic Reduction in Pesticide Use in the EU*. Vauk, Kiel, pp. 81-108.

Oskam, A.J. and Stefanou, S. (1997) The CAP and technological change. In: Ritson, Ch. and Harvey, D. (eds) *The Common Agricultural Policy*. CAB International, Wallingford, pp. 191-224.

Oskam, A.J. and Vijftigschild, R.A.N. (1999) Towards environmental indicators for pesticide impacts. In: Brouwer, F.M. and Crabtree, J.R. (eds) *Environmental Indicators and Agricultural Policy*. CAB International, Wallingford, pp. 157-176.

Oskam, A.J., Zeijts, H. van, Thijssen, G.J., Wossink, G.A.A. and Vijftigschild, R. (1992) *Pesticide Use and Pesticide Policy in the Netherlands: an Economic Analysis of Regulatory Levies in Agriculture*. Wageningen Economic Studies 26, Pudoc, Wageningen.

Oskam, A.J., Vijftigschild, R. and Graveland, C. (1998*). Additional EU Policy Instruments for Plant Protection Products*. Wageningen Pers, Wageningen.

Oude Lansink, A. and Peerlings, J. (1995) Farm-specific impacts of policy changes on pesticide use in Dutch arable farming. In: Oskam, A.J. and Vijftigschild, R.A.N. (eds) *Proceedings and Discussions of the Workshop on Pesticides*, 24 - 27 August 1995, Wageningen, pp. 414-428.

Oude Lansink, A. and Peerlings, J. (1996) Modelling the new EU cereals and oilseeds regime in the Netherlands. *European Review of Agricultural Economics* 23, 161-178.

Penrose, L.J., Thwaiet, W.G. and Bower, C.C. (1994) Rating index as a basis for decision-making on pesticide use reduction and for accreditation of fruit produced under IPM. *Crop Protection 12*, 146-152.

Produce Studies (1996) *Analysis of Agricultural Policy in Relation to the Use of Plant Protection Products*. The Produce Studies Group, Berkshire.

Rayment, M. (1995) *A Review of the 1992 CAP Arable Reforms*. Arable Policy Paper no.1, RSPB, Sandy, UK.

Reus, J.A.W.A. and Pak, G.A. (1993) An environmental yardstick for pesticides. *Mededelingen Faculteit Landbouwwetenschappen, Universiteit Gent*, 58/21, 249-255.

Reus, J.A.W.A., Weckseler, H.J. and Pak, G.A. (1994) *Towards a Future EC Pesticides Policy: An Inventory of Risks of Pesticide Use, Possible Solutions and Policy Instruments*. CLM paper 149, CLM, Utrecht.

Schou, J.S. (1998) The CAP reform and Danish agriculture - An integrated environmental and economic analysis. In: Napier, T. *et al.* (eds) *Soil and Water Conservation Policies: Successes and Failures*. Soil and Water Conservation Society, pp. 301-313.

Skinner, J.A., Lewis, K.A., Bardon, K.S., Tucker, P., Catt, J.A. and Chambers, B.J. (1997) An overview of the environmental impact of agriculture in the UK. *Journal of Environmental Management* 50(2), 111-128.

Vatn, A., Bakken, L.R., Lundeby, H., Romstad, E., Rörstad, P.K. and Vold, A. (1997) Regulating nonpoint-source pollution from agriculture: An integrating modelling analysis. *European Review of Agricultural Economics* 24, 207-229.

Whitby, M. (ed.) (1996) *The European Environment and CAP Reform: Policies and Prospects for Conservation.* CAB International, Wallingford.

Wossink, G.A.A. (1993) *Analysis of Future Agricultural Change.* Wageningen Economic Studies 27, Wageningen.

Wossink, G.A.A. and Renkema, J. (1994) Analysis of future change in Dutch arable farming: A farm economics approach. *European Review of Agricultural Economics* 21, 95-112.

Wossink, G.A.A., Van Kooten, G.C. and Peters, G.H. (eds) (1998) *Economics of Agrochemicals: an International Overview of Use Patterns, Technical and Institutional Determinants, Policies and Perspectives.* Ashgate Publishing Co., Aldershot.

Wu, J.J. and Segerson, K. (1995) The impact of policies and land characteristics on potential groundwater pollution in Wisconsin. *American Journal of Agricultural Economics* 77, 1033-1047.

WWF (1995) *Pesticide Reduction: Economic Instruments.* WWF, Gland, Switzerland.

Zeijts, H. van (ed.) (1999) *Economic Instruments for Nitrogen Control in European Agriculture.* CLM-report 409, Utrecht, Centre for Agriculture and Environment.

The Arable Crops Regime and Nitrogen Pollution Control

7

Floor Brouwer and Marga Hoogeveen

INTRODUCTION

The objectives of the 1992 Common Agricultural Policy (CAP) reform were to improve the competitiveness of European Union (EU) agriculture, to restore market balance and to stimulate less intensive production methods. Measures were taken to reduce surplus production, reduce price support (together with more targeted income support) and improve environmental soundness of agricultural production (CEC, 1993). One of the objectives of the reform of the arable crops regime was to enhance the competitive position of these crops. The share of cereals used as animal feed was foreseen to increase, as a result of the reductions in cereal intervention prices displacing high-protein cereal substitutes (e.g. soybean products).

Excretion is a necessary by-product of digestive and metabolic processes in animals. Nitrogen excretion in the EU is high because nitrogen is commonly fed far above requirements in the presently available least-cost formulated feed. The protein content of commonly used pig diets is 17% in Europe but may reach levels down to 14% in the USA. Such differences of feed affect excretion levels. Feed strategies for pigs and poultry could potentially contribute to reduce nitrogen surpluses. Such strategies include the use of feed supplements and the modification of feeding systems. The reform in 1992 of the arable crop regime significantly lowered the cost of reducing the protein content of compound feed. This enabled the pig and poultry sectors to apply nutritional management measures to reduce nitrogen output without compromising profitability throughout the production chain.

Assessments made so far on the potential economic and environmental benefits of agricultural policy reform in reducing nitrogen pollution from intensive livestock units remain limited. Some analyses have focused on feed strategies, and their potential contribution to reduce nitrogen pollution. Folmer *et al.* (1995), for example, assessed the potential beneficial effects on the environment, which may result from rebalancing, as cereals used as animal feed become cheaper. Dourmad *et al.* (1995) assessed the impact of CAP reform on practices for feeding pigs. A price reduction for cereals does favour lower protein contents in diets, and subsequently lower nitrogen excretion from the livestock. The average protein content of the diets of fattening pigs was reduced by 1-1.5% in Brittany in 1993 compared to the previous year. This is equivalent to a reduction in nitrogen content of pig manure of 0.2 to 0.4 kg nitrogen per animal. Brouwer *et al.* (1999) assessed the potential economic and environmental benefits of nutritional management in response to the reform of the CAP and the Nitrates Directive.

The present chapter investigates the role of agricultural policy in an effort to improve nutritional management practices in the EU, aimed to reduce nitrogen pollution from intensive livestock production units. The economic and environmental benefits are assessed in response to the reform of the CAP and the Nitrates Directive. The central part of the chapter examines the role of agricultural policy reform in reducing nitrogen pollution and subsequently contributing to meeting the requirements of the Nitrates Directive. It is argued that lower prices of cereals stimulate the use of cereals in compound feed, reducing the protein content of feed to grow pigs and poultry, and subsequently reducing nitrogen pollution. The effects of further reform of the CAP under Agenda 2000 are also evaluated.

We will first explore the consumption of animal feed in the EU, primarily focusing on raw material applied for the production of compound feed. Some main factors determining raw material consumption in compound feed will be investigated, including the prices of cereals and the world market prices of raw material (mainly high-protein products like soybean). Following this overview, we will investigate the possible impact of CAP measures on the use of cereals in compound feed and the protein content of compound, being a major factor in nitrogen excretion from livestock. Some concluding remarks are made in the final part of the chapter.

Emphasis is given in the chapter to the consumption of compound feed by pigs because of their major share in the excess amounts of manure which need to be disposed of and their critical role in meeting the requirements of the Nitrates Directive. Similar conclusions should apply to poultry production.

ANIMAL FEED CONSUMPTION IN THE EUROPEAN UNION

Compound feed producers provide feed, which meets nutritional requirements in terms of energy, proteins and vitamins. Feed formulae are generally calculated by computer to meet a set of nutritional constraints at the least possible cost per tonne. Least-cost formulae do not automatically correspond to the most environmentally friendly feeds. The main factor influencing the cost of a reduction of the protein content in feed is the difference between the grain and protein-based feedstuff prices (soyabean meal). This is due to the relatively high gearing of the cereal-grain to protein-source price ratio.

Compound feed is based on a wide variety of raw materials, including products such as soya, manioc, cereals, citrus products, maize and residual material from oil production. Total production of compound feed in the EU-12 in 1995 amounted to some 120 million tonnes. About a third was to feed pigs; about 28% for poultry, and the rest mainly for ruminant animals (Table 7.1).

Just four countries (Germany, Spain, France and The Netherlands) account for approximately 60% of the production of compound feed within the EU (Table 7.1), also covering the major part of its consumption.

Use of cereals in animal feed

The total consumption of animal feed includes the raw material fed directly to the livestock, and the compound feed delivered by feed companies. Cereals are an important component of both. Table 7.2 shows the total consumption of cereals as livestock feed in the EU. The figures presented are limited to the EU-12 because no detailed figures are available on the consumption of cereals by the new Member States. The use of cereals in animal feed in EU-12 amounted to some 90 million tonnes in 1994/95, which was some 7 million tonnes above the use in 1991/92. The figure for EU-15 was about 100 million tonnes. The European Cereal Trade Association (COCERAL) estimates a further increase to 108.0 million tonnes in EU-15 for 1996/97 (Agra Europe, 25 July 1997).

Table 7.1 Compound feed production in the EU in 1995 by Member State (in 1,000 tonnes)

Country	Cattle	Pigs	Poultry	Other	Total
Belgium	1,258	3,278	1,095	105	5,736
Denmark	1,545	3,255	660	151	5,611
Germany	8,419	5,900	4,009	602	18,930
Greece	1,502	620	650	50	2,822
Spain	3,710	6,040	4,050	1,457	15,257
France	4,556	6,343	8,767	2,025	21,691
Ireland	1,909	620	460	361	3,350
Italy	3,960	2,600	4,300	1,140	12,000
Netherlands	4,787	7,303	3,286	713	16,089
Austria	155	215	410	232	1,012
Portugal	1,023	1,383	1,307	217	3,930
Finland	507	324	212	115	1,158
Sweden	1,130	650	500	50	2,330
UK	4,400	2,450	4,000	1,000	11,850
EU-12	37,069	39,792	32,584	7,821	117,266
EU-15	38,861	40,981	33,706	8,218	121,766

The use of home-grown cereals (i.e. cereals grown on the farm) accounts for 45.5% of the total use of cereals in animal feed in the EU-12. The rest, amounting to about 49.5 million tonnes, is traded, almost 41 million tonnes of which are used to produce compound feed (see Table 7.3).

Variations between countries regarding the on-farm use of cereals are due to differences in agricultural production systems. The use of home-grown cereals for animal feeding is very important in Denmark, France and Germany, but is unimportant in Belgium and The Netherlands. In Denmark and Germany, about two-thirds of the cereals used for feed are grown on the farm. Thus, approximately 40% of the pig production in Denmark is on mixed holdings that include both livestock and crops. In Denmark, the use of home-grown cereals to feed animals has even increased since the early 1990s. This is probably due to the available equipment and knowledge of direct use of cereals in animal feed. Such skills do allow for substitution possibilities between home-produced and market-based sources of feed.

In general, the share of cereals in compound feed had fallen during the 1980s, reaching its lowest level (28.7%) in 1991 for the EU overall. In more recent years, however, it has shown an upward trend (Table 7.2). Reduction of intervention prices for cereals increased their competitiveness compared to substitutes. Since the reform of the arable crop regime in 1992 intervention prices have been reduced by about one-third to reach 100 green ECU per tonne by 1995/96. The effect has been to return to the amounts of cereal in compound feed of the early 1980s.

There are different national patterns here. In those countries (Denmark, France and Germany) with a high reliance on home-grown cereals, there is a medium share of cereals in compound feed which has been increasing steadily since the late 1980s. In countries with a medium reliance on home-grown cereals (Spain and Italy), there is a high share of cereals in compound feed which has been slowly decreasing since

Table 7.2 The use of cereals in compound feed (share of total compound feed, in %) by Member State; total consumption of cereals in compound feed in the EU (million tonnes) (1980-1995)

Country	1980	1986	1987	1990	1992	1993	1994	1995
Belgium	30.4	28.6	29.3	14.4	12.6	14.7	17.3	17.9
Denmark	34.3	27.4	27.5	27.9	26.6	28.6	33.4	38.6
Germany	27.0	23.7	19.9	21.3	25.4	25.9	28.4	29.8
Spain	65.1	63.1	64.0	48.3	58.7	62.5	59.4	58.3
France	45.6	41.3	33.4	32.2	30.7	32.5	37.2	38.8
Ireland	45.2	32.6	34.4	24.5	26.7	25.7	26.5	29.4
Italy	57.3	50.3	48.1	46.6	46.3	45.7	47.0	45.8
Netherlands	19.5	14.8	12.5	13.2	13.5	13.3	15.1	16.2
Austria	n.a.	n.a.	n.a.	n.a.	n.a.	n.a.	18.3	19.8
Portugal	62.3	34.8	27.8	23.3	25.5	27.0	28.7	30.4
Finland	n.a.	n.a.	n.a.	n.a.	35.8	36.0	35.1	38.3
Sweden	n.a.	n.a.	n.a.	n.a.	n.a.	n.a.	41.0	43.6
UK	51.7	40.6	40.0	35.8	32.4	32.1	33.7	34.7
EUR-12 [a]	42.2	35.6	32.9	30.1	30.5	31.2	33.4	34.6
EUR-15	-	-	-	-	-	-	33.4	34.7
Total	40.3	34.5	31.7	31.2	34.4	36.0	39.1	40.6

[a] Exclusive of Greece and Luxembourg.
Source: FEFAC (1995).

Table 7.3 Consumption of cereals in animal feed for some Member States in the EU in 1994/95

	Netherlands	Belgium	Denmark	France	Germany	Spain	Italy	EU-12
Cereals in animal feed (x 1,000 tonnes)	2,506.0	1,925.0	4,926.0	21,476.0	21,215.0	12,210.0	10,635.0	90,782.0
% change from 1991/92	15.9	4.6	2.0	15.7	19.5	-3.6	-5.2	9.3
Of which:								
- Home-grown cereals (x1,000 tonnes)	74.0	304.0	3,293.0	10,481.0	14,291.0	4,300.0	2,676.0	41,264.0
Share in total animal feed consumption (%)	3.0	15.8	66.8	48.8	67.3	35.2	25.2	45.5
Change from 1991/92 (%)	8.8	-9.0	40.7	0.6	17.1	-24.7	-8.1	4.7
- Marketed cereals (x 1,000 tonnes)	2,432.0	1,621.0	1,633.0	10,995.0	6,924.0	7,910.0	7,959.0	49,518.0
Share of total animal feed (%)	97.0	84.2	33.2	51.2	32.7	64.8	74.8	54.5
Change from 1991/92 (%)	16.1	7.6	-34.4	35.0	24.7	13.7	-4.2	13.6
Of which:								
- Wheat (x 1,000 tonnes)	1,022.0	712.0	2,586.0	9,902.0	7,625.0	1,664.0	1,041.0	31,367.0
in % of total animal feed	40.8	37.0	52.5	46.1	35.9	13.6	9.8	34.6
% change from 1991/92	52.1	43.0	67.8	44.9	44.0	12.9	-38.2	34.8

Source: Eurostat, Crop production 1-1997.

the early 1990s. Finally, The Netherlands Belgium and Luxembourg use only small amounts of cereals in animal feed, whether in compound feed or direct feeding on farm. Their position close to major harbours (Rotterdam, Antwerp, Ghent) which import raw materials for the compound feed industry, is a critical factor. Nevertheless, they have seen some increase in the proportion of cereals in compound feed since the early 1990s.

Differences exist between countries in the cereal composition (wheat, barley, maize and other cereals) of compound feed, due to agricultural production systems and the availability of raw material. At EU level almost 35% is wheat, 28% is barley and 27% is maize. The share of wheat is relatively high in The Netherlands, Denmark and Germany, but relatively low in Spain and Italy. Barley is an important raw material, which is widely applied in feed production in Spain, and maize is relatively more used in France and Italy. A substantial amount of rye is used in Germany to feed livestock. In respect of nitrogen content it is noteworthy that maize has a substantially lower protein content (8.7%) than wheat (11.9%) (CVB, 1996).

Import of raw material to produce compound feed

A large part (well over 40%) of the raw material requirements to produce compound feed is being imported from outside the EU - amounting to some 54 million tonnes in 1996. The four main countries of origin - accounting for about two-thirds of the raw material imports - are the US, Brazil, Argentina and Thailand (Table 7.4). The imported material is mainly soya products.

Table 7.4 Import of raw material for the production of compound feed in the EU from outside the EU in 1996, by commodity and country of origin (million tonnes)

Country of origin	Soya products	Manioc/tapioca	Cakes	Maize gluten feed	Total
USA	9.5	<0.1	0.2	2.8	15.1
Brazil	9.7	<0.1	<0.1	0.4	11.5
Argentina	5.6	<0.1	1.9	0.1	8.3
Thailand	<0.1	3.2	<0.1	<0.1	3.3
Other	1.2	0.3	4.4	<0.1	15.7
Total	26.0	3.5	6.5	3.3	53.9

Source: Eurostat, adaptation LEI-DLO.

The main countries importing raw material to produce compound feed are the Netherlands (15.1 million tonnes), Spain (6.9 million tonnes), Germany (6.5 million tonnes) and France (5.1 million tonnes). These countries already account for well over 60% of the total import from non-EU countries of raw material to produce compound feed (Table 7.5). More than half of the import of manioc and tapioca is going to The Netherlands.

The market conditions for raw material play a very significant role in global trade patterns. The price of soybean products may show large inter-annual variations. There are many factors which might contribute to an increase in soybean prices, including poor harvests in other parts of the world and the fluctuating exchange rates of the US dollar. The high prices for soybean products in late 1997 were partly due to lower fish catches in Latin America, which increased the demand for alternative sources of protein. The import of soybean (products) largely responds to such price variations (Figure 7.1).

Nitrogen content of imported raw material

The EU import of raw material to produce compound feed (exclusive of cereals and compound feed) from non-EU countries amounts to 53.9 million tonnes per annum. Total import (also including intra-trade) is some 70 million tonnes per annum. Such raw material includes high-protein material (soybean products with a nitrogen content of 3-7%), manioc and tapioca (nitrogen content of less than 1%). The import of raw material to produce compound feed in 1996 amounted to some 3 million tonnes of nitrogen, with more than 75% of it originating from outside the EU (Table 7.6). Measured in nitrogen, soya products include the major part of this import (1.9 million tonnes and around 60% of total import) because of their high protein content. The Netherlands accounts for more than 20% of the total import of feed (from inside as well as outside the EU) in the Member States.

CAP REFORM AND PROTEIN CONTENT OF COMPOUND FEED

The 1992 reform of the arable crop regime affected cereal prices and also contributed to price reductions of raw material to produce compound feed. A reduction of cereal prices, for example, results in a modification of the composition of compound feed. We will now assess the impact of CAP reform on the protein content of feed, in order to judge its contribution to a cost-effective reduction of nitrogen pollution from European livestock production systems. The focus is on pigs, because this sector largely contributes to the excess amounts of manure.

The analysis is based on the Cereal and Compound Feed Market Model (CCM) of LEI-DLO. CCM is a regionalised multi-commodity model of cereals and compound feed raw materials. The model only covers the original 12 Member States of the EU (Blom, 1995). The three Member States which only entered the EU in 1995 - Austria, Finland and Sweden - are not yet included in this model. As indicated in Table 7.1, these countries account for a very limited part of the compound feed market in the EU.

A distinction is made into six policy alternatives. They distinguish between three scenarios for agricultural policy and two types of nutritional management. The three scenarios for agricultural policy are as follows:

- the period before the reform of 1992 (*CAP-1988*);
- the period following reform of agricultural policy in 1992, taking conditions for 1994/95 (*CAP-1994*);

Table 7.5 Country of origin of the import of raw material for the production of compound feed by country of destination in 1996 (in million tonnes)

Country of destination	USA	Brazil	Argentina	Thailand	Total
Belgium	0.6	2.8	0.4	<0.1	3.3
Denmark	0.4	0.5	0.9	<0.1	2.6
Germany	2.2	1.2	0.6	0.1	6.5
Greece	0.1	<0.1	0.2	<0.1	0.6
Spain	1.9	1.8	1.0	0.7	6.9
France	0.7	2.8	0.4	<0.1	5.1
Ireland	0.5	0.1	0.1	<0.1	1.2
Italy	0.8	0.9	1.3	<0.1	4.7
Netherlands	5.4	2.7	2.1	1.9	15.1
Austria	<0.1	<0.1	<0.1	<0.1	0.2
Portugal	0.7	0.5	0.1	0.3	2.1
Finland	0.1	<0.1	<0.1	<0.1	0.2
Sweden	<0.1	0.1	<0.1	<0.1	0.5
UK	1.4	0.6	0.7	<0.1	4.9
EU-15	15.1	11.5	8.3	3.3	53.9

Source: Eurostat, adaptation LEI-DLO.

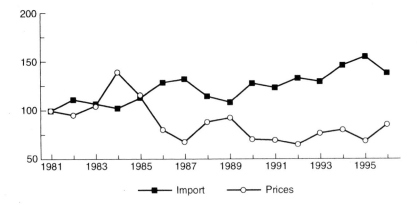

Figure 7.1   Relationship between import volumes and price of soy (bean) products from 1981 to 1996 (1981 = 100; prices in ECU per tonne) (Source: Eurostat, adaptation LEI-DLO)

Table 7.6 Import of feed concentrates (exclusive of cereals), expressed in 1,000 tonnes of nitrogen by Member State in 1996

Origin/destination	F	BLEU	NL	DE	IT	UK	EIR	DK	GR	PT	ES	SE	FI	AT	EU-15
France (F)	-	33.6	12.9	28.9	4.3	7.1	1.9	2.0	0.4	0.6	6.4	0.8	0.6	0.1	99.6
Belgium/Luxemb. (BLEU)	61.1	-	16.9	3.0	0.1	11.9	1.8	0.7	0.0	0.2	0.3	0.8	0.4	0.1	97.3
Netherlands (NL)	14.5	47.4	-	99.5	1.1	45.3	10.3	5.9	0.5	0.3	0.6	7.1	1.3	11.6	245.3
Germany (DE)	10.4	10.4	75.0	-	0.9	16.1	1.1	26.0	0.1	0.0	0.2	6.9	2.4	15.8	165.1
Italy (IT)	1.7	0.0	0.2	0.5	-	1.5	0.6	0.5	0.3	0.0	0.2	0.0	0.0	1.1	6.7
United Kingdom	3.3	1.6	3.1	1.0	0.1	-	6.8	4.9	0.4	0.1	0.2	1.5	0.3	0.0	23.2
Ireland (EIR)	0.3	1.8	0.4	0.0	0.2	4.7	-	0.0	0.0	0.0	0.0	0.0	0.0	0.0	7.4
Denmark (DK)	0.4	0.6	1.7	2.4	3.5	1.9	0.6	-	2.2	0.4	1.6	3.6	1.2	0.0	20.0
Greece (GR)	0.0	0.0	0.1	0.0	2.5	0.4	0.4	0.0	-	0.0	0.9	0.0	0.0	0.0	4.3
Portugal (PT)	0.1	0.0	0.0	0.1	0.0	0.0	0.1	0.0	0.0	-	1.8	0.0	0.0	0.0	2.1
Spain (ES)	4.3	0.4	0.2	0.1	0.8	0.9	0.5	0.2	1.0	3.0	-	0.1	0.0	0.0	11.5
Sweden (SE)	0.0	0.0	0.0	0.1	0.0	0.0	0.0	0.9	0.0	0.0	0.0	-	1.3	0.0	2.4
Finland (FI)	0.0	0.0	0.0	0.0	0.0	0.0	0.0	0.0	0.0	0.0	0.0	0.0	-	0.0	0.1
Austria (AT)	0.0	0.2	2.2	2.4	2.5	0.5	0.0	0.0	0.0	0.0	0.0	0.0	0.0	-	5.5
EU-15	95.9	96.0	112.8	137.9	14.0	89.7	29.2	41.3	4.8	4.6	12.2	20.9	7.4	28.7	695.5
From outside EU	282.3	155.2	558.4	311.5	237.1	215.8	36.8	135.7	29.6	79.5	308.1	23.2	12.6	7.1	2,392.7
Total	378.3	251.1	671.2	449.3	251.1	305.5	66.0	177.0	34.4	84.1	320.1	44.1	20.0	35.9	3,088.2

Source: Eurostat, adaptation LEI-DLO.

- a 15% reduction in the market prices for 1995/96, anticipating an additional re-form of agricultural policy around the end of the century (*CAP-2000*).

For nutritional management, the following two types were evaluated:

- *Current nutritional management*; based on least-cost formulation of diets using feeding stuffs and feed supplement available to commercial companies. It ig-nores nitrogen output.
- *Controlled nutritional management*; in which reduction of nitrogen output is taken into account in feed formulation. Feed prices may differ from current nu-tritional management, as dietary protein level is likely to be reduced to a technically proven limit, which does not compromise animal performance.

The model calculations presented focus on the compound feed market of Belgium and Luxembourg, Denmark, the northern and central part of Germany, the north-eastern part of Spain, the north-western part of France, the northern part of Italy, The Netherlands and east England. These regions cover more than two-thirds of the com-pound feed market to feed pigs.

The increasing amount of compound feed from 18.3 to 20.4 million tonnes from CAP-1988 to CAP-1994 is due to an increased consumption of the pig herd (see Tables 7.7 and 7.8). We assume the consumption of compound feed to remain un-changed in comparing CAP-1994 with CAP-2000.

Table 7.7 Prices and composition of pig feed used in concentration areas of pigs with current nutritional management

	CAP-1988	CAP-1994	CAP-2000
Compound feed (million tonnes) [a]	18.3	20.4	20.4
Share of cereals in compound feed (%)	23.2	31.3	44.5
Protein content of compound feed (%)	18.0	17.7	15.7
Costs of compound feed (ECU per tonne)	149.9	134.4	110.5

[a] Production of compound feed for sows and finishing pigs, exclusive protein con-centrate.
Source: CCM model.

CAP reform gives an incentive to lower protein diets in feed

Compared to the situation around the end of the 1980s (period of stabilisers) the re-form of CAP in 1992 induced a reduction of the intervention prices of cereals (CAP-1994 compared with CAP-1988). Following this policy change, the prices of com-pound feed fell as well. The costs of raw material on average were reduced by about 15 ECU per tonne (Table 7.7). The protein content of feed decreased slightly from

18.0 to 17.7%. This is caused by a changed price ratio between energy and protein crops. An opposite movement did also take place because of the substitution of low-protein raw material (tapioca) by raw material with a medium protein level (i.e. cereals). The share of cereals in compound feed for pigs is projected to increase with a further reduction of cereal prices. Meanwhile the protein content of pig feed would largely reduce as well. A substantial reduction of nitrogen content in feed would be achievable with CAP-2000, contributing to a large reduction of protein content in feed.

Protein content of compound feed is targeted with controlled nutritional management

The protein content of compound feed may be lowered by a combination of agricultural policy reform and nutritional management measures. Control of the nutritional management would allow control of the protein content of feed at lower costs, compared to the conditions of the late 1980s.

In conclusion, the reform in 1992 did provide an incentive to lower the protein content of compound feed. This is due to both the reductions of cereal prices and changes in the prices of other raw material. A reduction in the protein content of feed also induces a reduction on the excess of nitrogen from livestock manure. This is based on the assumption that the performance of livestock production will remain unchanged. A decrease in cereal intervention prices will stimulate an increase in the uptake of cereals in compound feed and further lower the protein content.

The extra costs of nutritional management measures need to be compared with their benefits for the environment. Benefits for the environment include the amount of protein which has not been fed and consequently has not been excreted by the animals. Table 7.8 shows the results of a controlled protein diet for pigs (controlled nutritional management) in all three scenarios of agricultural policy (CAP-1998, CAP-1994 and CAP-2000). Putting extra restrictions on the use of protein by definition lowers the protein content of compound feed. Such restrictions imply larger costs of compound feed (cf. Table 7.7). The prices of compound feed would increase by 6.6 ECU per tonne under CAP-1988, 4.7 ECU per tonne under CAP-1994, and

Table 7.8 Prices and composition of pig feed used in concentration areas of pigs with controlled nutritional management

	CAP-1988	CAP-1994	CAP-2000
Compound feed (million tonnes) [a]	18.3	20.4	20.4
Share of cereals in compound feed (%)	32.2	37.5	46.7
Protein content of compound feed (%)	14.4	14.4	14.4
Costs of compound feed (ECU per tonne)	156.5	139.1	113.2

[a] Production of compound feed for sows and finishing pigs, exclusive protein concentrate.
Source: CCM model.

2.7 ECU per tonne under CAP-2000. In the CCM model price changes occur when the demand and supply of raw materials change. Therefore it can be expected that as a consequence of the lower demand for protein - other influences (i.e. the world market situations) being equal - the prices of protein-rich feedstuffs will decrease.

The impact of nutritional management on the price and composition of compound feed is different under each of the agricultural scenarios. In the case of CAP-1988 the impact of controlled nutritional management on protein content of feed is larger than in the case of CAP-1994 and CAP-2000. CAP-2000 results in the smallest changes to nutritional management measures. The introduction of controlled nutritional management increases the share of cereals in compound feed, compared with the case of current nutritional management. Following the reform of the arable crop regime, the protein content is reduced from 18.0% to 15.7% (Table 7.7). In the case of CAP-2000, controlled nutritional management might be achievable at lower costs compared to the case of CAP-1988. In that case, the protein content is already lower than the situation in the late 1980s (Table 7.9).

Placing extra restrictions on the use of protein by definition lowers the protein content of compound feed, and the price of compound feed also increases. The total amount of protein shows a downward trend in the case of controlled nutritional management, but impacts differ across scenarios. The role of CAP reform in reducing the nitrogen content of feed exceeds that of nutritional management. In the case of CAP-1988 a 20% reduction of nitrogen in pig feed is achievable at an increase of 4% of the total costs of raw material to produce compound feed (Table 7.9). CAP-1994, however, already reduces costs of compound feed by some 10%. In that case a 20% reduction of nitrogen in pig feed is assessed to be achievable with costs of compound feed 7% below the conditions under CAP-1988. Comparison of CAP-1994 with CAP-1988 achieves a 2% reduction on the nitrogen-content of feed. It needs to be mentioned the CCM model did not allow calculating the marginal costs to reduce the nitrogen content of compound feed. Only the average costs within a range can be estimated. Comparing CAP-1994 with CAP-2000 shows that a further reduction of market prices of cereals reduces costs of compound feed by another 16% points. Meanwhile the nitrogen content of compound feed to grow pigs also is reduced by

Table 7.9 Indices of costs of compound feed (ECU per tonne) and nitrogen content of feed (kg N per tonne of compound feed) for pigs by agricultural policy and type of nutritional management

Nutritional management Agricultural Policy	Index of costs (ECU per tonne)		Index of N-content (kg N per tonne feed)	
	Current	Controlled	Current	Controlled
CAP-1988	100	104	100	80
CAP-1994	90	93	98	80
CAP-2000	74	76	87	80

Source: CCM model.

some 11% points. A reform of the arable crop regime beyond CAP-1994 therefore is expected to increase the application of low-protein feeds.

In the absence of specific regulatory mechanisms, the dietary protein level and related nitrogen output will remain uncontrolled and driven by the highly volatile world market price of protein sources. Hence, there is no certainty that feeding programmes will automatically be adjusted for lower nitrogen excretion without necessary initiatives. Therefore, world market conditions may play a very significant role in allowing nutritional management practices to go beyond current practice. Soybean meal prices would largely affect the protein content in compound feed. A reduction of soybean meal prices may lead to important dietary protein increases. This may also reduce the potential of nitrogen pollution control. In order to minimise the negative effects on nitrogen pollution of low prices of soya and protein-based raw material, environmental protection measures may be required to ensure low dietary protein levels in feed formulations.

CONCLUSIONS

This chapter has evaluated the relative economic and environmental benefits of the available options to reduce nitrogen pollution from intensive livestock production units in Europe within the context of the changing CAP regime and the Nitrates Directive. The main conclusions of the analysis are summarised in the following.

- First, high nitrogen excretion is a direct consequence of dietary protein being fed far above animal requirements. Protein levels in feed can be reduced by providing feed more closely allied to the animal's requirements without affecting its performance, resulting in a reduced nitrogen excretion by animals. This technology does not require large additional investments; it is flexible, reducing pollution at source, and can be adjusted to the nitrogen-reduction targets.
- Second, feed formulae are generally calculated at the least possible cost per tonne, with feed to be balanced according to the nutritional requirements of livestock among others, regarding energy and proteins. The reform of the arable crop regime reduced the costs of lowering the protein content of compound feed. The reform in 1992 facilitated nitrogen pollution control, but did not achieve it. However, appropriate incentives would allow inducing a large shift toward environmentaly friendly formulation practices. A further incentive to the 1992 reform of CAP may enhance the competitiveness of cereals in the compound feed market and subsequently stimulate low-protein feed and reduce nitrogen pollution from intensive livestock.
- Third, and perhaps most important, the recent decisions in the context of Agenda 2000 are expected to further provide for a wider application of low-protein feeds. However, the dietary protein level and related nitrogen output will remain uncontrolled and driven to a large extent by the highly volatile world market price of protein sources. Soybean prices play a very significant role in the nitrogen content of compound feed. These prices respond rapidly to changes in world market conditions, and the EU has a limited influence on such prices because of

its small share in global consumption of soya. Hence, there is no certainty that feeding programmes will automatically be adjusted for lower nitrogen excretion without necessary initiatives. Any beneficial effects of CAP on the nitrogen content of compound feed therefore may be jeopardised by the soya market.

- Finally, efforts to meet the requirements of the Nitrates Directive should go beyond the means of livestock manure application (good farming practices) and treatment. Preventive nutritional measures, which have immediate effects, and a minimum economical burden, rather than alternative curative measures, should be enforced in keeping with structural changes.

REFERENCES

Blom, J.C. (1995) *Een Geregionaliseerd Graan- en Mengvoedergrondstoffenmarktmodel voor de EU-12: Model en Toepassingen (A Regionalised Cereal and Compound Feed Market Model in the EU-12: Model and Applications).* Agricultural Economics Research Institute (LEI-DLO), Onderzoekverslag 134, The Hague.

Brouwer, F., Hellegers, P., Hoogeveen, M. and Luesink, H. (1999) *Managing Nitrogen Pollution from Intensive Livestock Production in the EU: Economic and Environmental Benefits of Reducing Nitrogen Pollution by Nutritional Management in Relation to the Changing CAP Regime and the Nitrates Directive.* Agricultural Economics Research Institute (LEI), Rapport 2.99.03, The Hague.

CEC (1993) *The Agricultural Situation in the Community 1992.* Commission of the European Communities, Brussels.

CEC (1997) *The Agricultural Situation in the Community 1996.* Commission of the European Communities, Brussels.

CVB (Centraal Veevoederbureau) (1996) Voedernormen landbouwhuisdieren en voederwaarde veevoeders. Lelystad, August 1996, CVB-Reeks, Vol. 20.

Dourmad, J.Y., Le Mouel, C. and Rainelli, P. (1995) Réduction des rejets azotés des porcs par la voie alimentaire: évaluation économique et influence des changements de la politique agricole commune. *INRA Productions Animales* 8(2), 135-144.

FEFAC (1995) Feed and Food. Statistical Yearbook - 1995. Brussels.

Folmer, C., Keyzer, M.A., Merbis, M.D., Stolwijk, H.J.J. and Veenendaal, P.J.J. (1995) *The Common Agricultural Policy Beyond the MacSharry Reform.* North-Holland Publishing Company, Series Contributions to Economic Analysis, Volume 230, Amsterdam.

The Arable Crops Regime and the Countryside Implications

8

Michael Winter

INTRODUCTION

The 1992 reforms to the Common Agricultural Policy (CAP) arguably found their most radical and far-reaching manifestation within the arable sector covering chiefly cereals, protein crops and oilseeds. The reform was radical because of the extent of the price cuts (for example, intervention prices for grain were to be cut by approximately 35% by 1995/96) and, through the introduction of compensatory payments, the corresponding decoupling of support from production. At the same time, the decision to make set-aside an integral part of the reform placed large-scale land diversion on the agenda of European agriculture for the first time. It is, therefore, of significance to assess the environmental outcome of the 1992 reforms in some detail. This chapter focuses on the impact of the 1992 arable reforms on the British countryside, placing this particular case in the broader context of the experience of other countries wherever possible.

The British case study draws on several sources, including two surveys of farmers' responses to the reforms. The first survey covered the period 1992 to 1995 and was based on interviews with 575 farmers in Great Britain undertaken in the autumn of 1995 and spring of 1996 (Winter and Gaskell, 1998a). The interviews focused on farm management responses to the 1992 CAP commodity reforms concentrating on the dairy, beef, sheep and arable sectors. The second survey was based on interviews in 1997 with 575 arable farmers in England and Wales (Andersons, 1997).

A BRIEF OVERVIEW OF THE ARABLE SECTORS 1992-1995

On 1 July 1993 the intervention prices for cereals were reduced for the first time (by 24.8% for common wheat and maize, 44.3% for durum wheat and 20.9% for other cereals). Further reductions of 2.7% and 7.4% occurred for all cereals in

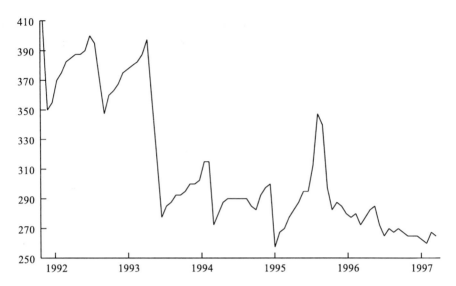

Figure 8.1   Monthly spot prices for milling wheat, Hamburg, July 1991 to June
                  1998 (DM per tonne)
Source: data supplied by UK Home Grown Cereals Authority.

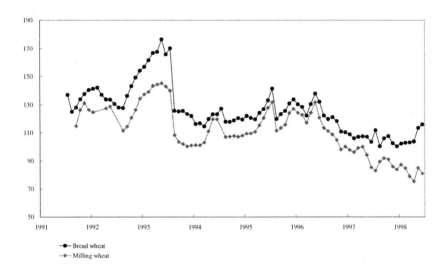

Figure 8.2   UK ex-farm spot prices - monthly averages, July 1991 to June 1998
                  (£ per tonne)
Source: data supplied by UK Home Grown Cereals Authority.

1994 and 1995 respectively. The reduction in intervention prices did not lead to a large reduction in market prices received by farmers in the UK, because of the trend in world supplies and prices combined with the devaluation of sterling after the UK's withdrawal from the European Exchange Rate Mechanism in September 1992. Consequently, British cereal farmers prospered during the immediate post-reform period, more so than many of their European counterparts, although in Spain, Italy and Greece prices were at or above those in the UK. Figure 8.1 shows that farm prices fell by 27% for milling wheat in Hamburg between May 1992 and May 1995 compared to a 6% fall in the UK during the same period (Figure 8.2). By 1996, European prices had rallied significantly with May 1996 Hamburg prices for milling wheat just 13% lower than in 1992, and in the UK, farm gate prices had returned to the same level as in the pre-reform markets in 1992. Of course, in real terms prices fell a little but UK cereal farmers were richly over-compensated by the Arable Area Payments Scheme (AAPS), and farm incomes in this sector increased in the period immediately after the reforms (Figure 8.3).

A large majority of arable farmers maintained or improved the profitability of their farms and there was an increase in turnover on four out of five farms as shown in Table 8.1. Amongst those interviewed in the 1997 study, three-quarters felt that the AAPS had increased profits (Andersons, 1997). The AAPS effect was considerably boosted by the weak pound during this period. Thus the green pound

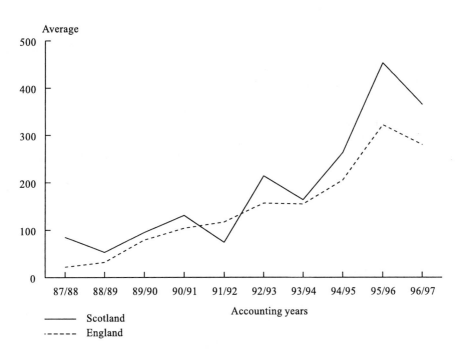

Figure 8.3   Cereal farmers: net farm income in England and Scotland, current
            prices (indices, 1989/90 – 1991/92 = 100)
Source: MAFF annual.

Table 8.1 Changes in total agricultural turnover 1991/92-1994/95 by major arable systems on surveyed British farms (percentage of total number of farms in each category)

Change in turnover	Cereal	Cropping	Mixed
Increase	82.1	80.5	80.0
No change	8.9	13.4	14.4
Decrease	9.0	6.1	5.6

Source: Winter and Gaskell (1998a).

Table 8.2 Crop subsidies (£ x 1,000) on UK cereal farms adjusted for green £ devaluations after September 1992

	1992	1993	1994	1995
Net farm income £ per farm	24.8	23.0	30.4	46.6
Crop subsidies £ per farm	9.1	26.3	31.0	38.1
Crop subsidies adjusted to Sept. 1992 green £ rate	9.1	22.0	26.4	29.8
Total devaluation effect £ per farm	-	4.3	4.6	8.3
Net farm income excluding devaluation effect £ per farm	24.8	18.7	25.8	38.3
Devaluation of green £ % since Sept. 1992	-	19.3	17.2	27.7

Source: MAFF data analysed by Andersons (1997).

devaluations amount to an average value of £4,000 per farm in 1993 and 1994 rising to over £8,000 in 1995 (Table 8.2). After 1995 the reverse situation came to prevail with the strengthening of sterling.

SET-ASIDE: HIGH HOPES AND LOST OPPORTUNITIES?

Under the terms of the 1992 reforms to the arable regimes (Regulation 1756/92), compensatory payments could be made under two schemes. The main scheme involved a complex range of differential payment rates for different crops and required farmers to set-aside a predetermined proportion of their arable land. Under the simplified scheme farmers could opt to receive a flat-rate payment equivalent to the cereals rate and avoid the set-aside requirement, but this option was only available to farmers claiming on a limited area of land (less than 15.62 ha in England). Given the size structure of English arable farming, it is not surprising that the overwhelming area of arable production came under the main scheme (96% of land, managed by 66% of producers). Consequently the total area of set-aside land has been considerable, accounting for just under 500,000 hectares in England alone in 1995/96, clearly posing important implications for farmed landscapes and for biodiversity. Table 8.3 shows the extent of set-aside within the European Union (EU) in 1996, showing a wide variation from country to country. This is, in the main, a consequence of the farm size structure. Countries with a high proportion of small arable holdings, most of which have opted for the simplified scheme, have low set-aside areas (for example Belgium, Greece, Ireland, The Netherlands and Portugal). In some countries, for example Sweden, the marginal nature of arable agriculture has encouraged some farmers into voluntary set-aside thus increasing the proportionate importance of set-aside. What is clear, though, is that set-aside, despite the publicity it attracted when first introduced, has occupied a relatively small area of all agricultural land (just over 5% of utilised agricultural land in the EU as a whole in 1996).

Evidence on the environmental implications of set-aside is mixed. It is accepted in northern Europe that there should be overall environmental benefits of set-aside at a field level, in terms of increased biodiversity, particularly overwintering birds, and that these benefits will be the greater the more long-term the set-aside (Clarke, 1992; Berg and Part, 1994). However, research on the actual agronomic and environmental implications of set-aside in the UK (Crabb *et al.*, 1998; Firbank, 1998) found no overall change in local bird populations since the introduction of set-aside, although Firbank (1998) speculates that perhaps set-aside has served to slow down rates of decline.

The use made of the different type of set-aside options is crucial. In the UK only a small number of farmers have used voluntary set-aside as a means of semi-permanent withdrawal of land from intensive arable production for more environmentally beneficial forms of land use. Farmers tend to see set-aside primarily in agricultural management terms, with the land treated as an agricultural break rather than as a potential site of environmental significance.

Long-term set-aside is not necessarily beneficial. For example, in Italy 36% of set-aside is voluntary and is concentrated in the hills of southern and central Italy, where its impact is generally seen as environmentally negative. It is associated with hastening a continuing decline of arable farming in the hills and mountains, contributing to the loss of 'traditional crops, landscape beauty, and the so-called 'mosaic' pattern made up of *ager* (arable), *saltus* (meadows and pasture) and *silva* (woodland) landscapes.' (Bordin *et al.*, 1998, p. 248).

INTENSITY OF PRODUCTION

Prior to the 1992 reforms, nitrogen fertiliser consumption in western Europe had declined by 13% between 1986 and 1992 as a result of on-farm improvements in nutrient efficiency (Williams, 1994). In England, physical applications of nitrogen to winter wheat had declined by 5.7% from 189.1 kg to 178.4 ha[-1] between 1985 and 1993 (Davidson and Asby, 1995).

The relatively attractive cereal prices and consequent rise in incomes on arable farms in the UK after 1992 interrupted these trends. The healthier than expected market prices up to 1996 actually encouraged a modest increase in the use of inputs. Figure 8.4 shows how the initial decline in fertiliser applications associated with the reforms in Great Britain was then reversed. It should be noted that these figures, which cover all categories of agricultural land, include set-aside, so it would be expected that the average application rate would decline after 1992, all other things being equal.

While the majority of farmers surveyed by Winter and Gaskell (1998a) had not changed the amount of nitrogen they applied to their major crops between 1991/92 and 1994/95, where changes had been made there was a general tendency to increase applications, as shown in Table 8.4. For example nearly a quarter of farmers with wheat had increased their fertiliser applications compared to just 11% who had reduced applications. This is consistent with the Cambridge University figures which show that winter wheat fertiliser applications per ha were static between 1993 and 1994 but increased from 740 kg ha^{-1} in 1994 to 790 kg (active ingredients = N 200, P58, K48) in 1995, this despite an increase in cost during the same period of 10% (Asby and Sturgess, 1997). Similarly,

Table 8.3 The scale of set-aside in EU agriculture, 1996

Country	Total UAA (1,000 ha)	Total base area (1,000 ha)	Total area set-aside (1,000 ha)	Set-aside as a % of total base area	Set-aside as a % of total UAA
Belgium	1,363	479	22	4.6	1.6
Denmark	2,712	2,018	262	13.0	9.7
Germany	17,308	10,156	1,472	14.5	8.5
Greece	5,163	1,492	17	1.1	0.3
Spain	28,929	9,220	1,481	16.1	5.2
France	30,343	13,526	1,874	13.9	6.2
Ireland	4,407	346	33	9.5	0.7
Italy	16,743	5,801	719	12.4	4.3
Luxembourg	127	43	2	4.7	1.6
Netherlands	1,963	437	20	4.6	1.0
Austria	3,479	1,203	125	10.4	3.6
Portugal	3,990	1,054	72	6.8	1.8
Finland	2,522	1,591	204	12.8	9.9
Sweden	3,438	1,737	322	18.5	9.4
UK	15,889	4,461	634	14.2	4.0
EU-15	138,376	53,561	7,259	13.6	5.2

Notes: UAA = Utilised Agricultural Area.
Set-aside is inclusive of 1992 set-aside and five-year set-aside.
Source: UAA figures for 1994: European Commission (1997,Table 3.5.2.2 page T/131). All other figures for 1995/96: European Commission (1997, Table 3.5.7.1 page T/162).

Andersons (1997) found that 8% of farmers farmed less intensively with the introduction of the 1992 schemes, 20% more intensively and 71% indicated no change.

The formal reduction in cereal prices under the 1992 reforms had been expected to lead to reduced intensity of production. In countries where the arable sector did not enjoy the temporary advantage of a weak currency that UK producers had, the picture was more complex. Table 8.5 shows how in France wheat increased in relative importance and nitrogen applications remained barely

Table 8.4 Changes in nitrogen application rates 1991/92-1994/95 on surveyed British farms (percentage of total number of farms having the particular crop)

Crop	Increase	No change	Decrease
Wheat	23.4	65.3	11.3
Winter barley	17.5	74.4	8.1
Spring barley	8.1	87.7	4.2
Oats	2.8	89.5	7.7
Maize	4.5	85.5	10.0
Winter oilseed rape	8.5	79.4	12.1

Source: Winter and Gaskell (1998a).

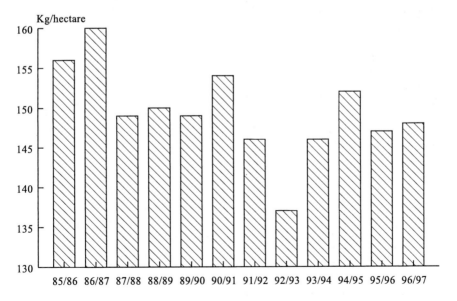

Figure 8.4   Nitrogen fertiliser usage in Great Britain (arable land)
Source: Fertilizer Manufacturers' Association (annual).

unchanged in the post-1992 framework, so that by 1996 wheat yields were at an historically high level with the area grown exceeding that under cultivation in 1990. A significant proportion of French farmers also responded to lower cereal prices by diversifying into other crops, notably fruit and vegetables. This, in its turn, brought about an increase in the use of plant protection and fertilisers as these crops are so demanding (Rainelli and Bonnieux, 1995). In Germany, only a little more than a third of surveyed farmers intended to reduce inputs following the 1992 reforms - unfortunately evidence on whether or not they did so is not provided (Kleinhanss, 1998). Evidence from Spain indicated likewise a likely initial reduction in fertiliser applications (Varela-Ortega and Sumpsi, 1998) but again there appears to be limited evidence as to whether these predictions were actually borne out. In Belgium, there is evidence of an upturn in nitrogen use on sandy loams in Flanders following legislation in 1991 to limit the spreading of animal manures. In contrast, elsewhere in the country on heavier soils applications seem to have declined a little in response to lower cereal prices (Bouquiaux *et al.*, 1998). Denmark offers perhaps the clearest evidence of a sustained decline in nitrogen use in the aftermath of the 1992 reforms but this had much more to do with legislative restrictions on use than the impact of lower prices per se (Linddal, 1998).

Table 8.5 Area of cereals, wheat yields and nitrogen use 1990-1996, France

	1990	1991	1992	1993	1994	1995	1996
Cereals area (1,000 ha)	9,060.6	9,226.1	9,338.5	8,541.2	8,166.9	8,293.0	8,829.3
Wheat area (1,000 ha)	4,737.5	4,635.0	4,630.8	4,262.2	4,314.9	4,485.2	4,741.3
Wheat yield (tonne ha^{-1})	6.6	6.9	6.6	6.6	6.8	6.6	7.3
Nitrogen (kg per fertilisable ha)	95.0	89.0	92.0	87.0	90.0	90.0	n/a

Source: Rainelli and Vermersch (1998).

There is evidence of declining inputs of plant protection products in The Netherlands (Dijk *et al.,* 1995; Brouwer and Van Berkum 1998), in France, Germany and the UK (Noé *et al.,* 1995) and in Denmark (SJI, 1995). Brouwer and Van Berkum suggest that factors other than the 1992 CAP reforms are of relevance in these trends including earlier CAP reform measures:

One important factor … relates to the reform of the CAP, already started in the 1980s. The stabiliser package for arable commodities reduces production growth and introduced the set-aside scheme. Moreover, a restrictive price policy led to real price decreases. Another factor influencing the sales of plant protection products is that contents of the active ingredients change so that farmers need less kilograms to protect crops against pests. Furthermore, the growing awareness of farmers of the negative environmental effects of excessive use of these products has affected the use of agrochemicals (Brouwer and Van Berkum, 1996, p. 74).

In The Netherlands, there is evidence of a continued decline in use of pesticides during the post-reform period as shown in Table 8.6 (Brouwer and Van Berkum, 1998) but at a lower rate than during the period immediately prior to the reform. It is very hard here to distinguish between the effects of the lower prices after 1992 and technological developments in pesticide use associated with cost-cutting initiatives linked to the development of integrated crop management, which has attracted particular interest in The Netherlands (Proost and Matteston, 1997). A similar story is told for Denmark (Linddal, 1998).

Table 8.6 Use of crop protection products on key arable crops in The Netherlands (kg of active ingredients per ha), 1990-1995

	Insecticides	Fungicides	Herbicides	Nematicides	Other	Total
Winter wheat						
1990	0.4	3.2	3.2	0.2	0.9	7.9
1992	0.3	2.1	2.8	0.0	0.8	6.0
1994	0.2	0.9	2.3	0.0	0.6	4.1
1995	0.2	0.8	2.4	0.0	0.6	4.0
Sugar beet						
1990	0.5	0.0	3.6	5.3	1.4	10.8
1992	0.4	0.0	3.3	2.0	1.6	7.3
1994	0.3	0.1	3.5	2.2	1.6	7.6
1995	0.2	0.0	3.5	0.7	1.9	6.3

Source: Brouwer and Van Berkum (1998, p. 172).

In one key respect the reforms have rendered moves towards a more rational use of agro-chemicals *less* likely within predominantly arable systems. This claim revolves around the complexity of the eligible land rules. In Britain, eligible land is land under eligible crops or temporary grass at 31 December 1991 as established and agreed by the farmer's IACS (Integrated Administration and Control System) submission. Eligible crops are those which are eligible to receive compensatory payments.[1] There are a number of aspects of these arrangements which

---

[1] Eligible crops: any cereals (including wheat and durum wheat, barley, oats, rye, triticale, sorghum, buckwheat, millet and canary seed, sweetcorn and maize for grain); oilseeds (rapeseed, sunflower seed and soya); proteins (peas for harvesting dry, field beans, sweet lupins); linseed; fodder maize. Ineligible crops: potatoes, turnips, swedes, kale, flax, linseed fibre varieties.

are environmentally unfavourable. By including temporary grass within the eligible crop area but not eligible for payment, farmers are encouraged to maximise their cropped area at the expense of temporary grass so as to maximise entitlement to arable area payments. Thus, the system encourages permanent cropping and fewer grass leys within rotations, thereby increasing dependency on chemical inputs. This is dramatically the case on traditional mixed farms with small arable areas operating sustainable long rotations where land may be cropped only every 5-10 years. The eligible rules 'capture' a particular area as arable and confine payments to that area of land. Thereafter, such farmers must either abandon their long rotations and intensify production on their eligible land or retain the long rotations and miss out on payments when cropping non-eligible land.

Thus, to sum up this section, taking fertiliser applications as a key indicator, there appears to have been no overall reduction in the intensity of arable production in the UK in the period from 1992 to 1996. On the contrary, in a significant proportion of cases, surveyed farmers intensified their application of fertilisers. We can infer a continuing trend of negative impacts, particularly with regard to biodiversity and natural resource protection. The evidence from other European countries is less clear-cut but there was certainly no general decline in production intensity. It can be concluded that the 1992 CAP reforms failed to halt intensification. A significant exception is the intensive livestock-arable systems of the Low Countries, where there is evidence of a decline in intensity following the 1992 reforms, albeit from extremely high levels. It is important to stress that many of the available data refer to the period prior to 1996 and do not take into account the significant reduction in cereal prices in 1997 and 1998 which may well have led to wider reductions in the applications of fertilisers and pesticides.

THE OPERATION OF THE ARABLE AREA PAYMENT SCHEME

We turn our attention now to the way in which the AAPS has been implemented and its direct impact on farming. Figure 8.5 shows general trends in arable crops in Great Britain between 1975 and 1995, derived from the agricultural census. Despite the introduction of compulsory set-aside, areas of both wheat and winter barley increased from 1993.

Just over one-quarter of the farmers in the Winter and Gaskell (1998a) study had introduced a new enterprise to their farms since 1992 and 16% had started one or more arable enterprises. One-third of these farmers said that the 1992 reform of the CAP had influenced their decision to introduce a new crop. This is an interesting finding given that it was not the intention of the arable reforms to alter cropping patterns. The likelihood of a farmer starting a new arable enterprise varied according to his or her farming system. Farmers running arable systems (i.e. cereal, cropping and mixed types) were most likely to have started a new crop. The most striking change was in the choice of break crops. Farmers had a tendency to introduce rape or peas as a break crop because of their favourable subsidy and, in some instances, because they could be grown on set-aside land.

Wheat was sometimes introduced as a new crop when a farm expanded but it rarely replaced an existing crop.

Few dairy and hill farmers had started a new crop on their farms. In many instances dairy farmers had already been involved in growing forage crops such as cereals or, more especially, maize. Significantly, 17% of lowland livestock farmers had introduced one or more new crops. A more detailed investigation of this group found that usually maize or, sometimes, wheat had been introduced, usually at the expense of grass leys, as a means of intensifying forage production for the livestock enterprises. The replacement of temporary grassland by maize or cereals is a major feature of the changes since 1992, greatly stimulated by the fact that the new regime allowed arable payments on maize whether it was used as a fodder crop or as a cereal. The environmental implications of this switch are serious with regard to potential pollution and soil erosion, decline in biodiversity and reduced access to farmland.

Significantly greater proportions of farmers in Great Britain increased rather than decreased their acreages of wheat, winter barley and maize between 1991/92 and 1994/95 (see Table 8.7). Evidence from other countries (Brouwer and Lowe, 1998) suggests that, despite the constraints of set-aside, a significant proportion of farmers across northern Europe managed likewise to increase their arable cropping area with potentially negative implications for the countryside, in terms of landscape, nutrient balance and wildlife.

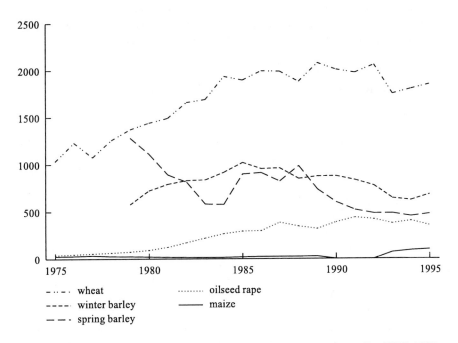

Figure 8.5   Arable crop areas Great Britain, June census results 1975-1995 (1,000 ha)

Table 8.7 Changes in arable crop area 1991/92–1994/95 by key crops on surveyed British farms (percentage of total number of farms having the
particular crop)

Crop	Increase	No change	Decrease	% of all arable farmers growing crop
Wheat	35.3	38.9	25.8	66.1
Winter barley	36.6	38.3	25.2	47.7
Spring barley	25.9	48.5	25.6	31.7
Maize	65.9	16.3	17.9	6.9
Potatoes	27.3	57.4	15.3	16.8
Sugar beet	18.8	70.4	10.8	13.7
Winter oilseed rape	26.8	41.0	32.2	20.1
Spring oilseed rape	33.2	33.1	33.6	4.0
Linseed	16.7	62.5	20.9	2.6

Source: Winter and Gaskell (1998a).

This was against the spirit of the 1992 reforms which had sought to freeze the
eligible land for arable area payments. But the eligibility rules allowed some latitude. It is true that some farmers with a high ratio of eligible crops to eligible land
had to reduce their areas of eligible crops (such as cereals) to fulfil set-aside requirements, and only in exceptional circumstances could eligible and ineligible
land be swapped. However, farmers with significant areas of temporary grassland
could increase their arable cropping. In addition, farmers were permitted to grow
arable crops on ineligible land although not permitted to claim AAPS payments
on those crops. This was an attractive option when market prices were expected to
be high. Farmers wishing to feed some of their crops to livestock had a strong incentive to maximise production of those crops on non-eligible land as part of their
forage area thus maximising their entitlement to AAPS payments on marketable
crops on their eligible land. Additionally, there was a strong incentive to grow
non-eligible crops on non-eligible land, for example fibre varieties of linseed in
1992 and 1993. With flax, there was a very strong subsidy-driven financial incentive to plough ineligible land. During 1994 and 1995 potato prices were
particularly buoyant in Britain, even resulting in permanent pasture, often close to
watercourses, being ploughed up for potatoes, causing particular concern about
threats to wildlife and pollution for agencies such as English Nature and the Environment Agency. Twenty seven per cent of the farmers in potatoes in 1994/95 had
increased their potato acreage since 1991/92, a very significant figure for such a
specialist and demanding crop (Table 8.7). The exclusion of, say, flax or potatoes
or swedes from the category of eligible crops meant that farmers who might otherwise have grown them within arable rotations were now unlikely to do so as
they would have missed out on the opportunity for arable payments on eligible
crops. The eligible area rules thus provided a strong stimulus for farmers to culti-
vate non-eligible land for these non-eligible crops.

Thus whilst it is true that the eligible area rules did help to place a limit on arable expansion in many instances, they were not so environmentally advantageous as suggested by some commentators (e.g. Brouwer and Van Berkum, 1996). Moreover, in some respects they were antithetical to the promotion of environmentally sustainable agriculture. For example, eligibility criteria and base area rules combined to discourage integrated mixed farming. The calculation of regional base areas would, at first sight, appear attractive from an environmental perspective, providing an apparently fixed limit to the base arable area. The total area on which claims were paid in any one regional base area (e.g. England) could not exceed the average area sown to eligible crops and set-aside in 1989-1991. If total claims within a regional base area exceed this historic level, all claims are reduced proportionately, and farmers are paid on fewer hectares than they have claimed for. The set-aside obligation is also increased (without compensation payment) the following year (penalty set-aside). The objective of this rule was to limit liability for payment from the EAGGF budget not to influence particular cropping decisions. Individual farmers had an incentive to maximise the area they sowed to crops eligible for payments, at the expense of any temporary grasslands which might be included within a holding's eligible area. Much as with the co-responsibility levies of pre-1992 CAP, the actions of any one farmer could not determine the eventual outcome at the regional level, and all farmers would have to bear the penalties if the regional base area was exceeded.

In lowland (i.e. non-LFA) Scotland there was an overshoot in each year following the 1992 reforms. In 1993, the lowland Scotland base area was exceeded by 5.4%, leading to a 0.54% cut in payments and an extra 0.54% penalty set-aside in 1994. In 1994, a 3.5% overshoot led to 0.7% penalty set-aside in 1995, and in 1995 a 3.7% overshoot resulted in 1.1% penalty set-aside in 1996. To date there has been no overshoot in England, except with regard to maize. This is not because farmers in England have been any less anxious than farmers in Scotland to maximise their AAPS returns. Rather it is because arable production within Scottish agriculture is more likely to take place in the context of mixed farming compared with the traditionally more specialist arable areas of England. Thus a higher proportion of Scottish eligible land was in temporary grassland than in England, where on many farms almost all of the eligible area was already under eligible crops during the base period. Such loss of temporary grasslands to all-arable systems of cropping has a negative landscape impact, negative access implications and may increase environmental problems from fertiliser and pesticide applications.

Given strong market signals coupled with generous AAPS payments, many farmers sought to maximise the proportion of their eligible area placed in eligible crops. The high cereal prices between 1993 and 1996 even induced some farmers to crop more than their eligible area. As Table 8.8 shows, more than a quarter of the surveyed farmers in Great Britain had cropped a greater area than their eligible area.

While for nearly a half the reverse was the case, this in itself is scarcely surprising given that temporary grasslands form part of eligible land and are very important in certain farming systems. What the figures appear to indicate is that

Table 8.8 The area of eligible crops planted compared with the land area registered under the Arable Area Payment Scheme on surveyed British
farms (percentage of total number of farms)

Area of eligible crops planted	England	Scotland	Wales	Great Britain
Greater than the REA	23.2	44.9	67.0	27.0
Equal to the REA (+/- 5%)	24.8	24.7	8.2	24.6
Less than the REA	52.0	30.4	24.8	48.4
Total	100.0	100.0	100.0	100.0

REA = Registered eligible area.
Chi-Square test significant at 5% level.
Source: Winter and Gaskell (1998a).

for a sizeable minority of farmers, the rules of eligibility did not prevent a continuing expansion of the arable area. These findings are consistent with the
Andersons (1997) study which found that 15% of farmers in England were growing crops on non-eligible land. The areas of land involved are not large.
According to the IACS returns, some 20,000 hectares of cereals were grown for
sale without the benefit of AAPS payments in England in 1996, representing 0.8%
of the area of cereal crops claimed under AAPS (source: unpublished data supplied by MAFF). While this is a small proportion of the total, it is likely that it is
concentrated on particular farms and therefore may have a significant local impact. As Table 8.9 shows, its greatest incidence was amongst dairy, mixed and
livestock farms. For such farmers, their eligible land was generally a relatively
small proportion of their total land area and some have sought to maximise the
benefits available from the AAPS. For most arable farmers, in contrast, the possibilities to expand beyond the eligible area would be very limited or non-existent.
Thus, the arable area rules, whilst clearly placing important limits on arable area
expansion, did not prevent some marginal expansion, which may have had significant negative environmental implications locally for landscapes, biodiversity
and natural resource protection.

In the UK, any further arable expansion is likely to be environmentallly unde-
sirable. In areas of Europe, though, where traditional arable agriculture has been
in decline, eligibility for arable payments may serve to encourage farmers to re
tain low intensity arable production, which has environmental benefits. For example, Beaufoy (1996) refers to the beneficial brake on the abandonment of arable
and arable fallow in Extremadura in Spain resulting from the introduction of arable area payments. It is likely that this has also occurred in other marginal
farming areas within Mediterranean countries, which have been experiencing a
decline in traditional low intensity arable systems over several decades with detrimental results for landscape and for wildlife species associated with mixed
arable farming (Diaz *et al.,* 1997; Suárez *et al.,* 1997). In these cases, environmental benefits may derive from the application of the arable area rules.

Table 8.9 The area of eligible crops planted compared with the land area registered under the Arable Area Payment Scheme by farming system on surveyed British farms (percentage of total number of farms in each category)

Farm type	Greater than REA	Equal to REA	Less than REA
Cereal	24.1	30.0	46.0
Cropping	22.3	27.8	49.9
Dairy	65.7	1.2	33.2
Non-LFA	40.3	21.3	38.4
Mixed	35.5	10.7	53.8
All farms	27.0	24.6	48.4

Insufficient data for LFA farms.
Source: Winter and Gaskell (1998a).

CASE STUDIES

The points made so far are best illustrated by individual case studies, two of which are reproduced below (from Winter and Gaskell, 1998a). Case Study 1 furnishes an example of a farm where the profitability of the arable sector led to maximum expansion within the constraints of the eligible area rules and a reduction in a livestock enterprise. In contrast, some farmers resisted the temptation to increase their area of eligible crops at the expense of grass leys as demonstrated in Case Study 2.

Case Study 1: An arable and dairy farm in southern England

Mr Corn farms approximately 500 ha, of which about half is in grassland for the dairy and sheep enterprises and the remainder in arable cropping, mostly cereals. A major change following the 1992 reforms was the decision to concentrate on the 300-cow dairy unit and the arable enterprise. In 1991/92 the farm carried 600 breeding ewes. By 1994/95 this had been reduced to rearing ewe lambs bought in the autumn, an enterprise that was discontinued in 1995/96, with some winter keep sold for sheep graziers. The sheep quota was sold and additional dairy quota purchased. As a result, the farmer maximised his cereal acreage on eligible land by removing grass leys from the eligible land rotation. In addition, he ploughed up additional non-eligible land for cereal and forage maize production for feeding to the dairy herd.

There were negative countryside implications from these farm management changes. In terms of landscape, the mosaic of arable and grassland was reduced in quality by the increase in arable cropping. In particular, there was a loss of some semi-permanent grassland with both landscape and wildlife habitat consequences. The nutrient balance was likely to have been adversely affected by the increase in cropping with possible consequences for surrounding watercourses. Increased maize cultivation posed a risk of autumn soil erosion.

Case Study 2: A traditional mixed farming estate in southern England

Mr Trad owns and farms 1,000 hectares in southern England, of which 350 is in arable crops, 500 in grassland and the remainder under woodland. Turnover: beef 40%, dairy 35%; arable 21%; sheep 4%.

The farmer is personally committed to conservation, lower inputs and mixed farming. Consequently, the scale of the arable enterprises had been reduced in the past two decades in stark contrast to what had happened on surrounding farms. Increasing the beef herd had been a strategy to decrease dependence on heavily fertilised arable cropping. But now Mr Trad found himself a victim of the BSE crisis and feared that in the long term he would have to reduce his cattle enterprise. His mixed farming pattern was under threat. He was considering the possibility of expanding sheep production but would need to buy quota in order to do so. The easy option would be to increase cereal production because much of his grassland was short-term ley on eligible land and therefore would qualify for AAPS payments if brought into cropping. Financially this would be hugely advantageous to Mr Trad. The environmental loss, though, would be considerable. This did not seem immediately likely due to the farmer's strong personal preferences for mixed low-input farming. But the future risk was very clear.

CONCLUSIONS

To sum up, given the strength of world market prices and of sterling, the Arable Area Payments Scheme did not, in the short term at least, lead to a substantial reduction in cereal and other crop prices in Britain. Consequently, there was not an extensification of production methods in the arable sector and so there was little or no environmental gain from reduced inputs or arable reversion. Indeed, on a significant minority of farms there was an intensification of production methods during the period studied (1992-1996). The eligible land rules did apply a brake on arable expansion and this was an environmental benefit, particularly in landscape terms, where it prevented the ploughing up of long-term grass leys and permanent grasslands. Given the profitability of arable enterprises relative to the livestock sector during the period in question, this limitation on arable expansion may well have had a very significant and positive environmental benefit in protecting grassland from arable expansion.

However, against this positive outcome, it is important to emphasise that the eligibility rules had their downside. Given the profitability of arable cropping, the scheme tended to encourage farmers to maximise the proportion of their eligible land that was cropped in any one year even if some of that land had been in short- or medium-term leys prior to 1992. Nor did the rules preclude all arable expansion. No conditionality rules were written into the arable reforms to prevent

Land use pattern	• AAPS eligibility rules prevented major expansion of arable area, although some farmers expanded their arable acreage outside their eligible area.
	• Some farmers substituted arable cropping, including forage maize, for short-term grass leys within eligible arable area.
	• Set-aside was initially a major change to land use pattern but a combination of the declining rate of set-aside and the cultivation of non-food crops diminished its possible environmental value.
Landscape features (including historic/ cultural features)	• Some protection accorded to features on or adjacent to set-aside, but only during the period the land was in set-aside.
	• The high profitabillity of arable farming and higher than expected market prices meant continuing threats to some landscape and historic features on farmland.
Wildlife habitats	• Protection of features on set-aside and eligible area rules meant that wildlife habitat on farms was probably under less threat than previously, apart from the risk from non-eligible cropping.
	• Under certain conditions, set-aside offered some modest wildlife gains.
	• The continuing dominance of winter cereals over spring cereals, resulting from higher than expected market prices, meant that anticipated improvements to arable habitats did not take place.
Natural resource protection	• Due to higher than expected market prices, the anticipated reductions in intensity did not take place on eligible land and fertiliser and pesticide inputs remained high.
	• Set-aside offered only a temporary respite for a small proportion of the farmed area.
	• Threats to watercourses increased, in some circumstances due to cultivation of non-eligible crops on non-eligible land in response to market stimuli.

Figure 8.6   Summary of the countryside implications of the post-1992 arable regime and arable market trends in Great Britain
Source: Winter and Gaskell (1998a).

cropping on non-eligible land because it was assumed that without compensation payments farmers would not grow crops on such land.

By effectively fossilising the arable area, the rules of the scheme fixed a pattern of arable agriculture that, after many years of arable expansion and overspecialisation, was neither ecologically sustainable nor desirable in landscape terms. In certain regions, such as eastern England, the eligible area rules served to preserve all-arable systems, and more generally they provided a powerful disincentive to a return to mixed or more integrated farming systems.

Figure 8.6 provides a summary of the major conclusions to be drawn on the impact of CAP arable policies on the British countryside in the years immediately after 1992. It has to be said that much of what happened was a result of unexpected market trends rather than the arable regime per se. However, the negative countryside consequences of the operation of the reformed arable regime strongly suggest the need for the application of environmental conditionality or cross-compliance to arable aid payments.

There is also a need for a wide-ranging review of the possibility of reforming the AAPS to make it easier to transfer eligible land, both within a holding and to other farms, so as to facilitate a return to mixed farming. Transfers would have to satisfy criteria that would promote integrated farming systems and prevent a net increase in the area of arable land. Indeed, there is probably a case for reducing the *total* area of eligible land through some form of 'tax' in any transfer arrangement.

Acknowledgements

This chapter draws extensively on work reported on more fully in Winter and Gaskell 1998a (see also Winter and Gaskell, 1998b). I am grateful to my colleague Peter Gaskell for his collaboration in the original research.

REFERENCES

Andersons (1997) *Economic Evaluation of the Arable Area Payments Scheme.* Report to MAFF by Andersons and the Department of Agricultural and Food Economics, University of Reading.

Asby, C. and Sturgess, I. (1997) *Economics of Wheat and Barley Production in Great Britain, 1995/96.* Special Studies in Agricultural Economics, Report No. 34, University of Cambridge.

Beaufoy, G. (1996) Extensive sheep farming in the steppes of La Serena, Spain. In: Mitchell, K. (ed.) *The Common Agricultural Policy and Environmental Practices.* European Forum on Nature Conservation and Pastoralism, pp. 34-48.

Berg, A. and Part, T. (1994) Abundance of breeding farmland birds on arable and set-aside fields at forest edges. *Ecography* 17, 147-152.

Bordin, A., Cesaro, L., Gatto, P. and Merlo, M. (1998) Italy. In: Brouwer, F. and Lowe, P. (eds) *CAP and the Rural Environment in Transition: A Panorama of National Perspectives.* Wageningen Pers, Wageningen, pp. 241-266.

Bouquiaux, J-M., Foguenne, M. and Lauwers, L. (1998) Belgium. In: Brouwer, F. and Lowe, P. (eds) *CAP and the Rural Environment in Transition: A Panorama of National Perspectives.* Wageningen Pers, Wageningen, pp. 143-166.

Brouwer, F. and Lowe, P. (eds) (1998) *CAP and the Rural Environment in Transition: A Panorama of National Perspectives.* Wageningen Pers, Wageningen.

Brouwer, F.M. and Van Berkum, S. (1996) *CAP and Environment in the European Union: Analysis of the Effects of the CAP on the Environment and Assessment of Existing Environmental Conditions in Policy.* Wageningen Pers, Wageningen.

Brouwer, F.M. and Van Berkum, S. (1998) The Netherlands. In: Brouwer, F. and Lowe, P. (eds) *CAP and the Rural Environment in Transition: A Panorama of National Perspectives.* Wageningen Pers, Wageningen, pp. 167-184.

Clarke, J. (ed.) (1992) *Set-Aside: Proceedings of a Symposium.* British Crop Protection Council, Farnham.

Crabb, J., Firbank, L.G., Winter, M., Parham, C. and Dauven, A. (1998) Set-aside landscapes: farmer perceptions and practices in England. *Landscape Research* 23(3), 237-254.

Davidson, G. and Asby, C. (1995) *UK Cereals 1993/94: The Impact of the CAP Reform on Production Economics and Marketing.* Special Studies in Agricultural Economics Report No. 28, University of Cambridge.

Diaz, M., Campos, P. and Pulido, F.J. (1997) The Spanish dehesas: a diversity in land-use and wildlife. In: Pain, D.J. and Pienkowski, M.W. (eds) *Farming and Birds in Europe.* Academic Press, London, pp. 178-209.

Dijk, J., Hoogeveen, M.W. and Haan, T. de (1995) *EU-landbouwbeleid en milieubelasting in graan- en grasteelt.* Agricultural Economics Research Institute, The Hague.

European Commission (1997) *The Agricultural Situation in the Community 1996 Report.* Office for Official Publications in the European Communities, Luxembourg.

Fertilizer Manufacturers' Association (annual) *British Survey of Fertilizer Practice.* Peterborough.

Firbank, L.G. (1998) *Agronomic and Environmental Evaluation of Set-Aside under the EC Arable Area Payments Scheme. Volume 1 Overview.* Report to MAFF. Institute of Terrestrial Ecology, Grange-over-Sands.

Kleinhanss, W. (1998) Germany. In: Brouwer, F. and Lowe, P. (eds) *CAP and the Rural Environment in Transition: A Panorama of National Perspectives.* Wageningen Pers, Wageningen, pp. 41-62.

Linddal, M. (1998) Denmark. In: Brouwer, F. and Lowe, P. (eds) *CAP and the Rural Environment in Transition: A Panorama of National Perspectives.* Wageningen Pers, Wageningen, pp. 185-197.

MAFF (annual) *Agriculture in the United Kingdom.* HMSO, London.

Noé, J.L., Wieting, K., de Bretagne, L., Dary, J.L. and Skylakakis, G. (1995) Crop protection products quantitative use patterns in the European Union. Paper presented to *Workshop on Pesticides*, Wageningen.

Proost, J. and Matteston, P. (1997) Integrated farming in the Netherlands: flirtation or solid change? *Outlook on Agriculture* 26(2), 87-94.

Rainelli, P. and Bonnieux, F. (1995) CAP and the environment in France. Unpublished paper.

Rainelli, P. and Vermersch, D. (1998) France. In: Brouwer, F. and Lowe, P. (eds) *CAP and the Rural Environment in Transition: A Panorama of National Perspectives.* Wageningen Pers, Wageningen, pp. 63-82.

SJI (1995) *The impact of price reductions on the use of inputs in the Danish arable sector,* Statens Jordrugsøkonomiske Institut, Copenhagen.

Suárez, F., Naveso, M.A. and de Juano, E. (1997) Farming in the drylands of Spain: birds of the pseudosteppes. In: Pain, D.J. and Pienkowski, M.W. (eds) *Farming and Birds in Europe.* Academic Press, London, pp. 297-330.

Varela-Ortega, C. and Sumpsi, J.M. (1998) Spain. In: Brouwer, F. and Lowe, P. (eds) *CAP and the Rural Environment in Transition: A Panorama of National Perspectives.* Wageningen Pers, Wageningen, pp. 201-240.

Williams, A.J. (1994) The Common Agricultural Policy and the general environmental policies concerned with agriculture in the European Community and their implications for fertilizer consumption. *Marine Pollution Bulletin* 29(6-12), 500-507.

Winter, M. and Gaskell, P. with Gasson, R. and Short, C. (1998a) *The Effects of the 1992 Reform of the Common Agricultural Policy on the Countryside of Great Britain,* 3 volumes. Countryside & Community Press and Countryside Commission, Cheltenham.

Winter, M. and Gaskell, P. (1998b) Agenda 2000 and CAP reform: is the environment being sidelined? *Land Use Policy* 15(3), 217-231.

The Wine Regime

9

Jordi Rosell and Lourdes Viladomiu

INTRODUCTION

The wine regime is one of the most complex market regimes within the Common Agricultural Policy (CAP), owing to the fact that it covers conventional market organisation measures as well as a significant number of technical matters, which vary both by country and by region. The wine sector has traditionally been the object of a substantial degree of public support and regulation, in the context of clearly interventionist policies at the national, as well as European Union (EU), level.

Since 1992, however, a far-reaching reform of the intervention mechanism has been under consideration. In July 1998 the European Commission presented a new proposal to regulate the Common Market Organisation of wine. The Commission's previous proposal had been rejected in 1994 by almost all Member States. The new proposal will probably be accepted without significant problems. In the intervening period the Commission had extensively revised its approach, and the situation on the wine markets had changed considerably (European Commission, 1998a).

The full details of the regulation were not known at the time of writing, but it is believed that it will not differ significantly from the proposal of 1 July 1998 (European Commission, 1998b). The new proposal is better suited to the complexity of the European wine sector and displays a better understanding of the specific social contexts of wine production. In addition, the environmental impacts of wine production are also considered, indicating the need to maintain beneficial effects and limit negative ones. The proposed reform leaves several questions open, and the most distorting elements remain.

The chapter aims to improve the understanding of the environmental implications of wine production. First some of the main features of the global trade in wine are described, followed by an analysis of the Common Market Organisation for wine. The principal characteristics of wine production that determine its environmental impact are then outlined. Finally, a detailed analysis of the impact of wine production on the environment, landscape and nature is presented based on a case study of La Mancha in the central part of Spain, which is the largest wine-producing area in the world (600,000 ha).

GLOBAL PRODUCTION, CONSUMPTION AND TRADE

The production and export of wine is concentrated in a limited number of countries, with the EU occupying a pivotal position globally, accounting for around 45% of the world area under vines, and around 60% of world production and consumption. It is also the leading importer and exporter, accounting for around 80% of global exports (FAO, 1997).

At the global level, the wine producing area has been contracting at a low rate since the end of the 1970s. This is accounted for by contractions in the traditional producer regions (the EU, the south-eastern part of Europe, North Africa, Argentina) not fully offset by the dynamic development of area and output in newer regions (the USA, South Africa and Chile in particular). Nevertheless, projections by FAO and the International Vine and Wine Office (*Office International de la Vigne et du Vin - OIV*) indicate a slightly rising trend in world production over the coming years, given the large number of vines planted in the mid-1990s (Sumpsi and Barceló, 1996). In the EU, in 1975, a ban on new planting was introduced as well as an incentive scheme to cease production: between 1976 and 1996, the area under vines fell from 4.5 to 3.4 million ha (1.4% or almost 56,000 ha year^{-1}), with particularly significant reductions occurring between 1991 and 1995.

Levels of consumption began to decline in the mid-1950s and have continued to do so until the present day in producer countries, with this decline being particularly marked in countries with a previously high level of consumption (France, Italy, Spain and Argentina) (Table 9.1). Conversely, consumption has been rising steadily in northern European countries as well as in some Asian countries (Pargny, 1997), although not enough to compensate for the decline amongst

Table 9.1 Consumption of table wine and quality wine (x 1,000 hl) [a]

	France		Italy		Spain	
	Table	Quality	Table	Quality	Table	Quality
1982-1985	31,755	11,607	36,747	5,086	13,695	3,830
1986-1989	27,450	12,759	31,067	5,502	10,005	5,276
1990-1993	22,475	13,856	28,066	5,695	8,488	6,896
1994-1996	19,656	15,378	26,029	7,302	6,523	7,190

[a] Community legislation classifies wine into two main categories: 'quality wines produced in specific regions' (also called 'quality wines psr'), and 'table wines'.
Source: Data from Eurostat in Polidori *et al.* (1997).

traditional consumers. Wine consumption in the EU in 1996 was 128 million hectolitres, an average of 34 litres per capita per annum; total wine consumption decreased by 10 million hl during the period from 1986 to 1996.

Over the last 20 years, EU production has fallen from an annual level of 210 million hl in the first half of the 1980s to an average annual production of 150 million hl in recent years (Table 9.2). Wine production is characterised by marked annual fluctuations, for example ranging between 152 and 165 million hl during the period 1993-1997.

The evolving pattern of world consumption and production has led to a growth in world trade. Nowadays, more than a quarter of global production is exported (FAO, 1997). Italy, France and Spain are the world's largest exporters, although in recent years there has been a growth in export from the newer producer countries (Chile, South Africa, the USA and Australia). European countries account for over 75% of world imports.

The first half of the 1990s showed a market imbalance between world production (an average of 271 million hl per annum) and world consumption (an average of 224 million hl per annum). The wine balance sheet changed in the period 1993-1995, especially in the EU (see Figure 9.1), owing in part to weather conditions. However, the world situation is generally characterised by excess production over demand.

Table 9.2 Production of wine, surface of vineyards and yields by Member State in EU countries (average 1993-1997)

	Area (1,000 ha)	Production (1,000 hl)	Yield (hl ha^{-1})
Austria	49	1,999	41
France	912	53,802	59
Germany	104	8,872	85
Greece	74	3,744	51
Italy	868	56,250	65
Luxembourg	1	150	118
Portugal	253	6,784	27
Spain	1,167	25,755	22
EU-15	3,428	157,488	46

THE COMMON MARKET ORGANISATION FOR WINE

The wine regime came into operation in 1970, being modelled on the French system of regulation but attenuating its planning and control aspects. There were no limitations on new plantings, and private storage aid and compulsory distillation measures were applied only in exceptional cases. Instead, a system of administrative control was adopted regarding the stocks and flows of wine, together with the

Wine Producers' Registry and different sets of rules governing quality wines and
table wines. Furthermore, as regards the addition of sugar to grape must, the codi-
fication of oenological practices carried over the system that existed prior to the
creation of the Common Market, permitting the practice in the wine producing
areas of northern Europe (Polidori *et al.*, 1997).

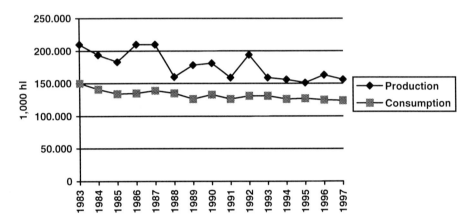

Figure 9.1   Production and consumption of wine in the EU between 1983 and
1997 (in 1,000 hl)
Source: European Commission (1998a).

This regime soon led to surpluses and a stand-off between France and Italy
referred to as the 'wine wars' (Montaigne, 1997). In 1975-76, a ban on new planting and an abandonment incentive scheme were introduced, but the surpluses
continued to increase, as well as the budgetary costs. With the forthcoming accession of Spain and Portugal to the EU, the European Council in session in Dublin
in 1984 decided on important reforms with the aim of limiting expenditure. A
price freeze was imposed, a policy of grubbing-up was reinforced and the compulsory distillation system confirmed. The measures were applicable to table
wines only, and were intended to curb the overproduction in this market.

Since 1987, the wine regime has been based mainly on EC Regulation
822/87. The Regulation has operated as a system of guaranteed prices, with
distillation to get rid of surpluses, application of customs duties and a system of
export 'refunds'. There is also aid for private storage, for grape must and its use
and for taking over the alcohol derived from compulsory distillation. The wine-
producing potential is regulated by a ban on new planting and premiums for
grubbing-up vineyards. These measures do not apply to quality wines, which are
the responsibility of each Member State to classify and control.

Over the past decade, expenditure on the wine regime has accounted for
between 2.5 and 4.7% of the total expenditure under FEOGA-Guarantee, while in

terms of final output wine production represents 5-6.5% of EU agricultural production. In recent years, the measures incurring the largest shares of the cost have been distillation and permanent abandonment premiums, accounting for about one-third each (Table 9.3).

Table 9.3 Expenditure on the wine regime between 1989-1991 and 1998 (%)

	1989-1991	1992-1994	1995-1997	1998 (estimation)
Export refunds	4.9	6.9	5.7	6.3
Stockage	4.7	4.0	4.3	6.8
Distillations	40.7	34.4	21.5	40.7
Buying-in of alcohol	21.0	16.1	19.2	19.7
Aids for use of must	11.1	11.4	15.4	18.4
Permanent abandon-ment premiums	17.6	27.2	34.1	8.1
Total	100	100	100	100
Total (MECU) (yearly average)	1,051.4	1,251.9	885.8	806.0

Source: European Commission (1998a).

In 1994, the Commission presented a proposal for reform of the wine sector which was poorly received by the Member States concerned. After 4 years of discussion, a new proposal was tabled in July 1998, which is likely to be approved. It envisages a budget increase of 25% on the previous proposal. The main aspects of the proposed reform are:

- *Maintenance and extension of the ban on new plantings*
The ban on new plantings is maintained until 2010, with some exceptions. The ban is extended to include new plantings of quality wines, which represents a fundamental change to the regime, which previously had largely controlled only table wine production. However, a number of new planting rights for quality wines are established, allowing Member States a degree of flexibility. Nonetheless, these replantation rights are conditional upon legalising the irregular plantings carried out in recent years.

- *Crisis distillation*
Compulsory, preventive, market support and complementary distillations come to an end and are replaced by crisis distillation. This would only be approved in exceptional cases, such as the occurrence of significant market imbalances. Prices are to be established by the Commission and approved by the Council. The Commission intends that crisis distillation should only operate in exceptional cases and that prices be sufficiently low as not to stimulate production. Although distillation was always intended as a disincentive to production, the producer response seems

to have been the opposite - raising output so as to maintain income (Castillo and Gil, 1995). Approximately 10% of the proposed budget for wine has been earmarked to cover the crisis distillation measures in 2001, but it is planned that this should decrease to 2.5% in 2004 (European Commission, 1998b).

- *Chapitalisation*

Chapitalisation (enrichment with sucrose) is permitted in those regions where it is traditionally practised. However, it is one of the sector's most controversial issues. Southern Member States, indeed, consider chapitalisation to be a major cause of surpluses as it allows for an increase of yields. Low alcohol content can be offset by the addition of beet sugar or concentrated grape must, but alcohol derived from sucrose costs around one-third of that derived from grape must. The new proposal for the wine regime authorises enrichment by sucrose and reduces payments for the use of grape must. These payments represented 13% of the budget over the last ten years, and a reduction to 11% is proposed for the period 2001-2005.

- *Permanent abandonment premiums*

Grubbing-up premiums are to be maintained, and each Member State must target payments to regions with structural surpluses, but it is going to be a much less significant measure. Having been nearly 30% of the total budget over the past ten years, it is proposed to be reduced to just 4% (European Commission, 1998b).

- *Reconversion measures*

The inclusion of reconversion measures in the wine regime represents an innovation which was well received, particularly by those within the sector. Each region can propose a regional reconversion programme, including plantation of new grape varieties, replanting of vines and improved techical processes. These reconversion measures are designed to improve competitiveness on the world market and to support sustainable cultivation practices (such as green cropping, reduced use of fertilisers and the adoption of techniques which reduce yields). They also are aimed to achieve beneficial effects on the environment, allowing for better protection of water, soil and landscape quality, as well as preserving and promoting biodiversity. Reconversion measures are intended to cover almost a third of the total budget (Table 9.4). Improved storage or promotion are not included but will be dealt with, *inter alia*, by rural development programmes.

- *Inclusion of quality wines in the CMO*

The traditional distinction between quality and table wines has increasingly less relevance for present-day European wine production. For example, almost all wines produced by Germany, Luxembourg, the UK and Austria are classified as quality wines, although this is not the case in other Member States.

Wine production has also benefited from agri-environmental policy and structural measures. The agri-environmental measures under Regulation 2078/92 included a series of schemes to aid vineyards: to maintain production on steep slopes, to convert to organic production and to maintain production in regions with high-risk of erosion. Particularly noteworthy are the German and Austrian schemes, which have the largest budgets of all the Member States. Certain wine producing areas in Italy and Greece also benefitted from the agri-environmental measures. The Structural Funds, especially in 'Objective 1' regions, have contributed to the modernisation of wine-making enterprises and in particular of co-operatives, as well as to the restructuring and modernisation of vine-growing holdings.

Table 9.4   Proposed expenditure for the wine regime, period 2001-2005 (in million ECUs)

Measures	2001	2002	2003	2004	2005
Export refunds	44	44	44	44	44
Stock	45.5	45.5	45.5	45.5	45.5
Aid for use of grape must	142.3	142.3	142.3	142.3	142.3
Permanent abandonment	45	45	45	45	45
Buying-in of alcohol	292.4	292.4	292.4	292.4	292.4
Crisis distillation	122	92	57.5	28.8	n.d.
Reconversion measures	379.8	422	443.2	443.2	443.2
Other	221.6	221.8	221.7	221.6	221.7
Total	1,292.6	1,305.0	1,291.6	1,262.8	1,234.1 [a]

[a] Not including crisis distillation.
Source: European Commission (1998b).

MAIN FEATURES OF WINE PRODUCTION

The environmental implications of wine production are complex and can be either harmful or beneficial. The type of impact depends on the production techniques applied, the location of production and the nature of the local ecosystem. It is against these variables that the suitability of the production system to the environmental conditions of a particular area can be judged. The principal characteristics of wine production in Europe are as follows:

- *Spatial concentration*

One of the main features of vine growing is its spatial concentration. In France, Spain and Italy, it takes place in particular regions, with ancillary processing activities located in the same area. In such regions, the business of wine production has deep roots, and the local economy is highly dependent on wine in terms of the growing of grapes, their transformation into wine and the eventual commercialisation of the final product. At the same time, vineyards form an important component of the landscape in wine-growing regions.

- *Adaptability to poor soil and adverse conditions of production*

Many wine regions suffer from poor soil quality and aridity, which means that they are unable to support a range of product types. In addition, these regions are often hilly, which reduces the options for economic diversification. In fact, European wine production is concentrated in the Less Favoured Areas (LFAs). For example, in Spain, while less than 50% of the total agrarian production occurs in LFAs, 80% of wine production occurs in these areas. Vineyards contribute to landscape conservation, as they prevent soil erosion and maintain the socio-economic fabric in regions without any economic alternative. The lack of alternative crops means that abandonment of viticulture implies increasing soil erosion and desertification.

- *Extensive versus intensive production systems*

Europe has a wide variety of vine production systems, particularly in terms of the level of input use. At one extreme are found the *intensive models* of certain German and Austrian regions, and at the other the *extensive models* of southern Europe (Laporte and Lifran, 1987). Thus Spain accounts for 34% of the total European vine-growing area but only 16% of total European wine production (Table 9.2), with an average yield over the last five years of 22 hl ha^{-1}. Germany, on the other hand, accounts for 3% of the European vine area but 5.6% of European production, with an average yield of 85 hl ha^{-1} - much higher than that of either France (59 hl ha^{-1}) or Italy (65 hl ha^{-1}) (European Commission, 1998a). This heterogeneity of both production systems and levels of input use brings forth a wide range of environmental impacts. Intensive systems can require heavy doses of inorganic inputs (up to 15 treatments in certain areas); this level of intensity causes serious deterioration in water and soil quality.

- *Perennial crop*

The vine is a perennial cultivar with an average lifetime of around 30 years, and even much longer in extensive regions of southern Europe. The long shelf-life of the final product makes it difficult to adjust annual production in response to market signals. However, the demand for particular types of wine and for specific varieties of grapes can change significantly over time. In recent years, for example, the demand for red wine has increased sharply at the expense of white wine, thereby reversing the trend of the 1980s. In responding to such fluctuations, it is important not to jeopardise the variety and diversity of European viniculture.

- *Differentiation*

Wine should be considered as a group of related products, with highly differentiated characteristics. Indeed, a range of definitions of wine are applied in the various production regions (in terms of oenological practices, alcohol content, acidity and so on). Differentiation also occurs in terms of the environment and the area in which a given wine is produced.

WINE PRODUCTION IN THE REGION OF LA MANCHA

The vine surface in Spain is around 1.2 million ha, or 4.7% of the UAA; annual production of wine over the past 10 years has ranged between 20 and 38 million hl, with an average of around 28.5 million hl. The extensive character of the Spanish vineyards, as implied by the generally low yields, reflects the fact that wine growing has traditionally taken place on land with a low fertility, with limited or no suitability to the production of other crops. Wine growing also takes place in areas with an arid or even semi-desert climate. In many areas, viniculture takes place under poor production conditions, both environmentally (sandy or stony soils, steep slopes, etc.) and practically (narrow spaces for machines to pass between rows, lack of infrastructure to provide occasional irrigation in times of severe drought, etc.) (Granda and García, 1987).

It should be noted that irrigation of vineyards was prohibited in Spain by the *Estatuto de la Viña, del Vino y de los Alcoholes* of 1970. Since most of the wine-growing areas in Spain are in arid regions, this prohibition implied that productivity remained low, and highly variable in terms of yields and quality. This reduced the competitiveness of the product. Nevertheless, these conditions allowed wine-growing to continue on marginal land where there is no alternative crop.

Wine accounts for around 5% of the value of final agricultural production in Spain (Ministerio de Agricultura, Pesca y Alimentación, 1996). This is around the average for the EU as a whole. Spain has a wide range of wine types. Due to the various micro-climates and diversity of soil qualities, there are numerous grape varieties, of both domestic and foreign origin (Viladomiu and Rosell, 1997).

In recent years, the Spanish wine sector has made efforts in two directions: on the one hand, quality improvement by means of the *Denominaciones de origen,* on the other, opening up external markets. The Spanish wine sector has made a big effort to change its traditional image of cheap, poor quality produce. There are now 53 *Denominaciones de origen,* half of which have been designated in the last ten years, and they cover around 50% of the total vine surface (Leza, 1998). These *Denominaciones de origen* are well known within their regions but do not have an international reputation, except for those of Rioja, Ribera del Duero, Malaga, Jerez and Penedes. Nonetheless, the average price of Spanish wine is amongst the lowest in Europe, which in addition to low yields means that gross production of the area under vines stands at around 675 ECU ha^{-1} compared to 10,400 ECU ha^{-1} for Germany (European Commission, 1998a).

There are numerous wine co-operatives, most of them relatively small operations hampered by overcapacity, and serious deficiencies in terms of organisation and management. They process around half of the production, but do not have as much market influence as this might suggest, as few of them undertake bottling or exporting. Indeed, many deal in low quality wines sold in containers or for distillation. Such inefficient enterprises are seen by many experts as the cause of the 'Spanish wine problem', since they have substantially raised output while continuing to perform unsatisfactorily in their task of distribution (Albisu and Martín,

1990). Low prices and organisational deficiencies account for the large volume of Spanish wine that is distilled.

In recent decades, Spain was one of the countries which experienced the greatest decline in wine consumption. The average consumption per capita currently is 38 litres. Spain is the third largest exporter, accounting for 10 to 15% of global trade in wine; the import of wine is relatively small. While still below the levels attained by France and Italy, Spanish exports have increased substantially since accession to the EU (FAO, 1996).

Although there are vineyards in almost every part of Spain, the largest areas are in the eastern (the Mediterranean arc) and southern parts of the peninsula. Viniculture is of particular significance in the vast Castilla-La Mancha region, in the eastern and southern part of Madrid (NUTS II; 79,230 km^2) (see Figure 9.2 for a detailed description of the La Mancha region). With more than 609,000 ha, it is the largest area of viniculture in the world and by far the most important region in Europe in terms of vine area, having almost twice the extent of the second largest region - Languedoc-Rousillon in France (Eurostat, 1997). The vine area of Castilla-La Mancha represents half of the total Spanish area and 16.5% of the European area.

Figure 9.2   La Mancha area

The major part of the Castilla-La Mancha vine area is located on the great Manchegan plains, the natural region of La Mancha, which themselves constitute the largest extent of flat land in Spain. In this vast region of around 32,263 km^2, viniculture occupies 16.5% of the total area and 22.5% of the UAA. Winter cereals (principally wheat and barley) are the other major crop in the region; viniculture, winter cereals and sheep production together have historically formed the basis of agricultural production.

Regional wine production shows large annual variation, mainly due to climatic conditions. The average total production is around 12 million hl, resulting from an average yield of 20 hl ha^{-1}, which is just below the average Spanish yield of 22 hl ha^{-1}. A key feature of Manchegan viniculture is its extensive, low-yielding nature, which has to do with climatic conditions (low levels of rainfall), poor soil quality and the density of planting. In this sense, Manchegan viniculture is a clear example of the *extensive model* of production (Laporte and Lifran, 1987). It also explains the low levels of fertiliser and agro-chemical use; indeed, for a few years now organic wine has been produced in the region without major variable costs.

Vineyards, even under extensive conditions of cultivation, require significant inputs of labour. The labour requirement in La Mancha averages 128 hours ha^{-1} year^{-1}, largely spent in the main periods of pruning and harvesting. This is more than 15 times the amount of labour required to grow winter cereals. In addition, viniculture has given rise to a dense network of agro-industrial activities in the region, the most important being the production of wine (with more than 1,500 wine producers) by both co-operatives and private producers, as well as others such as distilleries and the production of must. Other activities are also linked to vineyards and wine production, such as manufacturing (wood, metal, chemicals) and services (e.g. transport). As a result, the region of La Mancha, even though fundamentally agrarian in nature, has a population density which exceeds that of other rural areas in Spain, and even shows a slightly increasing trend.

In terms of the conservation of nature and natural resources, the combination of great expanses of vineyards, cultivation of winter cereals and the practice of sheep husbandry has been particularly fortuitous. This is especially the case with regard to the conservation of water resources in an area characterised by high levels of aridity, with annual precipitation levels of around 450 mm, spread unevenly over the seasons. Irrigated agriculture used to be limited to a few market gardens near towns and villages, and in river valleys. The majority of water supply came from aquifers and was abstracted by means of waterwheels, which brought the water up from a depth of just over five metres. There is a large aquifer in the central part of La Mancha (the Mancha Occidental aquifer) which extends for almost 5,500 km^2, as well as other smaller ones including the Campo de Montiel aquifer (around 2,500 km^2). Irrigation by surface water was less important, and the relatively flat relief and the pattern of rainfall produced a somewhat undeveloped water supply infrastructure. In the 1960s the construction of the Peñarroya Reservoir (fed by water from the Ruidera Lagoons, themselves created by overflow from the Campo de Montiel aquifer) allowed for the irrigation of around 7,000 ha of agricultural land using surface water. At the start of the 1970s, the irrigated area was almost 30,000 ha.

This model of agricultural activity allowed for the maintenance of the aquifers and of the wetland areas associated with them. Given the high levels of aridity which prevail in La Mancha, the existence of a number of sizeable wetland areas is noteworthy. The most important is Las Tablas de Daimiel, a wetland area of international importance formed by flooding of the Gigüela and Guadiana rivers together with the upwelling of groundwaters belonging to the Mancha Occidental aquifer through many wells, locally called 'ojos del Guadiana' ('Guadiana's eyes') (Alvarez Cobelas and Cirujano, 1996). Las Tablas de Daimiel was designated a National Park in 1973, and is covered by the RAMSAR Convention (List II) and designated a Special Protection Area (SPA) under the EC Birds Directive.

The objective of the designation of Las Tablas de Daimiel as a protected area was to stop the process of drainage. That process started around 1960 in order to dry the wetlands and to convert the new land into agricultural use. The designation occurred at a time of increasing interest in the protection of wetlands and their aquatic fauna. In 1969 the wetland of Doñana was also designated as a National Park and it and Daimiel are key points in the migration of an important number of European and northern African birds.

The National Institute for Nature Conservation (ICONA) is in charge of the conservation of Las Tablas de Daimiel. The park is 1,928 ha (of which 1,675 ha are wetlands), all publicly owned, and ICONA is also responsible for the management of the surrounding protection area (5,410 ha). The total of 7,338 ha represents just 1.3% of the surface of the Mancha Occidental aquifer that feeds it.

Equally noteworthy are the Lagunas de Ruidera, a network of 16 lakes formed by subsidence. They are joined by waterfalls and fed by water from the Campo de Montiel aquifer. Since 1980, they have been protected as a Natural Park. The level of protection is quite low in comparison with Las Tablas de Daimiel National Park and is limited to the actual area of the park (3,772 ha). There are also a considerable number of other small wetlands in La Mancha. Since 1980, all the wetlands of the region have been collectively listed as a UNESCO Biosphere Reserve under the name Mancha Humeda. They provide habitat for a large number of birds, and the marsh vegetation is of particular interest (Alvarez Cobelas and Cirujano, 1996). However, apart from Las Tablas de Daimiel National Park and Lagunas de Ruidera Natural Park, the remaining wetlands did not benefit from any specific management in line with their Biosphere Reserve status (Segura, 1994).

Unfortunately, the regional model of agriculture started to change at the end of the 1970s. Following the construction of wells to exploit underground water supplies, new land was put under irrigation beyond the immediate area of towns, villages and river valleys. These newly irrigated lands were mainly put down to three crops with high water demand: maize, sugar beet and forage crops (alfalfa). The irrigated area quadrupled between 1975 and 1990. This 'irrigation boom' led to the extensive grubbing-up of vineyards, and in some southernmost parts of the comarca (Campo de Montiel) the spread of irrigation led to a reduction in the area of forest (Velasco, 1998).

These changes in agriculture were due to the reform of agricultural policy and changing market conditions. The fall in Spanish wine consumption affected in particular the table wines of La Mancha, giving rise to growing surpluses. Following accession to the EC in 1986, Spain faced the application of EC Regulation 1442/88 to support the grubbing-up of vineyards. Between 1988 and 1997, almost 100,000 ha of vineyards were grubbed-up in Castilla-La Mancha as part of this programme (a fifth of the total area grubbed up in the EC). Meanwhile, by means of subsidy and soft credit, structural policy was providing an incentive for farm businesses to invest in machinery and the installation of irrigation. Indeed, the shift to irrigation has been the key factor in Spain's policy of agricultural modernisation throughout the twentieth century (Rosell *et al.*, 1995).

Changes in the agricultural sector, in favour of crops with high water requirements, had serious implications for the natural environment. The abstraction of underground water increased about fourfold, overstretching the capacity of the aquifers. The water level in the aquifers dropped markedly. The gravity of the problem became clear during the early 1980s when the Ojos del Guadiana dried up. This implied that the Mancha Occidental aquifer was no longer overflowing into the Tablas de Daimiel area. In the 1990s, the absence of water draining into the Ruidera Lagoons testified to the over-exploitation of Campo de Montiel aquifer also. The flooded areas, both in the Tablas de Daimiel National Park and in the other wetland sites of the area (La Mancha Humeda), contracted during a period of one decade from 22,000 ha to little more than 8,600 ha. The great expansion of irrigation in the area was also associated with an increase in the use of fertilisers, mainly nitrogen, which in turn has led to a deterioration in groundwater quality. According to a recent report (Instituto Tecnologico GeoMinero de España, 1997), at several places in Mancha Occidental and Campo de Montiel aquifers the nitrate content exceeds the limit of potability (50 mg l^{-1}). For this reason, both aquifers have been proposed to be declared as 'nitrate vulnerable zones' according to the Nitrates Directive (91/676/EEC). Over-explotation of the aquifers and deterioration in groundwater quality created serious problems in the area, given that more than 85% of the local population (300,000 inhabitants) depend on groundwater for their domestic water supply.

In 1987, and in compliance with the Spanish Water Law, the aquifers were declared over-exploited, which theoretically meant that abstraction should be controlled and reduced. In practice, most farmers disputed the legality of the over-exploitation order and continued to pump water for irrigation (Rosell and Viladomiu, 1998). Following the 1992 reform to the CAP, the opportunity was therefore taken to devise an agri-environmental scheme to induce farmers to reduce their water use. In view of the socio-economic impact of the imposed abstraction regime and the impact of over-exploitation on the Manchegan wetlands (especially the Tablas de Daimiel), the regional government sought compensation for farmers for the decrease in farm incomes under Regulation (EEC) 2078/92. The programme to reduce water abstraction from the Mancha Occidental and Campo de Montiel aquifers was approved by the European Commission in 1993 and became operational in that year.

The Scheme offers farmers the possibility of enrolling their irrigated land under a contract to reduce the volume of water used for irrigation over a period of 5 years, in exchange for a payment. Based on an average consumption on irrigated land of 4,920 m^3 ha^{-1} year^{-1}, the reduction can operate at three levels: 100%, 70% and 50%. The reduction of water consumption, to whatever level, includes the obligation not to exceed certain levels of fertiliser, pesticide and herbicide use. In terms of enrolment, the scheme has been a great success. In 1997, 2,624 farms participated, representing 85,834 ha. These high numbers suggest that initial enrolment forecasts are likely to be met. However, the scheme is particularly attractive to cereal growers who traditionally have not used much water, but not to farmers producing heavily irrigated crops who would need to convert to cereals to comply. The scheme is proving very expensive - 102 million ECU for the period 1993-1997 - and has only led to a slight improvement in the situation without really solving the problem (Rosell and Viladomiu, 1997).

CONCLUSIONS

The evolution of agriculture in La Mancha during the past three decades clearly shows the environmental repercussions which can result from changes in agricultural land use and, ultimately, in certain agricultural policies. The displacement of vineyards (as well as winter cereals and forests) by crops with high water requirements, driven as it has been by agricultural policies (subsidies to grub-up vineyards, subsidies for investing in machinery and irrigation, high prices for maize and forage crops), brought in its train serious changes to the Manchegan environment as well as to its most distinctive natural feature - the wetlands.

In this context, the 1994 proposal to reform the wine regime, which should have formed part of the MacSharry reforms, constituted a serious threat to the La Mancha region. In effect, the reform proposal introduced for each Member State a 'national reference production' (or national production quotas), calculated in such a way as to ensure a balance between supply and demand; Member States would be responsible for cutting back production (European Commission, 1994). Actual production during the reference year in Spain was 20% below the historical average production, and as a result around 250,000 ha of vines would have had to be grubbed-up. As such, the proposed reform included a compulsory grubbing-up scheme, upon which all other measures of support would be conditional; the likely impact of such a scheme in La Mancha is all too easy to imagine. Due to the characteristics of La Mancha wine production, it was likely that much of the Spanish grubbing-up programme would be concentrated in that region.

The new reform proposal for the wine regime presented by the Commission in July 1998 should be viewed as much more favourable to the environment, given that compulsory grubbing-up has been dropped (although some aid for voluntary grubbing-up remains). Instead, the new reform proposal includes a scheme to support the reconversion of vineyards with the aim of supporting the adjustment of production to the requirements of the market. This scheme, which is to receive almost a third of the budget for the wine regime for the period 2001-2005, could be well suited to a region like La Mancha where the production of low quality wine remains dominant.

On the other hand, the reform proposal does not include any payments to support the incomes of vinegrowers in low yield areas, nor does it consider the call from southern European countries to limit or prohibit the practice of chapitalisation. Moreover, despite Agenda 2000 pronouncements, the integration of environmental considerations into sectoral policies is strikingly absent from the new reform proposals save for a handful of general references to 'considerations relative to the environment' in the vineyard grubbing-up and reconversion schemes. The extent to which the reconversion programmes of the various European wine regions embrace environmental objectives remains to be seen.

The rapid spread of irrigated agriculture in La Mancha coincided with the designation of the Manchegan wetlands and lakes as protected areas. These two developments were incompatible. The protection of the wetland areas had to contend with the environmental impact of the spread of irrigation. This situation demonstrates the limitations of mere designation as a means of protection, without the integration and coordination of other policies.

More recently, an agri-environmental scheme has sought to halt and reverse the spread of crops with high water requirements. Nonetheless, the effect of the scheme remains insufficient to alter the dynamics of agricultural change in the region. The difference in profitability between dryland and irrigated agriculture continues to encourage transformation in favour of crops with high water needs. The sinking of new wells and the continued protests of the regional population asking for more water availability demonstrate the limitations of the course of action currently being pursued (Viladomiu and Rosell, 1996). In the same way, if the crops best suited to regional conditions continue to be subject to quotas or if aids to grub them up continue to be available, as is the case with vines and olives, then the options for development and even survival in regions like La Mancha are closed off. The events of recent decades in La Mancha constitute a strong argument in favour of the proclaimed integration of environmental objectives within agricultural policies.

Acknowledgements

This chapter has benefited from the results of two study projects: 'Regional guidelines to support sustainable land use by EC agri-environmental programmes' (EU Research Programme AIR3 CT94-1296) and 'Conservation and restoration of the wetlands of Mancha Humeda Biosfera Reserve' (CICYT Project HID 96-1298-C03-02).

REFERENCES

Albisu, L.M. and Martín, D. (1990) La publicidad de vino en España. In: Briz, J. (ed.) *Publicidad en el Sistema Agroalimentario*. Ediciones Mundi-Prensa, Madrid, pp. 101-122.

Alvarez Cobelas, M. and Cirujano, S. (1996) *Las Tablas de Daimiel*. Ecología acuática y sociedad, Dirección General de Conservación de la Naturaleza-CSIC, Madrid, 368pp.

Castillo, J.S. and Gil, F.J. (1995) Una producción mediterránea en el centro de la dinámica Norte-Sur en Europa: el vino. *Economía Agraria* 171, 161-190.

European Commission (1994) Proposal for a Council Regulation on reform of the common market organisation of the market in wine (COM(94)117 final). Commission of the European Communities, Brussels.

European Commission (1997) *Agenda 2000*. Office for Official Publications, Luxembourg.

European Commission (1998a) *Situation et Perspectives. Vin, PAC 2000*. Documents de travail, Commission of the European Communities, Direction Générale de l'Agriculture.

European Commission (1998b) Proposition de Règlement relative à l'Organisation Commune du Marché viti-vinicole (COM(98) 370). Commission of the European Communities, Brussels.

Eurostat (1997) *Regions. Statistical Yearbook*. Luxembourg.

FAO (1996) *Trade Yearbook*. Food and Agriculture Organization of the United Nations, Rome.

FAO (1997) *Production Yearbook*. Food and Agriculture Organization of the United Nations, Rome.

Granda, G. and García, J.L. (1987) El sector vitivinícola en Castilla y León y la CEE. Junta de Castilla y León, Valladolid.

Instituto Tecnologico GeoMinero de España (1997) Sintesis Hidrogeologica de Castilla-La Mancha, Madrid.

Laporte, J.R. and Lifran, R. (1987) Ampliación de la CEE y restructuración de la economía vitícola comunitaria. *Revista de Estudios Agrosociales* 139, 9-29.

Leza, L.F. (1998) Los vinos españoles con denominación de origen en el contexto de la Unión Europea. *Agricultura* 793, 624-628.

Ministerio de Agricultura, Pesca y Alimentación (1996) *Anuario de Estadística Agraria*. Madrid.

Montaigne, E. (1997) The Common Market Organization for wine: autopsy of a reform. In: Tracy, M. (ed.) *CAP Reform: the Southern Products*. Agricultural Policy Studies, Belgium, pp. 41-54.

Pargny, F. (1997) Quel avenir pour le vin? *Le MOCI*, no. 1, 289.

Polidori, R., Rocchi, B. and Stefani, G. (1997) Reform of CAP and the EU wine sector. In: Tracy, M. (ed.) *CAP Reform: the Southern Products*. Agricultural Policy Studies, Belgium, pp. 29-40.

Rosell, J. and Viladomiu, L. (1997) El Programa de Compensación de Rentas por reducción de regadíos en Mancha Occidental y Campo de Montiel. *Revista Española de Economía Agraria*.179, 331-350.

Rosell, J. and Viladomiu, L. (1998) Gestión del agua y política agroambiental. In: Cruces, J., Hernandez, J.M., Lopez, G. and Rosell, J. (eds) *De la Noria a la Bomba. Conflictos Sociales y Ambientales en la Cuenca Alta del Rio Guadiana*. Bakeaz, Bilbao, pp. 279-343.

Rosell, J., Álcantara, V. and Viladomiu, L. (1995) *Economic Appraisal and European Union Funds: a Study of Water Management and Irrigation in Spain*. Sociedad Española de Ornitologia/BirdLife España, Madrid, 38pp.

Segura, F. (1994) *An overview of the Spanish Network of Biosphere Reserves, 1994*. Comité Español del Programa MAB, Madrid, 155pp.

Sumpsi, J.M. and Barceló, L. (1996) *La Ronda Uruguay y el Sector Agroalimentario Español*. Ministerio de Agricultura, Pesca y Alimentación, Madrid.

Velasco, M. (1998) *Los Nuevos Regadios y Ruidera: un Modelo de Desarrollo Insostenible en el Campo de Montiel.* Instituto de Estudios Albacetenses de la Excma. Diputacion de Albacete, Albacete, 106 pp.

Viladomiu, L. and Rosell, J. (1996) *Preliminary Report About the Income Compensation Scheme in Mancha Occidental and Campo de Montiel (Castilla-La Mancha, Spain).* AIR 3 CT 94-1296 EU Research Project, 33 pp.

Viladomiu, L. and Rosell, J. (1997) The complexity of the CMO for wine: a view from Spain. In: Tracy, M. (ed.) *CAP Reform: the Southern Products.* Agricultural Policy Studies, Belgium, pp. 55-65.

The Olive Oil Regime

10

Guy Beaufoy

INTRODUCTION

Olive production is a major land use in southern Europe and has important environmental, social and economic influences. Furthermore, olives are one of the few Mediterranean crops with significant possibilities for expansion due to increasing world demand for olive oil. At present, olive oil production is heavily subsidised by the Common Agricultural Policy (CAP). This policy has led to intensification and increased output in the main producing regions, with generally negative consequences for the environment. To a lesser extent, it has helped to reduce the abandonment of small, traditional plantations in many marginal regions, thus preventing the loss of certain environmental and social values. As with many sectors typical of the Mediterranean regions, the olive sector generally receives little attention at the European level, particularly in the debate on the environmental role of agriculture. Very little work has been done on the environmental effects of different production systems, and only one publication was encountered which specifically considers the environmental effects of the CAP olive oil regime (Pain, 1994).

Currently, the CAP support regime for olive oil is being reviewed, which raises important questions about the rationale of this policy and its socio-economic and environmental effects. Various reform options are being discussed, including the introduction of a production-neutral aid per tree or per hectare to replace the current system of production support. Until now, the rather limited public debate on the reform of the CAP regime has been dominated by economic concerns, as Member States and producers fight to defend their current levels of subsidy, and the European Commission seeks a solution to the widespread fraud which plagues the existing support regime. Social and environmental arguments have been used from time to time, usually in defence of the status quo. However, such arguments have rarely been based on clear or objective data. Generally, there is a very limited understanding of the real environmental issues concerning olive production and this situation leads, in some instances, to the spread of environmental 'misinformation' and 'myths'.

This chapter identifies the key environmental issues concerning olive production and the CAP support regime and reviews and evaluates the main options for change. The chapter is based on a review of available literature and consultation with a number of experts and producers, principally in Spain (unless otherwise stated, the data refer to this country). It also draws on work undertaken in Italy for the European Forum for Nature Conservation and Pastoralism (EFNCP) and the Institute for European Environmental Policy (IEEP) (Petretti, 1995).

PRODUCTION PATTERNS IN THE EUROPEAN UNION (EU)

Olive farming is a major land use in Mediterranean regions, covering some 4.4 million ha in the EU Member States. The main areas of olive oil production are in Spain, with 2 million ha, and Italy, with 1.1 million ha. Greece and Portugal have around 0.9 and 0.4 million ha respectively, whilst France is a very much smaller producer, with 40,000 ha. Olive statistics vary considerably, depending on the source and the criteria used, and national data from the main producer countries are notoriously unreliable. The figures in Table 10.1 are from the International Olive Oil Council (IOOC) and relate to oil production. Plantations which produce table olives are not included. These cover a far smaller area in most countries, e.g. 130,000 ha in Spain.

Whilst olive plantations are found over most of the Mediterranean region, the greatest concentration of oil production is found in two Spanish provinces, Jaén and Córdoba in Andalucía, which between them account for over a third of EU output. The EU currently dominates the global market, producing over 70% of the world's olive oil. Tunisia, Turkey and Syria are the only other producers of significance, together accounting for over 20% of world production.

The area of land under olives in the EU countries has fluctuated considerably over recent decades, following different national and regional patterns. In Spain, olive plantations reached their greatest extent in the mid-1960s, when they were nearly 2.4 million ha, which was about three times the area covered only a hundred years previously. This expansion took place largely at the expense of Mediterranean woodlands, as a result of land privatisation (Parra, 1990). This was followed by the abandonment and subsidised grubbing-out of many older plantations in the 1970s and early 1980s. Some 300,000 ha were cleared in Andalucía under a government restructuring scheme started in 1972, to be replaced mostly by arable crops (MAPA, 1988).

The decline in the area of old olive plantations during the 1970s and 1980s due to a combination of abandonment and restructuring programmes seems to be common to other Mediterranean countries. In Italy, one estimate has put the decline in olive plantations due to abandonment at some 25-30,000 ha per annum from the mid-1980s to mid-1990s (Petretti, 1995). In Portugal, 30,000 ha of old plantations have been removed under the PEDAP (Specific Programme for the Development of Portuguese Agriculture) since 1986, which was funded by FEOGA with the aim of aiding the adaptation of Portuguese agriculture to EC market conditions. Only 7,000 ha were replanted with new trees (Fialho, 1996). It is reasonable to assume that the rate of abandonment in marginal areas would have been greater in the absence of the significant level of support for olive production provided by the CAP olive oil regime. However, no studies have been encountered of the extent to which the CAP regime has prevented abandonment.

Table 10.1 General data on olive area and production in EU (excluding table olives)

Member State	Olive area (ha)	Oil production (tonnes) [a]	Producers (no.)	Share of world olive oil output (%)
Spain	2,000,000	535,000	500,000	28
Italy	1,100,000	467,000	800,000	24
Greece	900,000	307,000	686,000	16
Portugal	340,000	35,000	70,000	2
France	40,000	2,000	20,000	<0.1
EU	4,380,000	1,346,000	2,076,000	70

[a] Annual average 1990/91 to 1995/96, IOOC estimates.
Source: Polidori *et al.* (1997).

At the same time as older plantations have been abandoned or grubbed-up in some regions, new plantations have been created, particularly in regions with a high concentration of commercially orientated producers, such as Andalucía and Badajoz in Spain. For producers in such areas, the CAP regime has provided a strong incentive to increase production, and new plantations have increased steadily since Spain's accession to the European Community in 1986. For example, in 1995 alone, 67,000 ha of olives were reportedly planted in Spain (López Sánchez-Cantalejo, 1996). Overall, there appears to be a tendency towards the gradual abandonment of traditional olive plantations in certain more marginal regions and the expansion of new plantations and intensification of management systems in regions with a comparative advantage.

PRODUCTION SYSTEMS

Olive production systems vary considerably. Tree density varies from as few as 40-50 stems per ha noted in some older plantations in Italy to as many as 300-400 stems per ha in the most intensive plantations of Spain. Management systems range from the highly labour-intensive to the highly mechanised. There seems, though, to be no widely used typology of olive production systems. This chapter uses a simple typology for production systems in Spain, which is adapted from that used by Petretti in Italy (Petretti, 1995). Three broad types are identified: traditional plantations, often with ancient trees and typically planted on terraces, which are managed with few or no chemical inputs; intermediate plantations which to some extent follow traditional patterns but are under semi-intensive management using large inputs of artificial fertilisers and pesticides, sometimes with irrigation; and modern plantations of smaller tree varieties, planted at higher densities and managed under an intensive and highly mechanised system, usually with irrigation (see Table 10.2). It must be emphasised that this typology represents a considerable simplification of real conditions. The reality on the ground is more complex. Nevertheless, the broad categories reflect the different environmental considerations associated with olive production and provide a basis for their examination. Thus, the traditional plantations have the highest natural value (biodiversity and landscape value) and generally represent a sustainable approach to exploiting marginal agricultural land. The semi-intensive and intensive systems are of much less natural value and have potentially much greater negative environmental impacts, particularly in the form of soil erosion and pesticide pollution.

Table 10.2 Typical characteristics of three broad plantation types in Spain

	Traditional	Semi-intensive	Intensive
Typical tree characteristics	Big and old	Smaller and young-er due to replanting	Dwarf varieties replanted re-gularly
Tree density (per ha)	80-150	150-200	200-400
Terraces with supporting walls	Common	Occasional	Rare
Understorey weed control	Harrowed occasionally	Harrowed repeatedly	Controlled with herbicides
Grazing	Rare in Spain but common in Italy	Rare	No
Chemical inputs	Very low	High	High
Irrigation	Not common	Increasing	Common
Typical yield (kg ha^{-1})	1,200	2,200	5,500
Consistency of annual yield	Very low	Low	High
Soil erosion	Usually low	Often very high	Medium
Biodiversity	High	Low	Very low
Landscape value	High	Low	Low
Other environment impacts	Fire preven-tion in margi-nal areas	Pesticide pollution, reservoirs for irriga-tion	Pesticide pollu-tion, reservoirs for irrigation

Traditional plantations

The trees in traditional plantations are mostly old and in some cases have been maintained in production for several hundred years by means of pruning and grafting of new stock. There is a great variety of traditional planting patterns, depending on local traditions, soil conditions, rainfall, etc. The most extensive plantations may have as few as 40 trees ha^{-1}, particularly in areas of low rainfall. In Grosseto province in Italy, the most extensively managed plantations have approximately 50 trees ha^{-1} (Petretti, 1995). On the other hand, in certain areas, traditional plantations may have over 200 trees ha^{-1}, for example, Sierra de Gata and Hurdes in Extremadura, Spain (Guerrero, 1994).

On steep slopes, the land is almost invariably terraced, which prevents soil erosion. Terraces are a characteristic landscape feature in many Mediterranean uplands, although some areas have already been lost to abandonment. In the past in Spain, the soil often was cultivated for the production of crops for on-farm use, such as barley, oats, chick-peas or vetch (Parra, 1990). Nowadays, the normal practice is simply to control spontaneous grasses and other vegetation by seasonal ploughing or harrowing, sometimes preceded by grazing or cutting. The frequency of tillage depends on the region and on the individual farmer. The control of vegetation in traditional olive plantations by sheep grazing is quite common in Italy but is rarely seen in Spain nowadays.

In recent years, traditional olive plantations have undergone a mixed process of abandonment in some cases and intensification in others. Abandonment goes hand in hand with the general social decline and depopulation of upland areas which have taken place in recent decades in Mediterranean regions. Olive plantation abandonment has taken place on a large scale in certain upland areas, for example in Grosseto Province in Italy, where it is associated with a loss of habitat and landscape diversity and an increase in the risk and destructiveness of forest fires (Petretti, 1995). Intensification may take different forms, such as the introduction of pesticides (particularly herbicides) and chemical fertilisers and more intensive soil cultivation to control weeds, leading to a general reduction in natural value and an increase in negative impacts, such as soil erosion.

Nevertheless, large areas of traditional olive plantations survive throughout the Mediterranean Member States particularly, but not exclusively, in more remote and upland areas. This type of production is low yielding and contributes a small proportion of total EU output. However, activities such as pruning and, especially, harvesting, provide a significant level of seasonal employment in areas which often depend on a mixture of part-time farming and tourism. Olive production therefore makes a significant contribution to the maintenance of rural communities in many marginal areas.

Semi-intensive plantations

Although relatively large areas of traditional plantations survive in most Mediterranean regions, the larger part of EU production comes from plantations which are managed under a semi-intensive system. This system dominates the landscape in provinces such as Jaén and Córdoba in Spain, which between them produce over a third of EU olive oil output. The trees are generally younger and planted closer together (80-220 ha^{-1}). Drip irrigation allows for higher densities in some cases. Artificial fertilisers and pesticides are used systematically and the land is cultivated repeatedly during the year to keep the soil free of competing weeds. Management systems are more intensive than in traditional plantations: the grass understorey is rarely allowed to become established whilst pesticide use reduces the diversity and richness of insect fauna. A large proportion of the plantations are found on slopes, which are often steep and generally have thin soils which are highly vulnerable to erosion. Flatter land with deeper soils traditionally has been reserved for annual crops. In Andalucía, many new plantations apparently have been established in recent decades by clearing natural woodland on slopes. Terraces are no longer constructed as they are very labour intensive and restrict the access of machinery.

Intensive plantations

The most modern plantations use smaller tree varieties planted at high densities. In the 1970s and 1980s, systems using very high densities were developed (up to 400 trees ha^{-1}), although the current tendency is to use a density of 200-225 trees ha^{-1} (Guerrero, 1994). These plantations are intensive and highly mechanised, often using seasonal irrigation and mechanical harvesting. Production is considerably higher than in the semi-intensive plantations described above.

Intensive plantations are inherently low in biodiversity due to the use of high levels of insecticides and herbicides, the absence of a grass understorey and the young age and small size of trees. Soil erosion generally is not so serious as in the semi-intensive type of plantation as they tend to be planted on better, flatter land, whilst modern production techniques usually involve reduced tillage and a greater use of herbicides to control competing vegetation.

ENVIRONMENTAL ISSUES ASSOCIATED WITH OLIVE PRODUCTION

This chapter limits itself to the principal environmental questions associated with the management of olive plantations: nature conservation and landscape, soil erosion and pesticide use. The question of water pollution resulting from the olive oil production process off the farm, which is a serious environmental problem in many olive producing regions, is not covered here.

Nature conservation and landscape

The value of olive plantations for nature conservation ('natural value') varies according to factors such as the level of pesticide use, whether a seasonal ground storey is allowed to become established beneath the trees, the age of the trees, etc. The intensive application of techniques intended to increase production (frequent tillage, heavy herbicide and insecticide use, etc.) has a detrimental effect on ground flora and insect populations and generally results in a reduction of natural value. Consequently, given the same conditions of soil and climate, low-yielding plantations tend to be of higher natural value than those from which high yields are achieved through intensive management practices. A possible exception is irrigation: introducing drip-irrigation to traditional plantations may lead to a significant increase and greater consistency in production without itself leading to a loss of natural values (P. Eden, personal communication). However, the development of irrigated production on a large scale may lead to considerable environmental impacts through the construction of reservoirs and the alteration of streams and rivers to provide water.

The structural diversity of traditionally managed olive plantations provides a variety of habitats. The older trees support a high diversity and density of insects which, together with the tree's fruit, provide an abundant supply of food (Parra, 1990). The low level of pesticide use allows a rich flora and insect fauna to flourish, which in turn provides a valuable food source for a variety of avifauna. Consequently, traditional olive plantations generally support a high diversity of wildlife, including reptiles, butterflies and other invertebrates, birds and small mammals. As well as many passerine species, typical nesting birds include hoopoe (*Upupa epops*), roller (*Coracias garrulus*) and scops owl (*Otus scops*) (Petretti, 1995). The little owl (*Athene noctua*) traditionally is associated with old olive plantations where it nests in the hollows of older trees and hunts insects, lizards and small mammals. The trunks of older trees are also used by mammals, such as the genet (*Genetta genetta*), and by reptiles. The spontaneous vegetation which develops between tillage can be of a high floral diversity, if sufficient time is allowed for it to develop. In a selection of plantations in western Andalucía, 75 plant species were recorded prior to the spring cultivation (Rodenas *et al.*, 1977).

In addition, the use of Mediterranean olive plantations as a food source by very large numbers of migrant passerine birds, both from northern and central Europe and from Africa, is well documented. Where pesticides are used intensively to control specific parasites, the overall insect population inevitably suffers and the trees' overall value as a food source for birds is reduced, although the olive fruit is still available to them.

Where olives form a part of a diverse land use system, for example in combination with pastures, arable cultivation and vineyards or where vines are grown between rows of olive trees, they are an important landscape feature and may add considerably to habitat diversity, particularly in countryside with few other trees. Traditional olive terraces are a characteristic of upland landscapes in many Mediterranean regions. However, in major oil-producing regions such as Andalucía, olive plantations tend to dominate the landscape, forming vast monocultures in which the trees themselves are the only form of vegetation for the greater part of the year. In this situation, landscape and habitat diversity are very limited.

The intermittent harvesting of olives which often precedes abandonment in marginal plantations should not be detrimental to their natural value; on the contrary, research has shown that in drought years when many plantations are not harvested, birds take particular advantage of the availability of olives, given the scarcity of wild fruits. One option for the management of non-viable plantations may be to abandon cultivation and introduce livestock to graze the spontaneous vegetation. This is reported to have happened in some less-productive plantations west of Sevilla in Spain (Rodenas *et al.*, 1977). The result is a more natural, *dehesa* type of landscape.

However, the total abandonment of olive plantations leads to scrub invasion through natural succession, and consequently to a more fundamental ecological change. Forest fires permitting, some form of woodland or Mediterranean *maquis* will result. Research undertaken in Grosseto province in Italy has shown that approximately 50% of olive plantations in the area are either marginal and threatened with abandonment or are already abandoned (Petretti, 1995). Abandoned olive plantations in this area develop into dense woodland within a period of 9-15 years. As a result, many species associated with the extensively managed olive plantations of the area decline or disappear, particularly reptiles, butterflies and birds. On the other hand, woodland species tend to benefit, for example mammals such as the porcupine (*Hystrix cristata*), pine marten (*Martes martes*) and wolf (*Canis canis*) (Petretti, 1995).

The natural woodland which results from abandonment may develop into a habitat of high natural value. However, it is important to consider the wider context of the landscape within which abandonment takes place. In upland regions which have suffered agricultural abandonment in recent decades, and in which landscapes are in danger of becoming dominated by forest and scrub, remaining olive plantations provide an important element of more open habitat which should benefit overall biodiversity (Petretti, 1995). They also provide effective firebreaks, a very important consideration in Mediterranean areas suffering from widespread abandonment and forest fires.

While abandonment has occurred in certain regions, expansion of olive production has happened elsewhere. Since the CAP regime began to be applied in Spain from 1986, there has been an expansion in olive plantations in the main producing areas. In parts of Andalucía this expansion has taken place at the expense of natural woodland, which is of high conservation value, because it contributes an element of diversity in landscapes already dominated by intensively managed olive plantations. There is also concern amongst conservationists in Spain about intensive olive plantations encroaching on arable land in areas of importance for steppeland bird communities. This has happened to some extent in Córdoba and Málaga in Andalucía (F. Cabello de Alba Jurado, personal communication).

Finally, several recent publications have emphasised the importance for nature conservation in Europe of traditional, low intensity farming systems (Beaufoy *et al.*, 1994). These agro-ecosystems and the wildlife species associated with them are threatened by a combination of abandonment and intensification in many parts of the EU. The same tendencies appear to apply to olive plantations.

Soil erosion

Soil erosion is a major problem associated with olive plantations. In Spain, dryland tree crops on slopes (principally olives) have been identified as the agricultural land use with the highest rates of soil erosion (Díaz Alvarez and Almorox Alonso, 1994).

Aggregate losses of topsoil from olive plantations in Andalucía have been calculated to reach as high as 80 tonne ha^{-1} per annum in some cases (Pastor and Castro, 1995), indicating a totally unsustainable farming system which is resulting in the widespread degradation of natural resources. Some 20% of all olive plantations in Spain are reported to have lost the upper soil horizon due to erosion.

The problem of erosion has been greatly exacerbated in recent decades by changes in practices associated with mechanisation (Díaz Alvarez and Almorox Alonso, 1994). It is now widespread practice for farmers to keep the soil of olive plantations bare of vegetation all the year round, by regular cultivation. Normally, this tillage is carried out up and down the slope, rather than following the contours. The most severe erosion takes place with the arrival of the autumn rains on bare soils which have been cultivated to a fine tilth by summer harrowing. According to experts, the frequent tillage which is widely practised is of doubtful agronomic value. Shallower and less frequent cultivation would be equally effective as a means of controlling ground vegetation and would reduce the soil's vulnerability to erosion. New minimum tillage and non-tillage systems have been developed which produce higher yields than conventional systems. These systems include techniques which greatly reduce erosion, particularly the maintenance of plant cover (crops such as barley or vetch, or spontaneous vegetation) on the strips of land between the lines of olive trees. Under such systems, the plant cover is normally eliminated with herbicides, rather than by mechanical cultivation. Unfortunately, from an environmental point of view, the herbicides which are used are mostly residual products of relatively high toxicity (for example, simazine and diuron) (Guerrero, 1994). More environmentally friendly options are available, such as glyphosate, but require far more frequent application as they are not residual. Mechanical mowing or grazing by livestock to control vegetation involve more complex management and generally result in lower yields, but are preferable from an environmental point of view. Various factors are cited in Spain which account for the fact that very few olive producers have adopted minimum or non-tillage cultivation systems:

- An ingrained mentality that land must be kept entirely free of weeds; also, most farmers have a tractor and harrow or plough and want to use them, and take pride in having a 'clean' olive plantation completely free of weeds.
- Changing to a non-tillage system would require the purchase of new machinery in many cases, for applying herbicides or for mowing spontaneous vegetation.
- Grazing generally is not a practical option as sheep often are no longer present in olive producing areas.
- Farm extension services in most regions are highly inadequate; consequently, alternative systems are not promoted to the great majority of farmers.

Finally, it is sometimes argued that olive plantations have an important function in protecting the soil from erosion, and that their abandonment should be prevented for this reason. This argument is misleading. While it is true that abandoned terraces in certain very dry areas (such as south-east Spain) are vulnerable to collapse which in some cases leads to serious erosion and landslips, this is a relatively localised problem. Erosion may also be a problem when plantations are grubbed-up prior to abandonment, as has occurred under various national and CAP grant schemes. However, such schemes are now much less common. Except in very adverse conditions, abandonment without grubbing-up tends to result in scrub invasion and the gradual development of natural woodland, which provides a high level of soil protection, so long as the vegetation is not destroyed by repeated fires. In general, soil erosion is a problem associated with the intensive cultivation of olive plantations, rather than with their abandonment.

Pesticide use

Apart from greatly reducing the natural value of olive plantations, excessive use of pesticides may create other environmental problems. For example, pesticide residues can arise in olive oil and in table olives (Civantos, 1995). Indiscriminate use of wide-spectrum products has led in some cases to explosions in coccid populations and other pests due to the removal of their natural enemies. Excessive herbicide treatment has caused die-back in the olive trees of some plantations. The creation of high levels of background pollution is a concern in regions where olive plantations cover extensive areas of land (Civantos, 1995). These problems have not been studied in detail yet.

MECHANISMS OF THE SUPPORT REGIME

This evaluation refers to the CAP olive oil regime prior to the interim reform of 1998. It focuses on the environmental considerations of different support mechanisms, although some reference is made to other issues (e.g. social) where these are considered relevant.

The current CAP olive oil regime incorporates no environmental elements. In fact, the support system is biased heavily in favour of the most intensive and competitive plantation types of least environmental benefit, at the expense of traditional systems of inherent natural value (see Table 10.3, illustrating the very different economic and production characteristics of three broad types of plantation in Spain). Ironically, most of the support goes to the producers who have the highest annual net incomes per ha (exclusive of subsidies).

While it is not realistic to expect to be able to design an EU regime which solves all of the environmental issues associated with the sector, it should be possible to establish a benign framework, within which Member States could then implement measures adapted to local conditions and needs. In particular, the following environmental issues need to be addressed:

- to prevent the abandonment of traditional olive plantations where these make a positive contribution to natural and landscape values;
- to reduce currently high levels of soil erosion in semi-intensive systems by promoting a change in management practices or, in extreme cases, in land use;

Table 10.3 Production characteristics of different plantation types in Spain, 1994

	Traditional upland [a]	Semi-intensive (dryland) [b]	Intensive dryland [b]	Intensive irrigated [b]
Trees per ha	150	180	200	300
Yield (kg olives ha[-1])	1,200	2,200	4,500	6,500
Direct costs (ECU ha[-1])	600	615	916	1,636
Total costs kg[-1] of oil produced (ECU)	4	3.2	1.7	1
Net income (without subsidy) (ECU ha[-1])	-114	276	907	997
Production subsidy (ECU ha[-1]) [c]	264	484	990	1,430
Gross income (subsidy included) (ECU ha[-1]) [d]	750	1,375	2,813	4,063
Net income (subsidy included) (ECU ha[-1])	150	760	1,897	2,427
Subsidy with tree payment of 4 ECU per tree (ECU ha[-1])	600	720	800	1,200

[a] Small upland plantation on terraces, Cáceres; author's estimates. [b] Plantations in the 'campiña' area of Córdoba; adapted from Guerrero (1994). [c] Calculated at 1.375 ECU kg[-1] of oil and average oil yield of 16%. [d] Calculated at 0.625 ECU kg[-1] of olives.

- to promote more environmentally sensitive production systems, particularly by reducing the use of high-toxicity pesticides;
- to prevent the further expansion of olive plantations onto valuable habitats and soils that are vulnerable to erosion and to control the spread of irrigation in areas with sensitive water resources.

An important additional issue is the distribution of financial resources. In order to address environmental (and social) needs, far greater resources must be made available for targeted measures with clear environmental (or social) objectives. Currently the great majority of support for olive farming is in the form of production subsidies. Meanwhile, agri-environmental schemes under Regulation 2078/92 have very limited resources (some 3% of the total FEOGA budget), the majority of which is absorbed by schemes in the northern and central regions of the EU (see Buller, Chapter 12).

Current regime (prior to the 1998 interim reform)

The olive oil regime involves an annual expenditure from FEOGA of some 2,000 million ECU on a complex range of 'classic' CAP support mechanisms. The key elements, prior to 1998, were:

- a minimum price for producers, maintained by a combination of import restrictions and intervention buying;
- an additional production subsidy per kilo of oil produced;
- a special support mechanism for small producers;
- an EU 'maximum guaranteed quantity' eligible for subsidy;
- a consumption subsidy, paid to the processing sector;
- export subsidies.

The consumption subsidy and small-producer mechanism were abolished under the 1998 interim reform of the regime. The system of production and consumption subsidies has been plagued by fraud on an alarming scale in all producing countries, as highlighted over the years by reports of the EU Court of Auditors (EC, 1997). Member States seem to have been unable to tackle this problem.

In recent years, the market price of olive oil has been well above the intervention price, so that intervention buying has not been necessary. Consequently, the most important mechanism has been the oil production subsidy, which provides producers with an additional income equivalent to approximately 80% of the intervention price. For example, in 1996/97 the intervention price in Spain was about 300 pesetas kg^{-1}, whilst the additional production subsidy amounted to approximately 240 pesetas kg^{-1}. Meanwhile, market prices during the year rose to well above 600 pesetas kg^{-1} (double the intervention price), partly due to the drought which resulted in shortages of supply. However, very large harvests in 1997/98 have brought market prices down to the level of the intervention price.

So-called large producers (with production levels which exceed 500 kg of oil per year) receive the production subsidy according to the quantity of olives converted into oil. As the main olive harvest occurs every 2 years, the amount received both from the market and in subsidy varies considerably from 1 year to the next. Under the pre-1998 regime, a different mechanism was applied to small producers (with production levels below 500 kg of oil per annum, typically holdings of less than 3 ha), for whom the subsidy was based on the average yield for the district or region during the previous 4 years, multiplied by the number of trees on the holding. Some 60-65% of EU producers received their production subsidy in this way (EC, 1997).

While it is not possible, without more comprehensive research, to ascertain the exact role of the CAP olive oil regime in defining production patterns and tendencies, certain conclusions can be drawn (see Table 10.3). In considering these effects, it is interesting to note that plantations yielding on average less than 1,500 kg of olives per ha cover approximately 55% of the olive area in Spain, while producing only 17% of total output (Unión de Pequeños Agricultores, 1998).

Production subsidy

Compared with traditional plantations, intensive systems produce very much higher yields and consequently benefit from a far higher level of production subsidy. The production subsidy for an intensive irrigated plantation is approximately *five times greater per ha* than the subsidy for a traditional, non-intensive plantation. However, the more productive plantations generate relatively high net returns per ha and, at 1994 prices, could produce a significant net income even without the production subsidy. Traditional plantations, on the other hand, appear to be barely viable under the present support system. Their costs per kg of production (four times higher than intensive irrigated plantations) reflect the high labour requirements of these systems (maintenance of terraces, difficulties of mechanisation, etc.), whilst their output is low. The features which lead to high labour costs (terraces, old trees, etc.) also form an integral part of their environmental value.

The production subsidy provides a strong incentive to maximise returns by intensifying production systems (see for example Fotopoulos *et al.*, 1997). Table 10.3 illustrates this effect in the case of irrigated and non-irrigated intensive plantations: with the production subsidy, conversion to irrigation results in a gain in net income of 530 ECU ha^{-1}, compared with a gain of only 90 ECU ha^{-1} without the production subsidy. The subsidy clearly provides a very strong incentive for converting to irrigated production. In certain areas (e.g. parts of Andalucía), this incentive also appears to be encouraging farmers to expand their plantations onto new land, sometimes at the expense of valuable natural habitats (Mediterranean scrub, extensive arable steppes) and onto steep slopes which are highly vulnerable to soil erosion.

Finally, the production subsidy promotes a high volume of output, but does not encourage quality. Especially in marginal areas, farmers are not concerned about quality because the subsidy constitutes their main income from olives and is paid at the same level regardless of quality.

Price support

As a result of high market prices, the price support mechanisms have played a relatively insignificant role in recent years, compared with the production subsidy. Consequently, they probably have not constituted an incentive towards intensification. If price support is set a level which provides a 'safety net' for producers when market prices fall dramatically (as was the case in 1997/98), then this is a valuable mechanism for helping to maintain traditional plantations which might otherwise be abandoned. Particularly given the large fluctuations in olive harvests and prices, there may be a strong case for keeping a system of basic price support and market intervention.

Small-producer mechanism

Under this mechanism, over 60% of EU olive producers received their subsidy according to the number of trees on the holding and in proportion to average historic yields in the district. Consequently, they benefited from a guaranteed level of subsidy each year, regardless of their own level of production. This is particularly important for traditional plantations which, because of their characteristics and management systems, have a strong tendency to produce a crop only every second year. Without the annual income provided by this small-producer mechanism, many traditional plantations probably would be much more vulnerable to abandonment. Although the mechanism itself is a sensible way of providing a continuity of income to more marginal plantations, the threshold of 500 kg for defining a small producer is arbitrary and vulnerable to fraud. It bears no relation to any specific social or environmental criteria and is thus a rather blunt instrument. The small-producer mechanism was abolished in the 1998 interim reform of the CAP regime.

Options for reform

In 1997, the European Commission (EC) produced a document (commonly known as the 'options paper') which assesses in general terms the need for a reform of the existing olive oil regime, and presents two broad options: a simplification of the current regime, involving the removal of the small-producer mechanism; or a more radical alternative, under which production subsidies would be replaced by a direct payment per tree (EC, 1997).

The limited reform option, involving the abolition of the small-producer mechanism whilst leaving the main production support system unchanged, could create the worst of both worlds from an environmental and social point of view. It would take away the important security for small producers currently provided by the tree-related annual subsidy, whilst maintaining the powerful incentive for larger producers for intensive, high output production. In fact, this was the option adopted in 1998 under the interim reform, which is to be reviewed before 2001.

Converting all or a part of the current production subsidy into a direct payment per tree (or per ha) would follow the principles of the 1992 reform of the CAP arable regime. In the longer term, this 'decoupling' will be necessary in order to comply with the trade liberalisation set in motion under the Uruguay Round of the GATT. Removing the production subsidy would reduce the incentive for intensification and thus could be considered broadly positive in environmental terms. Direct payments would provide an important continuity of income for more traditional producers, and thus may help to prevent abandonment (as with the small-producer mechanism).

However, a change to direct payments certainly would not solve all of the environmental issues associated with olive production. At the same time, it raises a number of specific environmental and social concerns that would have to be addressed by complementary mechanisms, such as:

- direct payment quotas to prevent the uncontrolled expansion of plantations;
- the modulation of support payments in relation to environmental and social conditions;
- 'cross-compliance' to prevent the payment of support to producers whose practices damage the environment, or to ensure the upkeep of plantations;
- 'agri-environment' incentive payments to promote particular systems and practices;
- regional strategies to address particular problems, such as soil erosion.

These options, and the various issues relating to tree and area payments, are discussed below. Perhaps the most widely expressed concern about the removal of production subsidies is the possibility of olive production becoming unviable, thus leading to a decline in harvesting and management and ultimately to the abandonment of olive plantations, which would lead to a decline in rural employment in olive-producing areas as well as having important environmental consequences. This question is discussed first.

Production-neutral support and the danger of management decline

Social concerns

The olive harvest (and other management, such as pruning) provides a very significant level of seasonal employment in some rural areas. It has been argued that the removal of price support and production subsidy might lead to a gradual collapse of this economy; farmers might simply take the direct payment and save themselves the costs of management and harvesting. It is uncertain whether this would be the outcome, and further investigations are needed so that the reform of the olive oil regime can be based on objective information.

However, such research should be more than a simple economic analysis of production costs versus returns. The viability of production depends greatly on the socio-economic context, which varies widely in the case of olives. In the main producing regions, olive farming usually involves a significant proportion of hired labour, as well as the other standard direct and indirect costs normally associated with agricultural production. For such systems, economic viability probably can be assessed quite easily. This chapter does not attempt such an analysis, although the figures in Table 10.3 suggest that there is no danger of abandonment or a decline in management in the more productive systems, which could produce a significant net return without any subsidy. At most, the change to direct payments may lead to a reduction in the intensity of management, thus potentially benefiting the environment.

If there is a danger of declining management and production, then it is surely not in the main producing regions, but in areas with more traditional, extensive systems, particularly where labour requirements are high due to the presence of terraces, older trees in irregular patterns, etc. Using conventional calculations of labour costs, it may seem that management and harvesting activities would cease to be economically viable in the absence of the production subsidy and price support. However, there are many additional factors which must be taken into account. For example, olive production in marginal districts generally is a part-time activity which depends principally on unpaid family labour. Furthermore, many marginal rural areas already suffer from considerable unemployment, so that casual labour often is available at a very low cost. So, whereas there may be a legitimate concern about the possible social effects in marginal areas of removing the production subsidy, the issue is more complex than it first appears and should be studied according to the different circumstances of each region.

Environmental concerns

In purely environmental terms, a decline in the management of plantations or in the harvesting of olives should not be a problem in itself; studies have shown the fruit of unharvested trees to be valuable for wildlife (especially overwintering birds) and the reduced level of disturbance might also be beneficial. Besides, the most extensive systems already receive very little management. However, within a few years, a totally unmanaged olive plantation will convert itself into scrub and ultimately woodland, thus changing completely its natural and landscape value. To prevent this process, probably the minimum management would be to maintain an open understorey, either through grazing, mowing or cultivation, and to remove the new shoots which grow from the base of the olive tree each year and undertake periodic pruning.

Mechanisms for maintaining plantations in production

The EC options paper (EC, 1997) suggests that all producers could be required to demonstrate that their olives have been processed, as previously happened with the small-producer mechanism, in order to avoid a decline in harvesting. Maintaining a basic level of price support for olive production is probably also important to ensure the continued viability of traditional systems. Another possibility would be to require the recipients of the aid to maintain their plantations in production by undertaking particular management tasks. This would be a form of environmental and social 'cross-compliance': the direct payment would be for recognisable olive plantations, not those invaded by scrub or woodland. The requirement could be simply to maintain the basic structure of an olive plantation with its key components (pruned trees and cultivated, grazed or mown understorey). If required to maintain their plantation in this way, most owners probably would continue to harvest, assuming prices remain above the intervention threshold of recent years.

Tree payments

From an environmental perspective, a support payment per tree potentially would be an improvement over the current production subsidy. Assuming that appropriate controls on new planting were implemented, it would remove the current incentive for intensifying and expanding production and would provide a basis on which to build environmental 'cross-compliance'. However, the conversion of a part of the production subsidy into an aid per tree could have some significant side-effects, given the fact that tree densities vary enormously. For example, national statistics show a range of average tree densities between a low of 54 trees ha^{-1} in Portugal up to 190 ha^{-1} in Greece (Guerrero, 1994). The density of trees ha^{-1} is dictated by many factors, including soil and climatic conditions, local planting traditions, harvesting techniques, availability of irrigation (permitting higher densities), etc.

Generally speaking, traditional plantations have a low density of large, old trees, whereas modern, more intensive plantations have a high density of small trees. A simple tree payment therefore would tend to provide a much higher level of aid to the most productive and intensive systems; these are the systems of least value for nature and landscape and which in many cases cause high levels of soil erosion. However, the picture is further complicated by the fact that traditional plantations in some regions also have a high density of trees. From an environmental perspective, there is a risk that the introduction of a tree payment might provide an incentive to a producer with an old, widely spaced plantation of 100 trees ha^{-1} to clear the existing trees and to replant with a more intensive system at, say, 250 trees ha^{-1}. On the other hand, some *additional* planting in very old and highly extensive plantations would be desirable in order to introduce a new generation of trees.

To avoid favouring modern, intensive plantations at the expense of traditional, more extensive plantations, a tree-density upper limit should be set above which the aid would not be paid (for example, 150 trees ha^{-1}). Tree payments should not be paid simply in proportion to the number of trees per ha as this would produce serious distortions which could be potentially detrimental to the environment. Overall, the tree-payment option appears to be fraught with potential problems and would run the risk of introducing a number of new distortions into the support system. There is no logical reason to pay farmers in proportion to tree density.

Area payments

An area payment system would avoid the distortions inherent in a tree payment system and would achieve more clearly the objective of decoupling support from production. There would be no incentive for clearing older trees to create new intensive plantations. To avoid claims from the owners of fields with only one or two trees, a minimum number per ha would be required; this should be low enough to allow very extensive, traditional plantations to benefit (e.g. 40-50 trees ha^{-1}). Overall, an area payment system would introduce fewer distortions and would provide a simpler basis for 'modulating' support in favour of plantations of environmental value and away from the most intensive, highly productive plantations. It could also be integrated, in the longer term, with the arable area payments scheme.

Direct payment 'quotas'

As with the direct payments in the arable and livestock sectors, a system of 'rights' or 'quotas' would need to be established in order to prevent fraud or an unlimited expansion of payments. It should be stressed that there appears to be considerable scope for expanding the area of olive production in the southern Member States of the EU; from a socio-economic point of view it would be wrong to fossilise existing land use patterns and prevent any further expansion. Olives represent one of the few Mediterranean crops with potential for expansion thanks to growing world demand.

From an environmental point of view, the main concern is to steer any future expansion of olive plantations onto the most appropriate land. This means principally avoiding steep slopes and land with particular conservation values, such as woodland, steppelands and wetlands. Following such an approach there would still be large areas of land currently under low-yielding arable cultivation, and with edapho-climatic conditions highly suitable for olive production, which could be planted with olives without environmental damage, and with potential environmental benefits, depending on the types of plantation and the management practices. There is also a need to take some plantations out of production, where extreme soil erosion cannot otherwise be avoided. A quota system therefore should be flexible enough to allow some redistribution.

Modulation of direct payments

The EC 1997 options paper for olive oil reform states that the modulation of tree payments should reflect variations in yield and particular environmental and social conditions. Modulation in proportion to yield may be necessary in the short term, in order to maintain approximately the current distribution of aid and so that the reform does not produce big 'winners' and 'losers'. However, a mechanism should be established in order to reduce, over time, the amount of aid going to the most productive plantations, in order to direct more support towards plantations of environmental value. Nevertheless, it must be recognised that there is a difficulty in defining precise criteria as indicators of these types of plantation. Insufficient work has been done in this area. Some possible approaches are mentioned below.

Ideally, environmental modulation should relate to particular management practices, or to the production of specific environmental benefits. For example, a higher payment could be offered for organic production. Many traditional olive plantations are managed with no or limited use of chemical inputs and therefore could benefit from such an aid. Although concrete data have not been encountered, organically managed plantations can be expected to be of considerably higher biodiversity than those which use pesticides systematically. Another useful environmental criterion for a higher level of support payment could be the maintenance of a permanent or semi-permanent grass understorey, cultivated no more than twice per year and managed without the use of residual herbicides.

One possible approach might be to allocate a higher payment for olives on terraces with retaining walls. This would help to correct the current situation under which a producer who is causing serious soil erosion by ploughing plantations on steep slopes receives the same support as a producer who is conserving the soil and landscape through the maintenance of traditional terraces. At the same time, such a bonus would aid producers in upland, marginal areas who have very limited production options and many of whom have plantations on terraces, which are also of significant landscape value and have higher costs.

Environmental 'cross-compliance'

Whichever support system is adopted for the olive sector, provision should be made for some basic form of environmental cross-compliance which would be applied to all producers benefiting from CAP support payments. Probably the best arrangement would be for the CAP regime to require Member States to implement this according to regional and local conditions. For example, in Spain aid should be withheld from producers who cause serious soil erosion, either by extending their plantations onto steep slopes or by using management practices, such as repeated tillage, which result in high erosion rates. Payments also should not be paid for new plantations which are created on land which is currently of high natural value, such as natural woodland, steppelands and wetlands. As an example of the conditions which might be set, Figure 10.1 gives details of the requirements that olive producers in Andalucía must meet in return for environmental area payments under Regulation 2078/92. Such criteria could be introduced by national or regional authorities as conditions which have to be met in order to be eligible for tree or area support payments. Other possible options would be to set maximum limits for pesticide usage.

Perhaps the biggest problem is how to ensure that the hundreds of thousands of farmers receiving olive support payments are complying, in practice, with cross-compliance requirements. Clearly to be credible, such a mechanism has to be applied effectively.

Replacement of systems involving repeated cultivation or the use of high toxicity herbicides with systems of reduced tillage or non-tillage, using the following options:

- on slopes of 10-15%, contour tillage with a maximum of two passes in spring and the creation of ditches to reduce the erosive effects of torrents;
- on slopes of > 15%, no tillage; herbicides may be used as an alternative.

Figure 10.1 Conditions for environmental payments under 2078/92 zonal programme for olives in Andalucía
Source: MAPA (1994).

'Agri-environment' incentives under Regulation 2078/92

Some schemes have been set up in Mediterranean regions under Regulation 2078/92 to provide additional aid for traditional olive plantations (Portugal) and to promote soil conservation measures (Spain) and organic production methods (several Member States). However, the impact of these measures is expected to be marginal as the financial resources available for Regulation 2078/92 are extremely limited. Unless a far greater budget is made available for such schemes, they will affect only a very small percentage of EU olive producers.

A greatly expanded agri-environment programme could provide an effective means of promoting olive production systems of high environmental value. This would allow the development of regionally appropriate schemes and avoid the need for a complex modulation of direct support payments. A mechanism should be introduced under the reform to gradually redirect resources away from untargeted direct payments, which should be seen as short-term compensation for the loss of production subsidy, and towards targeted schemes, such as Regulation 2078/92.

Regional strategies

Once an environmentally benign EU regime is established, Member States and regional authorities should develop strategies adapted to local conditions and needs. Such strategies should be based on a comprehensive study of olive production in the region, with the aim of identifying specific agronomic, social and environmental needs which could then be tackled with appropriate measures under a long-term programme. For example, the problem of soil erosion in olive plantations in Andalucía is so widespread that a long-term regional strategy with a range of measures is needed. This should include the promotion of different management systems, through effective advisory services, incentives, cross-compliance, etc., as well as more fundamental changes in land use in extreme cases. A study in the province of Jaén in 1975 identified 74,699 ha of olive plantations (17.2% of the provincial total) on land which should not be ploughed due to the high risk of erosion. It was recommended that 49,309 ha could be maintained in production under a non-tillage system whilst the remainder should be converted to a mixed management system, using the understorey for extensive grazing and forage crops (Aguilar Ruiz *et al.*, 1995). These proposals have not been implemented; the blanket support measures of the CAP have replaced such regional initiatives.

Research and information

Effective policies and strategies must be based on comprehensive, objective data, rather than in response to political pressures. In the case of olive production, such data is greatly lacking at present. Even the basic statistics on olive production are unreliable and widely disputed. There is practically no data available concerning the different types of plantation, their geographical distribution, economic viability, environmental values, etc. According to EC Regulations 75/154 and 2276/79, the producing Member States should have established comprehensive registers of olive plantations by 1981 (France and Italy), by 1988 (Greece) and by 1992 (Spain and Portugal), using aerial photography. Only Italy has come close to fulfilling this requirement. Greece, Spain and Portugal have made very little progress, a situation which is unacceptable and for which the Member States should be penalised. These registers should be developed as soon as possible and should include sufficient information to allow some categorisation of plantations (e.g. slope, presence of terraces, approximate age of trees, etc.).

CONCLUSIONS AND RECOMMENDATIONS

Certain types of olive plantation have important nature conservation and landscape functions. Traditionally managed plantations are of intrinsically high biodiversity and constitute an important element in the mixed farming landscape in Mediterranean regions. They also provide an essential source of winter food for large numbers of migratory birds. When planted on steep slopes, traditional plantations use terraces to prevent soil erosion. This is a sustainable system for the cultivation of land in upland areas which otherwise present very limited possibilities for agriculture. On the other hand, the more intensively managed olive monocultures which dominate the landscape of key production regions usually are of relatively low biodiversity and cause considerable degradation of natural resources through soil erosion on slopes. Recent and current trends show a regression of traditional production systems of high natural value and the expansion of intensive systems.

The CAP support regime for olives and olive oil should take account of the widely differing types of olive production existing within the EU, and of their very different environmental values. Although the European support regime cannot solve all of the environmental issues associated with the sector, it should be possible to establish a benign framework, within which national and regional authorities could implement measures adapted to local conditions and needs.

So far, the public debate and reported negotiations on the reform of the olive oil regime seem largely to ignore the environmental issues, partly because they are insufficiently understood. Commonly repeated arguments that the current regime of subsidies should be maintained because olive trees are 'important for the Mediterranean environment' are absurdly simplistic as well as misleading. At the same time, although the EC's proposal to replace production support with a system of tree payments makes some attempt to incorporate environmental concerns, there is also a danger that it would introduce new distortions and problems. The following key points emerge from this chapter on the different reform options:

- Currently, the great majority of CAP support for olive farming is in the form of production subsidies which favour the biggest producers and the most intensive production systems. This subsidy regime takes no account of environmental or social issues. In order to address these issues, far greater resources must be made available for targeted measures with clear objectives, such as the agri-environment schemes under Regulation 2078/92.
- Given that some form of 'decoupled' direct payment is likely to replace production support in time, a system of area payments would be less distorting, less problematic and a sounder basis for environmental integration than a system of tree payments.
- Although a change from production subsidies to direct payments should be broadly beneficial for the environment, there are several important environmental issues (as well as certain social issues) which a simple change to direct payments would not address, unless additional control and support mechanisms are built into the system.
- Initially, the justification for new direct payments would be to compensate producers for the loss of production subsidies. To fulfil this function, the payments would have to be modulated in proportion to regional yields. However, a mechanism should be established to reduce these compensation payments over time whilst transferring resources to targeted payments for production systems of environmental and social value.
- To help maintain small, traditional plantations in production, a basic level of price support and market intervention probably is necessary, in addition to direct payments. The upkeep of plantations also could be encouraged through

cross-compliance attached to direct payments.

- A limited reform, involving the abolition of the small-producer mechanism whilst leaving the main production support system unchanged, could create the worst of both worlds from an environmental and social point of view. It would take away the important security for small producers provided by the tree-related annual subsidy, whilst maintaining the powerful incentive for larger producers for intensive, high output production.
- It is essential to develop strategies at a regional level in order to address specific issues (social, agronomic and commercial as well as environmental). Measures should include research, advice and information as well as economic instruments. The CAP regime should be flexible enough to allow the development of regionally appropriate measures and a redistribution of agricultural land uses.
- There is an urgent need for real, objective data from the different regions concerned. An appropriate and effective reform will be difficult as long as there is no comprehensive information concerning the very different socio-economic and environmental circumstances existing in the various olive producing regions and districts of the EU.

REFERENCES

Aguilar Ruiz, J., Fernández García, J., Fernández Ondoño, E., de Haro Lozano, S., Marañés Corbacho, A. and Rodríguez Rebollo, T. (1995) *El Olivar Jiennense*. Universidad de Jaén, Jaén, Spain.

Beaufoy, G., Baldock, D. and Clark, J. (1994) *The Nature of Farming: Low-intensity Farming Systems in Nine European Countries*. Institute for European Environmental Policy, London.

Civantos, M. (1995) Development of integrated pest control in Spanish olive orchards. *Olivae* 59, 75-81. International Olive Oil Council, Madrid.

Díaz Alvarez, M.C. and Almorox Alonso, J. (1994) La erosión del suelo. In: Ramos, A. (ed.) *Agricultura y Medio Ambiente*. El Campo, BBV, Bilbao.

EC (1997) *Note to the Council of Ministers and to the European Parliament on the Olive and Olive Oil Sector (Including Economic, Cultural, Regional, Social and Environmental Aspects), the Current Common Market Organisation, the Need for a Reform and the Alternatives Envisaged.* COM(97)57 final. Commission of the European Communities, Brussels.

Fialho, M.M.R. (1996) *Olival & Azeite: Recuperar o Atraso*. III Congresso da Agricultura Alentejana, Évora, 1.2.1996.

Fotopoulos, C., Liodakis, G. and Tzouvelekas, V. (1997) The changing policy agenda for European agriculture: its implications for the Greek olive-oil sector. In: Tracy, M. (ed.) *CAP Reform: the Southern Products*. Agricultural Policy Studies, Belgium, pp. 79-84.

Guerrero, A. (1994) *Nueva Olivicultura*. Ediciones Mundi Prensa, Spain.

López Sánchez-Cantalejo, J. (1996) 1983-1995: trece años de mudanza en la agricultura española. *El Boletín* 31, Ministerio de Agricultura, Pesca y Alimentación, Madrid.

MAPA (1988) *El Olivar Español: Planes de Reestructuración y Reconversión*. Ministerio de Agricultura, Pesca y Alimentación, Madrid.

MAPA (1994) *Programa de Ayudas Para Fomentar Métodos de Producción Agraria Compatibles con las Exigencias de la Protección y la Conservación del Espacio Natural.* Ministerio de Agricultura, Pesca y Alimentación, Madrid.

Pain, D. (1994) *Case Studies of Farming and Birds in Europe: Olive Farming in Portugal.* Studies in European Agriculture and Environment Policy No. 9, RSPB/Birdlife International, Sandy, UK.

Parra, F. (1990) *La Dehesa y el Olivar.* Enciclopedia de la Naturaleza de España, editorial Debate/Adena-WWF España, Madrid.

Pastor, M. and Castro, J. (1995) Soil management systems and erosion. *Olivae* 59, 64-74. International Olive Oil Council, Madrid.

Petretti, F. (1995) The cultivation of olive trees in Grosseto. Unpublished report produced for the Institute for European Environmental Policy, London.

Polidori, R., Rocchi, B. and Stefani, G. (1997) Reform of the CMO for olive oil: current situation and future prospects. In: Tracy, M. (ed.) *CAP Reform: the Southern Products.* Agricultural Policy Studies, Belgium, pp. 67-78.

Rodenas Lario, M., Sancho Royo, F., Ramirez Díaz, L. and Gonzalez Bernaldez, F. (1977) Ecosistemas del area de influencia de Sevilla. In: Monografía 18. Doñana: *Prospección e Inventario de Ecosistemas.* ICONA, Madrid.

Unión de Pequeños Agricultores (1998) Resumen del informe de UPA sobre la reforma del sector olivarero. *La Tierra* 147, 8-15. UPA, Madrid.

Policies for Less Favoured Areas 11

Thomas Dax and Petra Hellegers

INTRODUCTION

Productivity and farm income vary greatly across regions within the European Union (EU). These longstanding interregional disparities led to the establishment of the Less Favoured Area (LFA) scheme in the 1970s. Over a long period it was the only significant structural measure of agricultural policy, but recent policy reforms have moved away from commodity market supports, towards direct payments and have increasingly emphasised the environmental implications of policy measures. With this thrust of present policy discussions in mind, this chapter will consider the rules governing the LFA scheme and its uptake in the Member States, as well as its implications for the environment, in particular with regard to low intensity farming systems.

The high coincidence of LFAs with High Nature Value (HNV) farming systems contribute to the assessment that farm development in LFAs is particularly relevant to the upkeep of the beneficial features of farming activity. The integration of HNV farming with the possible development of the LFA scheme is seen to be vital to addressing the notion of sustainable farming in the future Common Agricultural Policy (CAP) in an appropriate manner.

Origin of LFA support

For many decades European countries have addressed the problems of LFAs, and particularly those of mountain areas, through local sectoral policy programmes, mainly for forestry and agriculture (Barruet, 1995). At the end of the 1960s concern about the impacts of agricultural adjustment increased and the threats of a policy oriented solely towards productivity and the markets became visible. However, the idea to take up measures in disadvantaged areas to cope with depopulation and to preserve the economic, societal and landscape pattern of those areas did not find much interest in CAP debate (CEC, 1993, p. 7f.).

It was only when the UK negotiated entry to the EC, and laid down its condition to be able to continue to give special help to hill farming, that measures for

LFAs were taken at the EC level (Lowe *et al.*, 1998). In response to these negotiations, Directive 75/268 was introduced in 1975. This Directive provided support to farmers in certain agriculturally disadvantaged areas in order to achieve
'the continuation of farming and thereby maintaining a minimum population level
or conserving the countryside'.

From the very beginning, LFA policy was conceived as a structural policy
aimed at the prevention of land abandonment, to preserve the farming population
in those areas and conserve the countryside. In this respect, the LFA scheme was
one of the first measures to address environmentally beneficial farming systems,
at least indirectly. For the broader public the main relevance of the scheme was
that for the first time an explicitly regional approach in agricultural structural
policy was brought into play.

Delimitation of areas

The LFA scheme responds to the widely divergent regional situation of EU agriculture, with respect to both the socio-economic situation and natural
characteristics. It should set the framework for agricultural holdings in the LFAs
to benefit from direct payments and specific measures. The categories and the
criteria for the demarcation of the LFAs have been defined in EEC Directive
75/268 (Art. 3, para 3-5), later in Regulation 950/97 (Art. 23-25), and recently
integrated into Regulation 1257/1999 (Art. 13-21). A large number of more than
32 implementing Directives comprise the actual delimitation of the LFAs of each
Member State (CEC, 1997). There are three types of LFA:

- Mountain areas where altitude and slopes reduce the growing season and the
  scope for mechanisation. (High-latitude regions in Finland have also been
  included in this category.) These areas make up about 20% of the total
  Utilised Agricultural Area (UAA) (Article 3.3).
- Simple LFAs which are marked by poor soil conditions (low agricultural productivity), low agricultural income levels and low population densities or depopulation tendencies. These areas account for 34% of the UAA
  (Article 3.4).
- LFAs with 'specific handicaps' which are restricted to small areas with handicaps relating to the environment, landscape development or coastal areas
  and islands where agricultural activity should be preserved in order to maintain the countryside. About 2% of the UAA is classified under this type
  (Article 3.5).

The distribution of the three types of LFAs in the various EU Member States can
be seen from Table 11.1. It shows the particularly high share of mountain areas in
some Member States (Greece, Austria and Finland) and the predominance of simple LFAs in others (Luxembourg, Ireland, Portugal, Germany, UK and Spain).
The five largest Member States (Spain, France, Italy, Germany and UK), however, account for 75% of total LFA of EU-15.

The rising interest in the LFA scheme can be seen from the fact that there has always been pressure from Member States to increase the area of LFAs within their territory. The proportion of UAA designated as LFA therefore increased from 33% in 1975 to about 56% in 1996. This increase has been only partly due to the accession during this period of new Member States with high percentages of LFA (e.g. Greece, Portugal, Spain and later Austria and Finland). It was also due to real increases in LFA coverage in existing Member States (particularly Germany, France, Ireland, Italy and the UK) (Table 11.2).

Table 11.1   Utilised Agricultural Area (UAA) in LFAs (1996)

	In 1,000 ha (or % share of total UAA)				
Member State	Mountain areas Art. 3.3	Simple LFAs Art. 3.4	Specific handicaps Art. 3.5	Total LFA	Total UAA
Belgium	- (-)	273 (20)	- (-)	273 (20)	، 1,357
Denmark	- (-)	- (-)	- (-)	0 (0)	2,770
Germany [a]	336 (2)	7,987 (47)	199 (1)	8,522 (50)	17,012
Greece	3,914 (61)	964 (15)	402 (6)	5,280 (82)	6,408
Spain	7,503 (28)	11,343 (43)	700 (3)	19,546 (74)	26,330
France	5,284 (18)	7,809 (26)	804 (2)	13,897 (46)	30,011
Ireland	- (-)	3,456 (71)	12 (0)	3,468 (71)	4,892
Italy	5,218 (32)	3,405 (21)	218 (1)	8,841 (54)	16,496
Luxembourg	- (-)	122 (96)	3 (2)	124 (98)	127
Netherlands	- (-)	- (-)	111 (6)	111 (6)	2,011
Portugal	1,227 (31)	2,056 (51)	150 (4)	3,433 (86)	3,998
UK	- (-)	8,341 (45)	1 (0)	8,342 (45)	18,658
EU-12	23,482 (18)	45,756 (35)	2,599 (2)	71,836 (55)	130,070
Austria	2,047 (58)	208 (6)	164 (5)	2,419 (69)	3,524
Finland	1,407 (55)	536 (21)	220 (9)	2,164 (85)	2,549
Sweden	526 (14)	1,011 (28)	333 (9)	1,869 (51)	3,634
EU-15	27,462 (20)	47,511 (34)	3,316 (2)	78,288 (56)	139,777
Share of total LFA (%)	35	61	4	100	

[a] 16 Länder.
Source: CEC (1997, p. 54).

ROLE OF LFA WITHIN CAP AND OTHER POLICIES

Mountain areas comprise about 20% of total UAA in the EU-15. Some Member States, though, have a particularly high share of mountain areas, and their production patterns are dominated by LFA land use systems. The actual extent of mountain areas is much greater since such areas usually also have a high share of forest cover and unproductive areas.

The agricultural productivity of the LFAs is limited; the average production potential is about 60% that of 'normal areas' but down to about 50% in mountain areas (CEC, 1994). These disadvantages are reflected in the relative agricultural incomes. Land use in LFAs is characterised by a higher share of grassland and a lower share of arable. But the EU averages shown in Table 11.3 hide the much greater differences within and between Member States. Grassland production levels in the LFAs (including the mountain areas) of central northern Europe considerably exceed those of Southern European countries; whereas permanent crops are mainly concentrated in Southern European mountain areas and are nearly absent from the LFAs of Northern Europe (CEC 1993, p. 22).

The diversity of LFAs in the EU is even more striking when analysing the agricultural income disparities between LFA areas and non-LFAs. The differences within Member States are much smaller than those between 'northern' and 'southern' countries (Figure 11.1). The unfavourable income situation for southern countries generally and their LFAs in particular is more and more addressed by

Table 11.2 Agricultural area classified as LFA (as percentage of total UAA; Directive 75/268) .

Member State	Total UAA in 1990 (1,000 ha)	LFA as % of total UAA				
		1975	1981	1986	1991	1996
Belgium	1,357	19.8	21.2	21.9	21.9	20
Denmark	2,770	-	-	-	-	-
Germany	17,012	28.7	32.8	50.9	53.6	50
Greece	6,408	-	-	78.2	78.3	82
Spain	26,330	-	-	62.4	67.5	74
France	30,011	33.1	36.4	38.5	45.1	46
Ireland	4,892	51.2	55.4	58.0	71.4	71
Italy	16,496	37.7	42.3	51.1	51.9	54
Luxembourg	127	100.0	100.0	100.0	99.0	98
Netherlands	2,011	-	0.6	0.9	2.4	6
Portugal	3,998	-	-	75.6	75.6	86
UK	18,795	36.0	40.8	52.5	52.6	45
Austria	3,524					69
Finland	2,549					85
Sweden	3,634					51
EU-10	100,319	32.9	34.2			
EU-12	130,070			51.6	55.1	55
EU-15	139,777					56

Source: European Commission DGVI F.1; CEC (1993, p. 14); CEC (1994, p. 25); CEC (1997, p.54).

policy analysis in the south (e.g. Bazin and Roux, 1992; Frisio, 1997). Concern for the environmental impact of agricultural methods and the threat of land abandonment particularly in these countries will necessitate an increased awareness of the problem at the European level.

Extensive farming regions and regions with small-scale farming are most susceptible to marginalisation, with major environmental consequences (Baldock *et al.*, 1996). As mainstream CAP support is not oriented to these farming systems, expenditures per farm are especially low in small-scale farming regions and cannot suffice on their own to counteract marginalisation. At the same time, the widespread occurrence of low agricultural incomes and of less developed regional economies in LFAs (CEC, 1994) points to the need for a broader policy perspective. It underlines the requirement to integrate future rural policies in general and to adopt a common strategy across different policy sectors in order to combat the marginalisation tendencies in regional development. In particular, the income gap between normal areas and LFAs points to the need for specific and enhanced support for LFAs.

Table 11.3   The contribution of LFAs to EU agriculture

	LFAs as a whole	Mountain areas	Non-LFA
Share of utilised agricultural area (UAA) 1995 (EU-15, in %)	56.00	20.00	44.00
Share of standard gross margin (SGM) 1987 (EU-12, in %)	29.00	9.50	71.00
Production potential (SGM ha^{-1} 1987)	0.60	0.46	1.05
Agricultural income (1987-1989), index	59.00	50.00	100.00
Proportion in arable, 1987 (%)	42.30	40.90	64.60
Proportion in grassland, 1987 (%)	48.30	43.40	26.20
Proportion in permanent cultures, 1987 (%)	9.30	15.40	9.10

Source: CEC (1993); CEC (1997); Dax (1998c).

The land use of LFAs is largely characterised by the limits imposed by the naturally adverse conditions. Present farming systems have developed over many centuries and are usually well adjusted to the specific set of restrictions. To a large extent they have shaped much of the cultivated landscapes of Europe. The continuity of these farming systems is therefore seen as central to the preservation of these cultural landscapes and as a precondition to avoid erosion, desertification and land abandonment. In recent years there has also been growing interest in the relationship between LFA policies and nature conservation. The low intensity farming systems typically found in LFAs are associated with a diversity of wildlife and semi-natural habitats. Amongst conservationists there has also been increased understanding that species cannot be protected by site-specific measures alone, but depend on the integrity of ecological networks and sympatheric land uses in surrounding areas. However, the relevant conservation policies do not typically address land management and farming practices outside specifically designated areas. Some European studies have identified the conservation relevance of farming practices in LFAs. In an EU-wide study of high nature value regions the share of LFA in each of the 12 study areas exceeded 60% of total UAA (in 1989/90), with the exception of the Pindos Mountains in Greece and the Dutch Peatlands (Hellegers and Godeschalk, 1998).

The coincidence of LFAs and areas of nature conservation interest in the EU is shown in Figure 11.2. This overlap is particularly high for the mountain areas and in many cases for protected areas, too. Most of the farming systems in LFAs are low intensity ones. Although many organisations have focused on the environ-

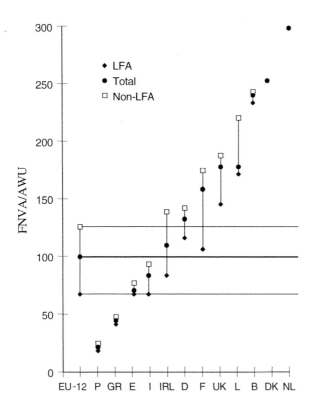

FNVA: Farm Net Value Added; AWU: Annual Work Unit.

Figure 11.1   Agricultural income disparities between LFA and non-LFA areas of the EU Member States, 1988 (Source: CEC, 1994, p. 64.)

mental aspects of intensive farming, the nature conservation aspects of less intensive systems must not be neglected. A study of low intensity farming systems
(Beaufoy *et al.*, 1994) concluded that 'it is ironic that many environmental initiatives on farmland tend to concentrate (often with little prospect of success) on

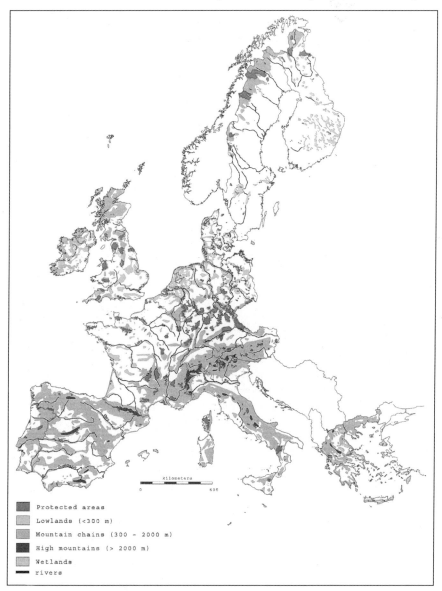

Figure 11.2 Areas of nature conservation
Commissioned by The Netherlands National Spatial Planning Agency, Ministry of
Housing, Spatial Planning and the Environment, as part of the project on Rural
Areas and Europe (DLO Winand Staring Centre for Integrated Land, Soil and
Water Research, Wageningen, the Netherlands). Source: Bethe (1997).

reversing actions that have been destructive, yet tend to ignore practices that are currently benign and could be sustained'. In this respect, most environmental initiatives tend to reward some farmers for their previously destructive activities but not others for their contribution to the maintenance of biodiversity, nature protection, cultural landscapes and socio-economic development (Bignal and McCracken, 1996). As low intensity farming systems are endangered by both abandonment and intensification, there is an urgent need to highlight the importance of LFAs also for nature conservation.

THE INSTRUMENTS OF LFA POLICY

The concept of LFA was based on the notion that these were areas of intrinsically low productivity where farming needed to be compensated for the disadvantages it faced to ensure that such areas did not become depopulated. Therefore, the main policy support was termed 'compensatory allowances' (see Table 11.4). This measure was granted per animal and/or per ha to offset natural handicaps. Other specific measures included:

- investment aids for farm modernisation which were set at a rate up to 10% higher than in 'normal' areas;
- aid for collective investment, which included the improvement of grassland and rough grazing.

Indirectly agricultural holdings in LFAs may also benefit from:

- A more favourable implementation of Common Market Organisations, such as a complementary premium to top up the Sheep Annual Premium, additional quotas and reserves for sheep and suckler cows. In addition, Member States may grant special reference quantities in the dairy sector.
- Various other measures which happen to have a greater impact on LFAs than elsewhere. The most important policy measures of this kind are the agri-environmental programmes under Regulation 2078/92 and direct aid for extensive farming under the CMOs. Moreover, the majority of LFAs are also eligible for measures of the regionalised Structural Funds programmes under Objectives 1, 5b and 6.

The criteria for eligibility for compensatory allowances specify that:

- applicants must have, in general, a holding of at least 3 ha UAA; and
- must continue farming for at least 5 years.

Aid granted has been provided so far on a Livestock Unit (LU) basis (for dairy cows, cattle, sheep, goats and equine animals) and may range from a minimum of 20.3 ECU per LU to 150 ECU per LU. There are additional restrictions intended to prevent funds being absorbed by too large or intensive holdings:

Table 11.4   Overview of LFA payments

	1976	1982	1986	1991	1994
Number of beneficiary holdings	339,735	683,105	880,681	1,147,600	1,056,419
Compensatory allowance payments:					
Total (million ECU)	269	509	782	1,060	1,381
Average (ECU per holding)	785	745	888	923	1,308
Number of LU (in 1,000)	7,330	9,496	15,333	16,827	15,998
Average (ECU per LU)	36.7	53.6	51.0	63.0	86.3

Source: Various CEC reports on situation of agriculture in the EC; CEC (1989, p. 86*f*); CEC (1997, p. 55).

- the premium is limited to 20 dairy cows per holding, except in mountain areas;
- the allowance can be granted for no more than 1.4 LU ha^{-1} of the total forage area of the holding;
- an additional aid per ha for crop production is calculated excluding the area for animal feed, durum wheat, fruit trees (apples, pears, peaches), vineyards and sugar beet and wheat area with yields higher than 2.5 tonnes ha^{-1};
- there is an upper limit for the EU contribution to individual holdings of 120 units (of LU or ha) and the full amount is only paid for the first 60 units. For the rest of the eligible units, half the premium is paid.

Co-financing is set in general at a rate of 25%, but for the following regions the following higher rates are fixed:

- 50% for non-Objective 1 regions in Italy and parts of Spain;
- 65% for Ireland and the Länder in Germany;
- 70% for Spain and Portugal;
- 75% for Greece and Objective 1 regions of Italy.

Although these co-financing rates show considerably higher levels for southern European countries, the uptake of compensatory allowances has been particularly weak there. The different implementation and use of the measure is reflected in the statistics of the uptake showing marked differences between Member States. In some countries like Italy, Germany and Spain the regional administrations are responsible for the running of the scheme and adapt it to local circumstances. Thus the national averages presented here have to be differentiated for the regions and type of LFA (mountain areas and other LFA). Whereas some 45% of all the farm holdings in LFAs benefit from the scheme (CEC, 1997, p. 55), the participation of holdings varies from between 84 and 99% in most north-

Table 11.5 Breakdown of LFA payments (1994)

Member State	Amount of allowances paid in 1994 Mio ECU	No. of beneficiary holdings	Average allowance per holding (ECU)	No. of LU granted (in 1,000)	Average amount per LU (ECU)	No. of ha granted	Amount per ha (ECU)	Share of beneficiary holdings in LFA (%)
Belgium	9	6,873	1,329	108	85	-	-	86
Denmark	-	-	-	-	-	-	-	-
Germany	500	231,277	2,163	3,199	93	2,689,563	75	87
Greece	99	190,262	521	1,170	61	424,472	65	66
Spain	85	187,059	454	1,417	36	960,351	35	34
France	297	139,435	2,127	4,205	70	6,658	327	54
Ireland	166	105,619	1,575	1,884	88	-	-	99
Italy	27	39,056	689	376	57	215,882	25	9
Luxembourg	11	2,515	4,437	53	113	46,437	111	84
Netherlands	3	3,901	884	33	104	33,670	102	98
Portugal	37	89,510	410	447	54	240,605	52	53
UK	147	60,912	2,419	3,106	47	-	-	95
EU-12	1,382	1,056,419	1,308	15,998	67	4,617,638	67	45

Source: CEC (1997, p. 55).

ern Member States to just 9% in Italy (Table 11.5). The main reason for the lower proportion of farmers receiving aid out of the total number of farmers in the LFAs in the countries of the south is inherent to the concept around which it has been built. The orientation of the compensatory allowances scheme on headage payments made it obviously more applicable in regions which focus on livestock production, including Ireland and the UK, but also Greece. In particular, the small structure of farms in the south, with many farms of a size beneath the eligibility threshold, excluded a large proportion from this payment. In spite of the fact that the minimum limits for the granting of aid in these countries has been lowered many farms are still not eligible, e.g. in Italy, where 29% of farms are less than one hectare in size. Moreover, the ha payments for crop production which exclude areas devoted to forage crops, grapes, wheat, etc. disfavour the application of the scheme in regions where permanent cultures and arable land are a significant proportion of land use. The difference is most outstanding between mountain areas in the north and the south. Whereas in the north arable land and permanent cropping are almost absent from mountain areas (and have limited relevance in other LFAs), it is a marked feature of land use in the southern LFAs. Another reason for a lower commitment of southern Member States can be found in the process of allocating Objective 5a budgetary resources by each Member State. The focus can be chosen between the three types of action: modernisation of holdings (Reg. 2328/91), improvement of processing and marketing structures (Reg. 866/90), and compensatory allowances. In the past, Member States in the south with great structural handicaps preferred to spend their Objective 5a resources on modernisation schemes and the improvement of processing and marketing structures.

The different priorities identified by Member States lead to considerable differences in farmer participation which are not to be explained by structural differences alone. In particular, variations in the implementation of the scheme by Member States and regions greatly affect the uptake and budget spent on the measure. Whereas some countries do not modulate the payment according to the size of the holding, in others provisions exist to differentiate grants according to type of production, number of productive units, the stocking rate, maximum payments or the income of the farmer. Partly as a consequence, the average payment per beneficiary holding ranges between Member States from 4,400 ECU in Luxembourg to 410 ECU in Portugal (see Table 11.5). Member States with the majority of allowances paid, like Germany, France, Ireland and the UK, have an average of more or less 2,000 ECU, whereas payments in the four southern countries (Portugal, Spain, Italy and Greece) only reach about 400-700 ECU per holding. Thus the payments are concentrated in the four cited north-western European countries which account for about 80% of the allowances paid (1994), even though they hold only 48% of the total LFA (EU-12).

Direct income support in the context of general agricultural policy plays an important role in maintaining the viability of farming in LFAs (over and above the compensatory allowances). The value of such support measures often exceeds the net farm income for a holding. The sheepmeat and goatmeat regime and the beef regime are of particular importance in the many LFAs where livestock grazing dominates. They have helped to prevent a decline in the area grazed and may have prolonged the distribution of sheep and goat and beef cattle on holdings in marginal areas (Baldock *et al.*, 1996).

ENVIRONMENTAL ASSESSMENT AND NEW POLICY ELEMENTS

Whereas initially LFA policy did not reflect policy concern over the impact of agriculture on the environment, but was conceived as a measure to compensate for handicaps and to mitigate income gaps, the subsequent reorientation of the CAP and the rise of environmental policy itself started to address this issue directly. Since 1989 Member States have had the possibility to put environmental conditions on the payment of compensatory allowances. Such limits, however, were only applied in the UK and in a fairly rudimentary way. As agri-environmental measures have been developed, in particular since the CAP reform of 1992, attention has also focused on the beneficial effects on the environment of the LFA scheme (European Economy, 1997). These lie particularly in the maintenance of important habitats, both on cultivated and grazed land, and in features such as hedgerows, ponds and trees which historically were integrated with the farming system.

With the end of traditional farming methods and a switch to more harmful farming practices also in parts of some LFAs, landscape degradation and a reduction in biodiversity might take place, and thus the continuation of sympathetic farming activities in these areas is extremely relevant. The continuation of farm management, particularly in mountain areas, thus plays a central role in rural development, as it acts as a prerequisite and basic activity for other sectors, such as tourism, and the maintenance of infrastructure facilities (OECD, 1998). From European-wide comparative work on the integration of the environment into mountain farming (EUROMONTANA, 1998) it could be seen that tendencies of farming in European mountain areas are somewhat divergent. Whereas farming in countries that are heavily committed to the use and integration of their mountain areas in the national economy has tended to stabilise in most areas, other countries are displaying stronger tendencies either to intensification or to land abandonment. In general, these processes are the results of a long-term evolution and can hardly be monitored and evaluated over a short time-scale and within a limited geographical area.

Assessment is made even more difficult since abandonment and intensification phenomena often appear simultaneously within one region. Moreover, they can also occur within a local community or even a single farm holding. The divergence within a given area adds to the complexity of processes and makes it difficult to attribute a straightforward positive or negative overall impact (Dax and Wiesinger, 1998).

Given the interrelation of land use and regional economy, the regional context has to be taken into account when assigning value to farming practices and changes in farming practices in mountain areas. What may be regarded as a positive effect for the environment in many southern European regions (e.g.

afforestation), might be seen as negative in central and northern European regions (with a high forest cover), causing a reduction in biodiversity and the disappearance of cultural landscapes. Likewise, Baldock and Mitchell report that 'there is a strong case for seeking to limit grazing pressure from subsidized livestock where this is causing environmental damages', but 'there are also areas where habitat value is deteriorating as a result of under-grazing' (Baldock and Mitchell, 1995, p. 58).

In addition to the requirement for a contextual interpretation, and the assessment of the simultaneous occurrence of land abandonment and intensification tendencies, non-agricultural sectors have acquired a leading role as the driving forces for farm household decisions. A thorough understanding of the socio-economic development and integration of farm households into the general economy (e.g. via pluriactivity and education) reveals its impact on the continuation of the provision of environmental goods (Arkleton Trust, 1992).

The evolution of extensive farming systems in the LFAs has attracted attention because the shift to other more harmful practices could lead to considerable degradation of the environment and the cultural landscape. These changes are especially serious as they very often involve the irreversible disappearance of valuable elements of the environment. Moreover, the negative impact on the natural environment often induces a further weakening of the socio-economic situation of the region and is thus detrimental to a sustainable regional development.

The reform of the CAP in 1992 for the arable and beef sector represented a significant shift in the nature of the support provided, from price support measures to more direct subsidies through the provision of direct payments. The changes were assumed to strengthen incentives towards a decrease in input factors and to induce an improvement in environmental performance in general. The effect on farming in the LFAs is likely to have been diluted as these areas are rather characterised by small-scale and/or low intensity farming. The introduction of the agri-environmental Regulation was of much greater significance. With many schemes only coming into operation in 1996 the specific effect on LFAs has not yet been analysed in greater detail. To give an example where there has been considerable and wide-ranging impact on farm income, the implementation and uptake of the Austrian scheme can be mentioned. In that case, although the scheme represents a horizontal programme for all farmers, the ecologically more demanding measures are concentrated on LFAs, and in particular in mountain areas: e.g. 94% of organic farming support is given to farmers in LFAs, even though the number of holdings and UAA in LFAs represents about two-thirds of Austrian agriculture.

MAINTAINING HIGH NATURE VALUE FARMING SYSTEMS

A series of recent studies have evoked the existence of high nature value (HNV) farming systems in Europe and their beneficial role for nature conservation and biodiversity (Baldock and Beaufoy, 1993; Beaufoy *et al.*, 1994; Hellegers and

Godeschalk, 1998). They have also highlighted the imminent threat to those farming patterns by impending marginalisation processes in the regions where they occur, which are mainly LFAs.

In general, HNV farming systems are referred to as low intensity farming systems with highly diverse habitat types (Baldock and Beaufoy, 1993), though there may also be high intensive farming systems with rich natural potential, like the polders in The Netherlands. The main categories of farmland with HNV features are (Hellegers and Godeschalk, 1998, p. 21):

- semi-natural grasslands (permanent, and with hardly any use of fertilisers);
- important areas for breeding and migratory birds;

Table 11.6   Farming systems which are likely to be of high nature conservation value

Farming system or practice	Distribution in Europe
Grazing and mowing of semi-natural dry grassland	Parts of south Italy, Spain, south Portugal, France, England, Germany
Grazing and mowing of lowland wet grassland	Parts of Ireland, Netherlands, France, UK
Grazing of moorland and heaths	Large areas of UK uplands and Ireland and smaller areas in other regions
Grazing of high (e.g. Alpine) mountain wooded agro-pastoral	Pyrenees, Cantabria, Alps, etc.
Grazing of Iberian dehesa and montado wooded agro-pastoral	Large areas of west and south-west Iberian Peninsula systems
Grazing of Mediterranean scrub (maquis, matorral, etc.)	Large areas of Spain, southern France, Greece, Italy
Grazing of coastal marshes	Part of Netherlands, UK, France, Spain, Portugal
Grazing and traditional silviculture of forests and woodlands	Mainly upland/mountain areas in the south of the Community
Arable cultivation and grazing of 'pseudo' steppes	Mainly Spain, also parts of Portugal, Italy, Greece
Management (including replacement planting) of perennial/tree crops, especially olives and orchards	Olives in Spain, Portugal, Italy, Greece. Orchards in Normandy, Provence, southern Germany, Italy
Maintenance of bocage landscapes and others rich in semi-natural features, as part of livestock and mixed farming	Parts of northern France, Britain, Ireland, Portugal, Spain, Italy, Greece
Mixed, low intensity arable land use	Especially in southern Europe: Portugal, parts of Spain, Italy and Greece

Note: the regional assessment of this table does not cover the new entrants to the EU in 1995 (Austria, Finland, Sweden) nor the central and eastern European countries where low intensity systems are also of great importance.
Source: Baldock and Beaufoy (1993).

- areas with many 'natural' features, like hedgerows, small woodlands, ponds, etc.;
- dehesas/montados (which are agro-forestry systems with rotation of arable and livestock production under trees in Iberia);
- low intensity arable and perennial crops.

In such areas, an appropriate land management is required to maintain the existing biodiversity. Marginalisation with ensuing land abandonment that is not properly managed might lead to a great loss of biodiversity. However, agricultural policy requires achievement of a balance between the provision of support for traditional husbandry and cultivation practices, and an approach relying more on ecologically sound processes.

Given the great variety of low intensity farming systems, analyses on the characteristics, types and distribution of such systems were needed. Within a comparative study, focusing on the countries with the main occurrence of these systems, Beaufoy *et al.* (1994) developed such a typology. A rough categorisation of the main extensive farming systems and practices most often found in LFAs in Europe is given in Table 11.6. These farming systems tend to be HNV and specifically contribute to the maintenance of the cultural landscape.

An estimation of the area of farmland under low intensity farming systems (Beaufoy *et al.*, 1994) shows that southern Europe has both the largest number of farming types, and the greatest area of land under low intensity farming. Given that the more intensively farmed Member States in north-western Europe are missing in these calculations, the actual share of UAA under low intensity systems in Europe might be somewhat lower than the average of the study (38%) (Table 11.7). Although the estimated areas are preliminary and indicative, the figures reveal the outstanding importance of low intensity farmland on the Iberian peninsula and for many other European regions, particularly LFAs. As incomes

Table 11.7 Farmland under low intensity farming systems

Country	Agricultural area under low intensity systems (in million ha)	Share of UAA under low intensity systems (in %)
Greece	6	61
Spain	25	82
France	8	25
Ireland	2	35
Italy	7	31
Portugal	3	60
UK	2	11
Hungary	2	23
Poland	3	14
Total	56	38

Source: Bignal and McCracken (1996).

from HNV farming systems tend to be low and the future of farming is threatened by marginalisation, special emphasis on the issue is necessitated. A particular consequence of the decline of these farming practices would be negative changes to cultural landscapes and biodiversity. Labour-intensive grazing and cultivation systems and the maintenance of valuable features of landscapes might be endangered, which may result in the encroachment of scrub and woodland, leading to a loss of environmental value. As a consequence HNV is no longer understood as an automatic by-product of agricultural activity, but its preservation comes to be highly dependent on direct payments, such as the compensatory allowances.

REFORM OF LFA POLICY

LFA payments are one of the core measures of the 'Regulation on support for Rural Development' from the EAGGF agreed as part of the 1999 CAP reforms. Given the high awareness of environmental problems and the requirement to better address the beneficial role of farming in LFAs there has been significant change in the basis of LFA payments.

These apply particularly to the overall objective to develop an instrument in favour of the preservation of low intensity farming methods. The compensatory allowance will remain the prime instrument but will have to be calculated on a per ha basis. This will disfavour more intensive livestock holders and regions. As a comparative analysis across a great part of EU Member States has shown (Hellegers and Godeschalk, 1998, p. 78ff.), many areas will not experience serious problems in staying eligible for support since livestock density is for a majority of farmers beneath 1.0 LU ha^{-1}, except for dairy farmers. But the overall assessment is that the reduction of the density threshold for the compensatory allowances will have greatest effects for farmers with problems of overgrazing and tendencies towards intensification of livestock production which occur in certain LFAs (Dax, 1998a).

The new regulation also envisages the requirement to define production methods compatible with environmental objectives and the maintenance of natural resources. This rule will play a major role when additionally NATURA-2000 areas will have to be integrated (Bennnett, 1997; a provision to increase the more flexibly disposable third category of LFAs, the 'small areas', from 4 to 10% is intended to give the necessary room for manoeuvre).

Discussions about the integration of areas with environmentally specific conditions revealed that support for this new type of area should remain separate from the kind of areas classified up to now. The decisions on Agenda 2000 of the Berlin Meeting in March 1999 regarding this issue made clear that there shall be two categories for complementary payments in the future, although the same rules shall apply to them, and the size for areas with environmentally specific conditions will be linked to the category of 'small areas' and limited to 10% of the national area. The new strand of the scheme focused on the application of EU regulations with respect to environmental prescriptions for farm management in specific areas (in general, NATURA-2000 areas). The recent decisions thus tried to alleviate arising conflicts between farmer-oriented support and 'environmental-ists'. Given the considerable overlap of LFA with the new category of areas with environmental prescriptions it will be decisive for the future of LFA support to continue and deepen the debate about the impact of all types of LFAs for the achievement of environmental objectives and, particularly, the preservation of natural resources.

The vivid debate on rural and regional development within the last decade has, to a large extent, also incorporated the beneficial role of agriculture in LFAs, and particularly mountain areas (Dax, 1998b). Analysis has recently focused on the positive impact that 'rural amenities' might play for rural development, thus highlighting the importance of harnessing the benefits stemming from rural re-sources (OECD, 1994). For the preservation of HNV systems within LFAs it will be of central importance that regional development programmes adopt this view-point. This means that the development of farming methods, as shown by this example, cannot be left to agricultural policy alone, but must relate to regional development processes also.

In conceiving the environmental sensitivity of mountain areas and other LFAs not only as a handicap to agricultural production but also as a rural development asset (Dax, 1998c) it seems appropriate to address rural amenities too. Targeting of support must not be limited to LFA payments and agri-environmental schemes, but be extended to the set of measures for agricultural and forestry and general ru-ral development. A special recognition of the environmental sensitivity in mountain areas and other LFAs through the Structural Funds Regulation could also enhance initiatives at the local and regional level.

REFERENCES

Arkleton Trust (1992) *Farm Household Adjustment in Western Europe 1987-1991*. Final report on the research programme on farm structures and pluriactivity, Oxford, 313 pp.

Baldock, D. and Beaufoy, G. (1993) *Nature Conservation and New Directions in the EC Common Agricultural Policy*. Institute for European Environmental Policy (IEEP), London/Arnhem.

Baldock, D. and Mitchell, K. (1995) *Cross-compliance within the Common Agricultural Policy: A Review of Options for Landscape and Nature Conservation*. Institute for European Environmental Policy (IEEP), London, 83pp.

Baldock, D., Beaufoy, G., Brouwer, F. and Godeschalk, F. (1996) *Farming at the Margins - Abandonment or Redeployment of Agricultural Land in Europe*. Institute for European Environmental Policy (IEEP) and Agricultural Economics Research In-stitute (LEI-DLO), London and The Hague, 202pp.

Barruet, J. (1995) Politique de la Montagne, L'enjeu européen et transfrontalier. In: Bar-ruet, J. (ed.) *Montagne, Laboratoire de la Diversité*. Cemagref, Grenoble, pp. 227-238.

Bazin, G. and Roux, B. (eds) (1992) *Les Facteurs de Résistance à la Marginalisation dans les Zones de Montagne et Défavorisées Méditerranéennes Communautaires (MEDEF-network)*. Paris.

Beaufoy, G., Baldock, D. and Clark, J. (1994) *The Nature of Farming: Low Intensity Farming Systems in Nine European Countries.* Institute for European Environmental Policy (IEEP), London, 66pp.

Bennett, G. (ed.) (1997) *Agriculture and Natura 2000, EU Expert Seminar.* Apeldoorn, The Netherlands.

Bethe, F.H. (ed.) (1997) *Rural Areas and Europe, Processes in Rural Land Use and the Effects on Nature and Landscape.* Ministry of Housing, Spatial Planning and the Environment, National Spatial Planning Agency, The Hague, 72pp.

Bignal, E.M. and McCracken, D.I. (1996) The ecological resources of European farmland. In: Whitby, M. (ed.) *The European Environment and CAP Reform: Policies and Prospects for Conservation.* CAB International, Wallingford.

CEC (1989) *Les Exploitations Agricoles des Zones Défavorisées et de Montagne de la Communauté*, Brussels/Luxembourg, 129pp.

CEC (1993) *Support for Farms in Mountain, Hill and Less-Favoured Areas.* Green Europe 2/93, Brussels/Luxembourg, 80pp.

CEC (1994) *The Agricultural Income Situation in Less Favoured Areas of the EC.* Brussels/Luxembourg, 208pp.

CEC (1997) *CAP 2000, Working Document, Rural Developments.* Commission of the European Communities Directorate-General for Agriculture, Brussels, 75pp.

Council Directive 75/268/EEC of 28 April 1975 on mountain and hill farming and farming in certain less-favoured areas. *Official Journal* L128, 19/5/1975.

Council Regulation (EC) No 950/97 of 20 May 1997 on improving the efficiency of agricultural structures. *Official Journal* L142, 2/6/1997.

Council Regulation (EC) No 1257/99 of 17 May 1999 on support for rural development from the European Agricultural Guidance and Guarantee Fund (EAGGF) and amending and repealing certain Regulations. *Official Journal* L160, 26/6/1999.

Dax, T. (1998a) *Räumliche Entwicklung im Berggebiet und Benachteiligten Gebiet Österreichs.* Facts and Features Nr. 18 der Bundesanstalt f. Bergbauernfragen, Vienna, 80pp.

Dax, T. (1998b) Die Probleme der Berggebiete in Europa – Eine vergessene Dimension in der EU-Regionalpolitik. In: Agrarbündnis (Hg.): *Landwirtschaft* 98, Der kritische Agrarbericht, Kassel, pp.153-159.

Dax, T. (1998c) Responding to ecological sensitivity of LFAs: a challenge to rural development. Paper presented at the *Conference on Biodiversity and a Sustainable Countryside*, London, 9-10 March.

Dax, T. and Wiesinger, G. (1998) *Mountain Farming and the Environment: Towards Integration; Perspectives for Mountain Policies in Central and Eastern Alps.* Bundesanstalt für Bergbauernfragen, research report no. 44, Vienna, 177pp.

EUROMONTANA (1998) *L'integration des Préoccupations Environnnementales dans l'Agriculture de Montagne.* Etude réalisée pour la Commission Européenne DG XI, Paris/Brussels, 105pp.

European Economy (1997) *Towards a Common Agricultural and Rural Policy for Europe.* Directorate-General for Economic and Financial Affairs, Reports and Studies, Vol. 5, Brussels.

Frisio, D. (1997) Presentazione di primi risultati della ricerca, relazione al seminario 'Quale futuro per l'agricoltura montana in Europa: politiche di sviluppo, di sostegno o di tutela?' Milan, 7 October 1997.

Hellegers, P.J.G.J. and Godeschalk, F.E. (1998) *Farming in High Nature Value Regions: The Role of Agricultural Policy in Maintaining HNV Farming Systems in Europe.* Agricultural Economics Research Institute (LEI-DLO), Onderzoekverslag 165, The Hague, 130pp.

Lowe, P., Hubbard, L., Moxey, A., Ward, N., Whitby, M. and Winter, M. (1998) United Kingdom. In: Brouwer, F. and Lowe, P. (eds) *CAP and the Rural Environment in Transition: A Panorama of National Perspectives*. Wageningen Pers, Wageningen, pp. 103-140.

OECD (1994) *The Contribution of Amenities to Rural Development*. Organisation for Economic Co-operation and Development, Paris, 87 pp.

OECD (1998) *Rural Amenity in Austria, A Case Study of Cultural Landscape*. Organisation for Economic Co-operation and Development, Paris, 115pp.

The Agri-environmental Measures (2078/92)

12

Henry Buller

INTRODUCTION

Regulation 2078/92 was approved in May 1992 as one of three 'accompanying measures' of the reformed Common Agricultural Policy (CAP) (the other two being Regulation 2079/92, the early retirement scheme, and Regulation 2080/92 allowing for on-farm forestry aid). Under the terms of the Regulation, Member States were required to submit 5-year programmes for co-fundable schemes of agri-environmental aid, to be made available to farmers either within defined sensitive zones or across entire agricultural territories (CEC, 1992). Although agri-environmental legislation within the European Union (EU) has a number of precedents (notably, Article 19 and later 21 of Regulation 797/85), three things distinguished the 1992 policy from its predecessors. First, although farmer participation in individual schemes remains wholly voluntary, Regulation 2078/92, unlike its predecessors, imposed an obligation on all Member States to draw up agri-environmental programmes within a specified time frame.

Second, post-1992 agri-environmental policy has been co-financed by the Guarantee Section of the EAGGF (FEOGA) and not, as was the case under previous legislation, by the Guidance Section. This shift was significant not only because it represented a move to associate agri-environmental measures with the mainstream of CAP commodity management (Baldock and Lowe, 1996) but also because, initially at least, EAGGF Guarantee imposed no immediate budgetary ceilings (unlike EAGGF Guidance), enabling agri-environmental programmes to take an increasing share of the CAP budget. The amount of money spent on it and the other accompanying measures did grow very fast against the classic areas of EAGGF Guarantee expenditure (Table 12.1). Even so, after 5 years of operation, EU agri-environmental policy still only accounted for around 4% of the total CAP budget.

Third, Regulation 2078/92, and the 127 programmes and 2,200 distinct measures that have been approved by the Commission following its application, covers a far wider range of agricultural activities and strategies of agricultural development than previous attempts to reconcile agriculture and environment have done. The Regulation encourages the adoption of environmentally beneficial productive methods, including organic farming, non-organic farming with environmental improvements and the maintenance of existing low intensity systems. It also advocates the management of farmland for non-productive purposes, whether to address agricultural pollution, biodiversity loss, landscape protection and management or the maintenance of traditional practices. In a small number of countries, notably France and Ireland, the Regulation has led to the development of integrated and whole-farm sustainable plans.

Table 12.1   Percentage change in EAGGF Guarantee expenditure; 1993-1994 and 1994-1995 by expenditure category (%)

	1993-1994	1994-1995
Crop sector	2.7	5.06
Animal sector	-16.8	5.35
Accompanying measures	121.2	69.7
Other aids	-44.0	-56.3
Total EAGGF Guarantee	-4.6	+4.6

Source: CEC (1996a).

With the passage of time since the adoption of these programmes and measures, the attention of policy makers, environmental and social scientists, the environmental lobby and the agricultural community has shifted from initial concerns over implementation and uptake, finance and procedure towards assessment and evaluation. Has Regulation 2078/92 achieved its stated goals and are these goals still relevant? Since its introduction in 1992, the place and role of EU agri-environmental policy has shifted. From being an accompanying measure to the 1992 CAP reform at its inception, agri-environmental policy has, in recent years, been placed at the forefront of a far wider debate concerning the future of rural, and not just agricultural, Europe. The Cork Declaration of 1996 gave particular emphasis to the role of environmentally friendly forms of agriculture in contributing to sustainable rural economic development (CEC, 1996b). The European Commission's own Agenda 2000 proposals (CEC, 1998a) sought to place agri-environmental schemes at the centre of a new rural policy strand within the CAP.

The intention here, in the words of the current Agriculture Commissioner, is to resolve 'the dichotomy between agriculture and rural areas' (Fischler, 1998). As such, the goals of EU agri-environmental policy are evolving and current scheme objectives need to be considered against this changing agri-environmental agenda.

This chapter presents a review of the impact of Regulation 2078/92 within the EU and seeks to assess the effectiveness of European agri-environmental policy in reconciling agricultural (and rural) development and the maintenance or improvement of environmental quality.

Identifying the objectives

Any evaluation task is wholly dependent upon the correct identification of the initial policy goals. Before going anywhere, we need to tease out the objectives of Regulation 2078/92 and of the agri-environmental measures that it has engendered. This is no easy assignment for these objectives are complex. They might be seen as a means of reducing agricultural pollution or more broadly as a genuine attempt to create a policy framework for more sustainable forms of farming - a post-Maastricht, post-Rio environmental agenda which has gradually taken on board a growing number of policy domains, including agriculture. Alternatively, the 'greening' of the CAP might be seen in a more cynical or perhaps realistic light, as a means of maintaining funding and subsidy levels within European agriculture in the face of the increasing globalisation and liberalisation of agricultural trade. It might equally form part of a broader re-definition process with respect to the 'multifunctionality' and indeed the 'originality' of European rural and agricultural space particularly in the light of the increasingly metropolisation of European society. In this process, there is perhaps a necessary legitimation of those traditional forms of agriculture that have long been marginalised by the more productivist model promoted by the CAP. Finally, the objectives of agri-environmental policy might be interpreted as an attempt by the Northern European states to impose their rural agenda on the Southern European States and thereby preserve both their hegemonic position with respect to European agricultural policy making and their position as leading agricultural producers.

Article 1 of Regulation 2078 lays down clearly the three goals of EU agri-environmental policy: to accompany the changes to be introduced under the market organisation rules; to contribute to the achievement of the Community's policy objectives regarding agriculture and the environment; and to contribute to providing an appropriate income for farmers (CEC, 1992). The Regulation's goals and objectives need to be seen, therefore, in the light of a series of different concerns. As an accompanying measure to CAP reform, agri-environmental aid schemes cannot be disassociated from the basic objective of that reform, to reduce agricultural over-production within the EU and thereby lessen the overall costs of agricultural support both by encouraging agricultural extensification and through the introduction of compensatory rather than production-related support mechanisms. The agri-environmental Regulation also constitutes an agricultural policy response both to growing concern over the environmental consequences of intensification, notably with respect to water pollution, biodiversity loss and landscape change, and to the need to comply with the Maastricht Treaty, which requires that EU environmental policy be integrated into all other EU policies, and with the 5th Environmental Action Plan (CEC, 1997a). As such, a second fundamental goal has been to reduce agricultural pollution and landscape degradation. A related, but very different objective, has been to support and maintain extensive farming practices not only against intensification but also, particularly in upland and peripheral regions, against agricultural decline and withdrawal. In contrast to the preceding goals dealing with the regions and problems of agricultural over-production, this third agenda relates specifically to regions and problems of agricultural under-production.

These general concerns are not only very different, but they relate to very divergent rural territories, agricultural production systems and rural/agricultural traditions. It is not surprising, therefore, to find that Member States have tended to interpret the Regulation in widely differing ways. As a Commission official has remarked:

> It is not said in the Regulation that it is exclusively an environmental instrument. It is certainly not exclusively an income aid instrument for farmers. It is something which has several objectives, and I think this is very important to keep in mind when you talk about this scheme and when you apply it (Priebe, 1997, p. xiv).

In the UK, both the Ministry of Agriculture and Department of the Environment clearly see the agri-environmental Regulation chiefly in terms of wildlife and landscape protection (MAFF, 1993, 1997) linked also to the reduction in production and, thereby, reduced expenditure on CAP market regimes. As the British House of Commons Agriculture Committee recently put it:

> Although agri-environmental payments may become a *de facto* means of some modest income support to farmers in scenic and less productive regions of the countryside, their rationale should not be to compensate farmers for reductions in production support but to recognise positive efforts made by farmers to protect and enhance the environment (House of Commons, 1997, paragraph 112).

Such a view is not universally shared within the EU. Commenting on the response of the Mediterranean states to Regulation 2078/92, Moyano and Garrigo (1998) are unequivocal in their assertion that these states see agri-environmental aid schemes primarily as 'a complementary source of revenue for their farmers', in certain cases using them to finance farm structural changes that have little or nothing to do with environmental maintenance or improvement. In France, the most important agri-environmental scheme in terms of both budget and uptake, the grassland premium or *Prime à l'herbe*, is widely regarded as being as much a compensation for grass-based beef and milk farmers who do not benefit from fodder maize payments as a means of maintaining extensive husbandry practices (Fruit and Lompech, 1997). These differences in national interpretation are important to the current discussion for two reasons; first, they determine the strategies of implementation adopted by individual Member States, including the selection of aid schemes, institutional arrangements and patterns of negotiation and, second, they have major consequences for the evaluation and assessment of scheme performance.

Implementation strategies

The agri-environmental Regulation requires Member States to draw up and implement national agri-environmental programmes comprising any number of national, regional or local schemes. Although states can draw up their own agri-environmental policies, programmes submitted to the Commission for co-funding (up to 50%, or 75% in Objective 1 regions) should in principle respond to all of the seven basic objectives specified in the Regulation (Table 12.2).

Table 12.2 Scheme objectives eligible for aid under Article 2 of Regulation 2078/92

Fundable objective

a   to reduce substantially the use of fertilisers and/or plant protection products, or to maintain the reductions already made; or to introduce or continue with organic farming;

b   to change, by means other than those referred to in (a), to more extensive forms of crop, including forage production; or to maintain extensive production methods introduced in the past; or to convert arable land into extensive grassland;

c   to reduce the number of sheep and cattle per forage area;

d   to use other farming practices compatible with the requirements of protection of the environment and natural resources, as well as maintenance of the countryside and landscape; or to rear animals or local breeds in danger of extinction;

e   to ensure the upkeep of abandoned farmland or woodlands;

f   to set aside farmland for at least 20 years with a view to use it for purpose of the environment, in particular for the establishment of biotope reserves of natural parks or for the protection of hydrological systems;

g   to manage land for public access and leisure activities.

Source: CEC (1992).

The Regulation, however, permits a wide variety of implementation strategies. While agri-environmental schemes introduced following Regulation 797/85 were, for the most part, targeted zonal measures applied to particular sensitive areas (generally following the British ESA model), schemes established following Regulation 2078/92 vary considerably both between Member States and within regions (IEEP, 1998; Buller, 2000). In general, one might distinguish four broad 'models'. Most Member States continue to apply targeted or zonal measures focusing on specific landscape types, natural regions or farming systems, and being made available only to eligible farmers operating within the selected zone (for example, the British and Danish ESA or the French *Opérations locales*). Some states, however, have also adopted 'wide and shallow' or horizontal schemes which, subject to certain eligibility criteria, operate over the whole national or regional territory (for example, the French *Prime à l'herbe* or the Swedish grassland measure). A third model is the broad regulatory framework which is often composed of an initial basic payment with additional 'add-on' aid schemes involving further constraints and, consequently, higher payments (for example, the Austrian ÖPUL programme, the MEKA programme in Baden-Württemberg in Germany, the Irish REPS programme and the Finnish GAEPS). Finally, and in virtually all Member States, there exist a number of highly specific aid schemes linked to the conversion to and maintenance of organic farming, the protection of threatened stock breeds and the training or farmers. Not targeted in a spatial sense, these schemes are characterised by highly variable budgetary allocations.

From the range of scheme types identified by the Regulation (see Table 12.2), the European Commission has recently proposed a five-point regrouping based upon stated scheme objectives. Though a certain amount of care needs to be employed in the interpretation of these figures, an examination of the estimated proportion of the total agri-environmental budgets (EAGGF and national contribution) allocated by each Member State to each of these five categories (Table 12.3), gives us an idea of the extent of national variation in agri-environmental policy implementation.

Table 12.3   Type of agri-environment measure as proportion of total budget per Member State (based upon expenditure for programmes approved by March 1996) in %

Country	Organic farming	Farming with environmental improvements	Maintenance of low intensity systems	Non-productive land management	Training and demonstration projects
Austria	17	59	21	3	0
Belgium	20	58	5	14	3
Denmark	24	46	16	14	0
Finland	5	42	42	7	5
France	3	15	79	3	1
Germany	1	56	21	21	1
Greece	14	35	0	50	0
Ireland	2	49	21	24	4
Italy	23	43	22	10	2
Luxembourg	1	39	56	3	0
Netherlands	2	32	0	0	66
Portugal	4	18	68	6	4
Spain	4	35	15	42	4
Sweden	15	6	71	1	7
UK	2	53	30	14	0
EU-15	8	41	35	14	3

Source: CEC (1997b).

While Austria, Belgium, Denmark, Sweden and particularly Italy accord a large proportion of their 2078 budgets to supporting organic farming, expenditure in this category in virtually all other states is very low (generally under 5% of the total), though in a number of cases it is subject to strong sub-national variation. Similarly, training and demonstration projects, with the single exception of The Netherlands where this has been particularly emphasised, are generally given a very low priority in national programmes. Undoubtedly the bulk of spending under Regulation 2078 is concentrated, firstly, on measures to actively improve environmental quality and, secondly, on those that seek to maintain recognised environmentally friendly agricultural practices. The difference between the two is often blurred. Tier 1 payments under British ESA schemes, for example, might in practice fall under the second type while higher tiers promote actual changes in farm management. Nevertheless, taking these two types of measure and the management of non-productive land, a pattern clearly emerges (Figure 12.1). One group of Member States (Austria, Belgium, Denmark, Germany, Ireland and the UK) can be distinguished by the significant proportion of their agri-environmental budgets allocated to schemes that seek to improve environmental quality. A second group (Finland, France, Luxembourg, Portugal and Sweden) is identified by the priority they give to the maintenance of low intensity systems while a third group (Greece and Spain) is concerned essentially with the management of non-productive and very extensively used agricultural land.

There are thus strong variations in agri-environmental spending both between countries and between general scheme types. It is significant that the two scheme types potentially most likely to have a long-term positive impact upon the relationship between agriculture and the environment - the promotion of organic farming and the training of farmers in environmentally sensitive agricultural techniques - are generally given the lowest priority by Member States. Equally revealing is the geographical grouping of scheme priorities, with the more intensively farmed and predominately lowland Member States differing from the southern and Scandinavian States (with the exception of Italy) where extensive grass-based systems predominate. A more detailed examination of the different aid schemes proposed by Member States (Buller *et al.*, 2000) reveals that, although the broad category of measures seeking to actively improve environmental quality accounts for a significant proportion of the budgets allocated, relatively few agri-environmental schemes (both in number and in terms of targeted areas) are directly concerned with reducing farm pollution. The vast bulk focus rather on improved landscape and wildlife management. This is something to which we return below.

The total expenditure on EU agri-environmental policy for the 5-year period, 1993-1997, was around 6,244 million ECU (MECU), of which some 3,787 MECU came from the EAGGF, the remainder coming from national sources. Due to a stronger than expected response to the initial Regulation from certain Member States, the late start of some national programmes and the entry in 1995 of an additional three states, the total EAGGF contribution has grown substantially and has well exceeded the initial estimate of around 2,256 MECU for the first 5 years. Just for the one year 1997, EAGGF expenditure on agri-environmental schemes amounted 1,556 MECU, which was more than three times that of 2 years earlier when only nine national programmes had been fully operational (CEC, 1997b).

The bulk of the EAGGF expenditure to date has gone to three Member States, Germany, Austria and France, which together accounted for 59% of the total for 1993-1997 (Table 12.4). However in the case of France in particular and Germany to a lesser extent such large agri-environmental budgets need to be seen in the context of equally large utilisable agricultural areas (UAA). In the year 1997, for example, EAGGF expenditure on French agri-environmental programmes amounted to 144 MECU for a country with a UAA of some 30.3 million hectares, yielding a broad expenditure ratio of 4.7 ECU ha^{-1}. A similar calculation

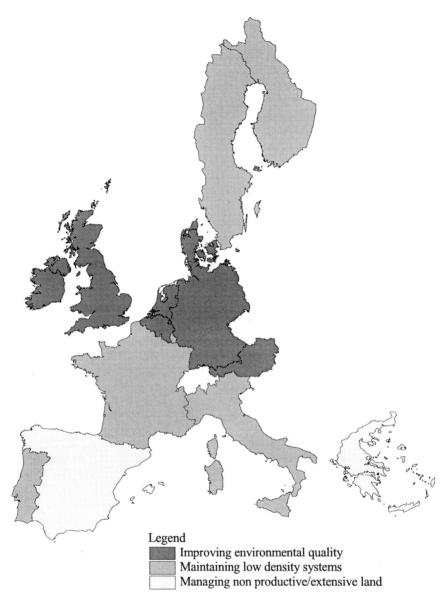

Figure 12.1 Dominant orientation of agri-environmental budget spending by
Member States (1997)
Source: Derived from EU budgetary statistics, CEC (1997b).

for Germany would yield a ratio of 17.7 ECU ha^{-1}. By way of contrast, countries
such as Austria, Finland and Sweden, with far smaller UAAs, yield ratios of 76,
53 and 25 respectively. Spain, the UK, Denmark and Greece turn in figures
around 2 ECU ha^{-1}.

To understand these variations, we need to look more closely at how agri-environmental aid schemes under Regulation 2078 have been taken up by farmers in the different Member States of the EU.

Table 12.4 Total expenditure under Regulation 2078/92 and EAGGF contribution, by Member State for the period 1993-1997

Member State	Total expenditure (EAGGF and national contribution) (MECU)	EAGGF contribution (MECU)	Proportion of total EU-15 EAGGF agri-environment budget (%)
Austria	1,553	806	21.30
Belgium	6	3	0.08
Denmark	38	19	0.50
Finland	798	399	10.54
France	1,018	509	13.44
Germany	1,294	918	24.24
Greece	15	11	0.29
Italy	714	432	11.43
Ireland	217	163	4.30
Luxembourg	9	4	0.11
Netherlands	49	25	0.66
Portugal	197	148	3.92
Spain	167	125	3.30
Sweden	252	126	3.33
UK	192	98	2.59
Total	6,519	3,787	100.00

Source: Buller (2000) from CEC (1997b).

Assessing the impact

EU agri-environment policy really began with Article 19 of the agricultural structures Regulation 797/85, later followed by Regulations 1760/87 permitting EU co-financing for agri-environmental schemes, 2328/91 relating to extensification and 866/90 relating to organic farming support. While Article 19 of Regulation 797/85 represented, in many respects, an important departure in EU agricultural policy (Whitby and Lowe, 1994; Baldock and Lowe, 1996), its implementation was largely restricted to six of the 12 Member States (Table 12.5). By 1990, some 900,000 ha of UAA had been placed under agreement (see Whitby, 1996; Potter, 1998; Buller *et al.*, 2000 for a more detailed analysis). By the time Regulation 2078/92 appeared, therefore, certain Member States had a prior experience not only in designing and implementing agri-environmental schemes but also in recognising and acting upon an agri-environmental agenda. Other states did not have this experience, a distinction that has subsequently marked the implementation and uptake of the more recent legislation (Buller, 2000).

Table 12.5   Land designated (ha) under Article 19 of Regulation 797/85 in 1990

Country	Area designated	Area under contract	No. of participants
Denmark	127,970	28,060	3,459
France	114,620	36,620	nd
Germany	2,560,000	291,646	40,780
Ireland	1,140	nd	nd
Italy	944,430	229,259	6,038
Luxembourg	2,800	610	4
Netherlands	75,800	26,815	5,013
UK	740,930	282,351	4,997
Total	4,567,690	895,461	60,291

Source: Potter (1998); CEC (1991).

As has been said above, the adoption of agri-environmental programmes under Regulation 2078/92 was mandatory for Member States. By April 1997, some 1.3 million contracts had been signed under 2078/92 schemes, representing approximately 18% of EU farms (CEC, 1997b) and covering 22.6 million ha, approximately 17% of the total UAA of the EU (Table 12.6). Clearly, this is a substantial advance over the initial round of schemes introduced after Regulation 797/85. At a national scale, uptake varies from over 73% of UAA under 2078/92 in Austria, around 20% in France to around 8% in the UK. For such areas of farmland, major changes in land use and agricultural practice are, for the duration of the contract period, no longer permitted. At one level, this represents a considerable step forward both in the drive towards a more comprehensive management of the environmental effects of agricultural activities and in the recognition of the importance and the richness of European agricultural landscape diversity. In addition, a significant number of European farmers are benefiting from some level of income support, in exchange for the maintenance of environmental goods.

However, a detailed examination of the pattern of implementation and uptake, both in terms of the numbers of contracts and the areas concerned, reveals a strong spatial concentration within a small number of states. Not only do Germany, France and Austria collectively account for well over half the total number of contracts, they also include 64% of the total area currently under agreement. The Mediterranean states (excluding France), though they have well over half the number of farms within the EU, account for only 16% of all contracts and 9% of the total agreement area. Some care needs to be taken, however, with these figures. First, it is notorious that Member States within the EU do not adopt the same definition of 'agricultural area'. Common rough grazing land, for example, is considered as part of the UAA in certain states but not in others. Second, it must be remembered that the bulk of the Mediterranean states (as well as Austria, Sweden and Finland) are less far into their agri-environmental programmes than Germany, France and the UK. Finally, care needs to be taken in the definition of contracts. In the majority of states, farmers enter single schemes. However, in those states or regions where cumulative framework instruments operate (such as, for example, the Austrian ÖPUL scheme or the MEKA scheme in Baden-

Württemberg), farmers may enter a range of schemes. In such areas, the number of contracts can greatly exceed the number of farmers (as is the case in Baden-Württemberg).

Table 12.6   Take-up of aid schemes under Regulation 2078/92 at mid-1997 (1996 for Italy)

Member State	Total number of contracts	No. of contracts as % of total farms	No. of contracts as % of all	Total area under contract (ha)	Proportion of total UAA under contract (%)
Austria	168,804	75.90	12.50	2,500,000	72.9
Belgium	1,242	1.70	0.09	17,000	1.2
Denmark	8,193	11.80	0.60	94,000	3.4
Finland	91,509	a)	6.80	2,000,000 [b)]	91.2 [b)]
France	177,695	24.10	13.20	5,725,000	20.2
Germany	554,836	a)	41.20	6,353,000	37.0
Greece	1,839	0.20	0.10	12,000	0.3
Italy	63,841	2.50	4.70	977,000	6.6
Ireland	23,855	15.50	1.70	801,000	18.5
Luxembourg	1,922	60.00	0.10	97,000	76.9
Netherlands	5,854	5.10	0.40	31,000	1.5
Portugal	125,479	27.80	9.30	606,000	15.4
Spain	29,599	2.30	2.10	532,000	2.1
Sweden	68,969	77.60	5.10	1,561,000	51.0
UK	21,482	9.16	1.60	1,322,000	8.1
Total	1,345,119	18.30	100.00	22,628,000	16.5

[a)] Impossible to determine with any accuracy as many farms hold multiple contracts. [b)] These figures must be considered as a considerable over-estimation for the reason given above.
Source: Eurostat (1996, 1998); CEC (1997b); Buller (2000).

At a sub-national level, the uneven distribution of 2078/92 implementation is even more apparent (Buller, 2000). The German Länder of Bayern and Baden-Württemberg, Austria, the northern regions of Italy and the southern regions of France (including the Massif Central) all display high levels of contracted agricultural land (Figure 12.2). In France, this reflects above all the geography of the *Prime à l'herbe* scheme. With some 95,000 contracts currently operational covering 5.5 million ha, this single national scheme, which seeks to maintain extensive practices, represents 67% of all agri-environmental contracts engaged in France and around 90% of all currently contracted land (Alphandéry and Bourliaud, 1998; Buller and Brives, 2000).

By contrast, southern regions and the principal arable regions of northern Europe show much lower levels of contracted land. As I have argued elsewhere (Buller, 1998a), the current distribution of agricultural land under agri-environ-

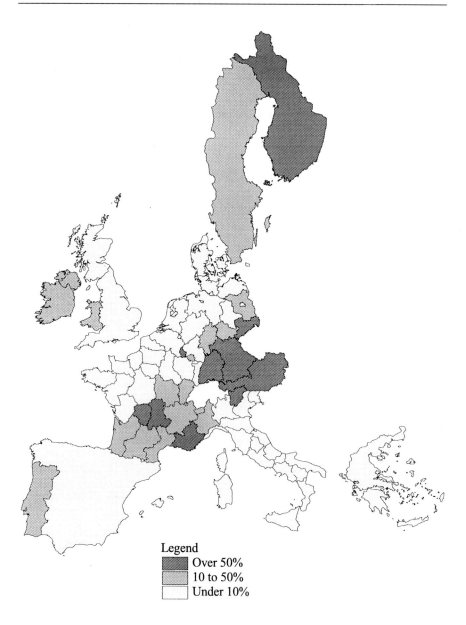

Figure 12.2   Proportion of the UAA currently under contract, data for 1997 (except Italy, 1996)
Source: Unpublished statistics from EU and individual Member States and Eurostat (1998).

mental agreement within Europe follows closely that of upland and essentially pastoral Less Favoured Areas. While such regions clearly constitute, in many cases, areas of high nature value (Baldock *et al.*, 1996), the similar spatial configuration of agri-environment policy and a policy designed to compensate farmers for the constraints imposed by natural handicaps reinforces the notion, once again, that the primary thrust of the former is maintaining existing environmentally friendly practices. That these should take place largely in areas where agricultural intensification is constrained by the physical environment, implies a polarisation of European agriculture-environmental relations that Regulation 2078/92, by itself, will do little to resolve.

Assessments of the uptake of agri-environmental schemes based solely upon the contracted area or the number of contracts, however, fail to take into account the variable intensity of the different measures on offer. Grassland schemes such as those which operate in Germany, Sweden and France may cover large proportions of agricultural land but, in general, they require few if any changes to existing farm practices and, as a result, offer farmers low premiums.

The distribution of current agri-environmental expenditure thus exhibits a slightly different pattern (Table 12.7). Schemes that require changes in agricultural techniques (reducing inputs, lowering livestock densities, adopting organic methods) offer, in general, higher premiums than those that target the maintenance of existing extensive practices. Higher average per hectare payments are thus recorded in Denmark, The Netherlands, Sweden and Ireland. In three of these states, a large proportion of the total agri-environmental budget is given over to active environmental improvement and/or pollution reduction, particularly in Denmark and The Netherlands (see Table 12.3 above). The Swedish example is more complex. The high currency value combined with a number of high premium schemes targeting low intensity systems (notably the meadows scheme

Table 12.7   Average annual expenditure by area under Regulation 2078/92 (situation at mid-1997 except Italy, 1996) (ECU ha^{-1})

Member State	Payment ha^{-1} under contract (area schemes only)
Austria	140
Belgium	84
Denmark	186
Finland	124
France	42
Germany	89
Greece	na
Italy	na
Ireland	147
Luxembourg	90
Netherlands	260
Portugal	137
Spain	81
Sweden	156
UK	55
Total	117

Source: CEC (1997b); Buller (2000).

which pays on average around 300 ECU ha^{-1} on around 4,000 ha of agreement land) explain the generally high payment rates (Carlsen and Hansund, 2000). By comparison, for states like France and Germany, as well as Spain and the UK, where low-paying maintenance schemes are most common (including lower tier ESA payments in Britain), average payments are substantially less.

Evaluating the results

Undoubtedly, one of the most formidable challenges facing EU agri-environmental policy is how to evaluate its effects. Partly, this derives from the lack of precise targets and the potential confusion of objectives arising out of the original Regulation. Indeed, concern over the emerging difficulties surrounding the assessment of schemes and the adoption of appropriate monitoring procedures prompted the establishment of Regulation 746/96 requiring all Member States to submit evaluation strategies to the Commission. Not only are agri-environmental policies intrinsically difficult to evaluate due to the almost universal absence of suitable baseline data, but the wide variety of implementation strategies and scheme types adopted at the national and sub-national level has led to a similar diversity of scheme goals. As a consequence, cross-national comparison of scheme effectiveness is highly problematic. Many of the 'classic' indicators of agri-environmental measures, such as those proposed by the OECD (OECD, 1997) are often inappropriate for a policy that is, in the manner in which it has frequently been implemented, essentially concerned with maintaining existing practices and environmental conditions rather than actively improving them (Biehl *et al.*, 1999). Furthermore, in addressing changes in environmental states, they remain too crude a means of assessing a policy that addresses itself primarily at agricultural management practices.

Agri-environmental schemes offer a series of different levels for evaluation (Figure 12.3) running from the simple assessment of scheme uptake to a potentially more involved evaluation of the ability of scheme operation to lead to long-term structural changes in the use and flow of environmental knowledge. To date, however, most monitoring and evaluation exercises have tended to focus upon the first (for example, Siikamäki, 1996; CEC, 1997b) and the second of these (Bird-Life International, 1996; MAFF, 1996; IPEE, 1997). Regulation 746/96, while it requires Member States to supply monitoring reports, does not bind them to a single methodology. Given the major variations in data availability, in the mechanisms for its collection and in Member State engagement, a clear picture of the effects and indeed the effectiveness of agri-environmental schemes is proving extremely difficult to secure (IEEP, 1998). Indeed, one might ultimately question the appropriateness of a macro-scale comparative evaluation exercise which fails to account for the very diversity of agricultural, environmental and rural contexts that the Commission itself is at pains to defend.

Given its relatively long experience in agri-environmental policy, it is not unsurprising that the UK has, to date, one of the most developed approaches to monitoring and evaluation (Whitby, 1994). A series of Monitoring Reports have been produced for the UK Environmentally Sensitive Areas, building, in some cases, on over 10 years of agreements (MAFF, 1996). The Countryside Steward-ship scheme for England has also had a monitoring component built into it since

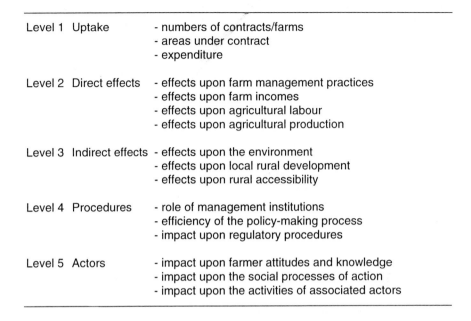

Level 1	Uptake	- numbers of contracts/farms - areas under contract - expenditure
Level 2	Direct effects	- effects upon farm management practices - effects upon farm incomes - effects upon agricultural labour - effects upon agricultural production
Level 3	Indirect effects	- effects upon the environment - effects upon local rural development - effects upon rural accessibility
Level 4	Procedures	- role of management institutions - efficiency of the policy-making process - impact upon regulatory procedures
Level 5	Actors	- impact upon farmer attitudes and knowledge - impact upon the social processes of action - impact upon the activities of associated actors

Figure 12.3   Parameters for the evaluation of agri-environmental scheme effectiveness

its introduction in 1991 (Countryside Commission, 1996). Elsewhere, Denmark, The Netherlands and, more recently, Germany (though in a few Länder only) might equally be characterised by the importance given to scheme monitoring and evaluation. Environmental evaluation of the German MEKA scheme (Zeddies, 1996), the Austrian ÖPUL programme (Dietrich, 1997) and the French Input Reduction scheme (Ministère de l'Agriculture, 1997) have all shown that these schemes or their relevant components do have a demonstrably positive effect. In all, those schemes that reduce or interdict the use of chemical inputs are held to have the highest ecological effectiveness. However, these are often focused studies. Beyond these limited examples, effective scheme monitoring has often been very meagre (for examples, see Buller *et al.*, 2000), rarely extending beyond the extrapolation of intended environmental effects from uptake figures. Broad schemes that might initially appear as eminently monitorable, such as those that seek falls in agricultural pollution through reductions in input use, often yield little in the way of concrete and measurable results. Changes in water quality are rarely immediate while the geographical spread of agreement land is of vital importance in the effectiveness of the measure. Similarly, organic farming techniques, while they might be considered environmentally beneficial in terms of the reduced use of chemical inputs, may have a limited impact if they are restricted to a small number of isolated farms within an otherwise intensively farmed region. Effectiveness, in terms of reduced input use, might be observable on the farm, but within the context of the wider environment it may be negligible.

Furthermore, reduced forage yields frequently associated with organic production techniques sometimes necessitate the expansion of forage crop area leading to the removal of hedgerows and landscape features.

The difficulties of monitoring the effectiveness of agri-envronmental schemes can be seen in the French example. Within the national programme available for the 1993-1997 period (Boisson and Buller, 1996; Buller and Brives, 2000) were three specific measures that sought, via indirect means, to reduce environmental pollution through the reduction of stocking densities, the conversion of arable land to grassland and 20-year set-aside. Not only has their environmental impact been judged as low, but the failure to obtain baseline data at their inception has frustrated attempts to go beyond the simple analysis of uptake (Table 12.8). For 1998 and the new 5-year French agri-environmental programme, the long-term set-aside measure has been abandoned and the others considerably altered to focus more closely upon targeted sensitive areas.

Table 12.8   Evaluation of three regional measures under the French agri-environmental programme, 1993-1997

Scheme	On-farm changes (1997) resulting from scheme introduction	Environmental impact
Reduction of stock density	23,271 LU removed	Extremely difficult to determine as contracting farms are rarely concentrated in the more sensitive areas
Conversion of arable land to pasture land (particularly in sensitive zones, such as along rivers)	16,689 ha engaged	Not measurable as no baseline figures for environmental quality
Long-term set-aside	495 ha engaged	Extremely slight

Source: Ministère de l'Agriculture/ISARA (1998).

Yet, lack of information on environmental change or the absence of clearly discernible environmental improvements should not, in itself, be taken as a sign of scheme ineffectiveness. This is particularly so for those schemes which seek to protect cultural landscapes and semi-natural grassland habitats through the maintenance of existing low intensity agricultural practices. For such schemes, which as we have seen dominate the agri-environmental programmes of most Member States, evaluation is perhaps more closely allied to assessments of uptake. Thus, we might claim that, for the 5 years of contract duration, the 5.3 million ha of species-rich grasslands under the French *Prime à l'herbe* coupled with the 400,000 ha of protected grassland under the *Opérations locales*, in addition to some 1.3 million ha of German grasslands (Grafen and Schramek, 2000), some 100,000 ha of common pasture land in the Spanish Asturias region (Peco *et al.*,

2000) and the bulk of Tier 1 land under the UK ESAs, are protected from un-wanted intensification, afforestation or abandonment. Such schemes represent to some extent the strength and the weakness of agri-environmental policy to date; their strength is that they have undoubtedly been successful in attracting partici-pants and, consequently, in achieving their principal goal, to protect landscapes and maintain traditional practices. Their weakness is two-fold. On the one hand, they often demand little in return from participating farmers and as such attract the criticism that they are, in effect, merely another form of subsidy. On the other hand, they lead to the increasingly frequent critique that agri-environmental pol-icy, as it has been implemented since 1992, is not addressing the central issue of agricultural pollution.

In responding to these charges, two lines of argument might be advanced. The first concerns the relationship of the agri-environmental measures to the CAP. To what extent should these measures exist to directly counterbalance the more environmentally damaging effects of other forms of CAP subsidy? If that is to be their *raison d'être*, then not only might they fall foul of the polluter-pays principle but they would also render somewhat incoherent the concept of 'accom-paniment'. Is there not a greater legitimacy and indeed returns in using such measures to protect farm systems of high ecological or landscape value and the agricultural and rural communities that they sustain; and in employing different means - whether they be regulatory (normative), conditional (cross-compliance), negotiative (for example, the French PMPOA or DEXEL schemes) or fiscal (pollution permits) - to address the more specific problem of agricultural pollu-tion (see, for example, Kleinhanss, 1998)?

A second response relates back to the different levels of scheme effect (see above, Figure 12.3) particularly those concerning procedures and actors. What is possibly of greater long-term significance than the fact that some 22.6 million hectares of European farmland have been placed within agri-environmental schemes, is that well over a million European farmers are engaged, to greater and lesser degrees, in the drive towards reconciling the needs of agricultural produc-tion with those of environmental sustainability and that agricultural extension services and the traditional vehicles of production policy diffusion are adapting to these new demands (see, for example, Brives, 1998). We might hope that the changes engendered here by the implementation of agri-environmental policy will extend beyond the simple limits of remunerated contractualisation.

CONCLUSIONS

Three things are clear from the first 5 years of post CAP-reform agri-environmental policy. First, at the European level, its implementation does repre-sent a degree of genuine environmental and agricultural policy integration (CEC, 1998b). However, agri-environmental policy remains nonetheless marginal to the central thrust of the CAP with agri-environmental schemes continuing to repre-sent a very small part of the total CAP budget when compared with price supports, compensatory payments to farmers and market intervention. Unless that balance is changed in favour of agri-environmental policy, there seems little point in extending schemes and measures to include more farmers.

The second, albeit tentative, concluding point we can make is that more durable gains will be longer in coming. The majority of farmers participate in schemes because it is to their economic advantage to do so, particularly in those cases where schemes represent supplementary income sources. While this is not necessarily incompatible with environmental protection, it suggests that building real shifts in the attitudes and behaviour of farmers and other actors needs to be a stronger component of scheme development than has hitherto been the case. Nevertheless, agri-environmental policy does act as a crucial source of legitimation for those farmers who, for a variety of reasons, have not followed the otherwise ubiquitous model of agricultural intensification and modernisation. In such cases, and indeed for agri-environmental policy in general, schemes are, we might suggest, not only more likely to be successful in achieving their varied goals but also more pertinent to rural and agricultural concerns if they are anchored in a coherent territorial and, thereby farm/rural system (Buller, 1998b).

Third, and perhaps most important, the environmental gains of agri-environmental policy are not always clear. On the one hand, the policy has manifestly failed to adequately address the issue of pesticide and nitrate pollution largely because farmers in the areas affected do not generally participate in schemes (where such schemes are available) as this would likely lead to a drop in farm income. On the other hand, the dominance by schemes that seek merely to maintain extensive production methods raises a number of questions regarding both their real objectives (supporting marginal farms, compensating grass-based farming systems for their ineligibility for arable and forage crop payments) and the evaluation of their effectiveness. This polarisation, between those farm systems benefiting from voluntary agri-environmental support and those systems for which, increasingly, regulatory means to reduce pollution are being demanded, points to a failure to genuinely integrate environmental and agricultural policy objectives.

Ultimately, agri-environmental policy needs to be seen as the first step, rather than the conclusion, of an integration process. The need for cross-compliance, the establishment of aid ceilings and the imposition of a base level, minimum standard of acceptable agricultural practice without premiums are perhaps the next stage, as are the development and extension of whole-farm sustainable development plans on the French and Irish models. However, it is becoming increasingly apparent that the integration of agricultural and environmental policy is ultimately not enough. A further policy domain, that of rural development, needs to be added. Perhaps this is the key to future policy evolution. Environmental policy and agricultural policy continue to be seen, by many farmers, as mutually opposing, the one, to some extent, 'controlling' the other. Their integration within a broader policy context of rural development policy should strengthen the role of environmental management and protection as a legitimate form of economic diversification.

REFERENCES

Alphandéry, P. and Bourliaud, J. (1998) L'agri-environnement, une production d'avenir. *Etudes Rurales* 141/142, 21-44.

Baldock, D. and Lowe, P. (1996) The development of European agri-environmental policy. In: Whitby, M. (ed.) *The European Environment and CAP Reform.* CAB International, Wallingford, pp. 8-25.

Baldock, D., Beaufoy, G., Brouwer, F. and Godeschalk, F. (1996) *Farming at the Margins - Abandonment or Redeployment of Agricultural Land in Europe.* Institute for European Environmental Policy (IEEP) and Agricultural Economics Research Institute (LEI-DLO), London/The Hague.

Biehl, D., Schramek, J., Buller, H. and Wilson, G. (eds) (1999) *Implementation and Effectiveness of Agri-environmental Schemes Established under Regulation 2078/92.* Final Report of Research Programme FAIR1-CT95-274. DGVI of the European Commission, Brussels.

BirdLife International (1996) *Nature Conservation Benefits of Plans under the Agri-environment Regulation.* Birdlife International, Sandy, UK.

Boisson, J-M. and Buller, H. (1996) Agri-environmental policy in France. In: Whitby, M. (ed.) *The European Environment and CAP Reform.* CAB International, Wallingford, pp. 105-130.

Brives, H. (1998) L'environnement: Nouveau pré-carré des Chambres d'Agriculture? *Ruralia* 2, 73-84.

Buller, H. (1998a) Agrarianism, environmentalism and European agri-environmental policy: towards a new "territorialisation" of European farm systems. Paper to the *Royal Geographical Society/Institute of British Geographers Annual Conference,* Guildford, January, 1998.

Buller, H. (1998b) Green boxing the CAP. Paper to *the Third Environmental Summer Workshop,* Centre Robert Schuman, Florence, July, Italy.

Buller, H. (2000) Patterns of implementation. In: Buller, H., Wilson, G. and Höll, A. (eds) *European Agri-environmental Policy.* Ashgate, Basingstoke.

Buller, H. and Brives, H. (2000) Farm production and rural product: the French experience with agri-environmental measures. In: Buller, H., Wilson, G. and Höll, A. (eds) *European Agri-environmental Policy.* Ashgate, Basingstoke, in press.

Buller, H., Wilson, G. and Höll, A. (eds) (2000) *European Agri-environmental Policy.* Ashgate, Basingstoke.

Carlsen, H. and Hansund, K.-P. (2000) Sweden: agri-environmental policy and the production of landscape qualities. In: Buller, H., Wilson, G. and Höll, A. (eds) *European Agri-environmental Policy.* Ashgate, Basingstoke.

CEC (1991) *The Agricultural Situation in the Community,* 1990 Report. Office for Official Publications of the European Communities, Luxembourg.

CEC (1992) *Council Regulation 2078/92 on the Introduction and Maintenance of Agricultural Production Methods Compatible with the Requirements of the Preservation of the Environment and the Management of the Countryside.* Office for Official Publications of the European Communities, Luxembourg.

CEC (1996a) *EAGGF/FEOGA Annual Report, 1995.* Office for Official Publications of the European Communities, Luxembourg.

CEC (1996b) *The Cork Declaration: A Living Countryside.* European Conference on Rural Development, November 1996, Cork, Ireland. Office for Official Publications of the European Communities, Luxembourg.

CEC (1997a) *Agriculture and the Environment.* CAP Working Notes, Office for the Official Publications of the European Communities, Luxembourg.

CEC (1997b) *Report from the Commission to the Council and Parlement on the Application of Council Regulation 2078/92,* COM (97) 620 Final, CEC, DGVI, Brussels.

CEC (1998a) *Agenda 2000: Propositions of the Commission.* Office for Official Publications of the European Communities, Luxembourg.

CEC (1998b) *Evaluation of Agri-Environmental Programmes.* European Commission Directorate General DGVI Working Document, VI/7655/98, Commission of the European Communities, Brussels.

Countryside Commission (1996) *Countryside Stewardship Monitoring and Evaluation*: Fourth Report. Countryside Commission, Cheltenham.

Dietrich, M. (1997) Ökologische Evaluierung des Umweltprogrammes. *Der Förderdienst* 1, 10-11.

Eurostat (1996) *Farm Structure Survey.* Office for Official Publications of the European Communities, Luxembourg.

Eurostat (1998) *Agriculture Statistical Yearbook.* Office for Official Publications of the European Communities, Luxembourg.

Fischler, F. (1998) *Nature Friendly Farming: Agricultural Perspectives in Rural Areas.* Speech of the Agriculture Commissioner of the European Commission to the Informal Council of Agriculture Ministers, St Wolfgang, 20 September, 1998.

Fruit, J.-P. and Lompech, M. (1997) Les politiques agri-environnementales dans l'espace français. *Information Géographique* 1977-2, 65-74.

Grafen, A. and Schramek, J. (2000) Germany: complex agri-environmental policy in a federal system. In: Buller, H., Wilson, G. and Höll, A. (eds) *European Agri-environmental Policy.* Ashgate, Basingstoke.

House of Commons (1997) *Environmentally Sensitive Areas and other Schemes under the Agri-Environmental Regulation.* Second Report of the Agriculture Committee, HMSO, London.

IEEP (1998) *Assessment of the Environmental Impact of Certain Agricultural Measures.* Final Report, Institute for European Environmental Policy, London.

IPEE (1997) *Evaluation des Effets des Mesures d'Accompagnement de la Réforme de la PAC* - Tome 1. Institut pour une Politique Européenne de l'Environnement, Paris.

Kleinhanss, W. (1998) Austria. In: Brouwer, F. and Lowe, P. (eds) *CAP and the Rural Environment in Transition.* Wageningen Pers, Wageningen, pp. 305-322.

MAFF (1993) *Agriculture and England's Environment.* Ministry of Agriculture, Fisheries and Food, HMSO, London.

MAFF (1996) *Monitoring Report, South Downs ESA.* Ministry of Agriculture, Fisheries and Food, HMSO, London.

MAFF (1997) Evidence of the Ministry of Agriculture Fisheries and Food to Agriculture Committee of the House of Commons. In: *House of Commons, Environmentally Sensitive Areas and other Schemes under the Agri-Environmental Regulation*, Second Report of the Agriculture Committee, HMSO, London.

Ministère de l'Agriculture (1997) *Evaluation Nationale de la Mesure Agri-environnementale 'Réduction d'Intrants'.* Ministère de l'Agriculture, Paris.

Ministère de l'Agriculture/ISARA (1998) *Evaluation des Measures Agri-environnementales.* Ministère de l'Agriculture, Paris.

Moyano, E. and Garrigo, F. (1998) Acteurs Sociaux et Politique Agri-environnementale en Europe. *Le Courrier de l'Environnement* 33, 106-114.

OECD (1997) *Environmental Indicators for Agriculture.* Organisation for Economic Co-operation and Development, Paris.

Peco, B., Suárez, F., Oñate, J., Malo, J. and Aguirre, J. (2000) Spain: first tentative steps towards an agri-environmental programme. In: Buller, H., Wilson, G. and Höll, A. (eds) *European Agri-environmental Policy.* Ashgate, Basingstoke.

Potter, C. (1998) *Against the Grain: Agri-environmental Reform in the United States and the European Union.* CAB International, Wallingford.

Priebe, J. (1997) Evidence to the House of Commons Agriculture Committee. In: *House of Commons, Environmentally Sensitive Areas and other Schemes under the Agri-Environmental Regulation*, Second Report of the Agriculture Committee, HMSO, London.

Siikamäki, J. (1996) *Finnish Agri-Environmental Programme in Practice: Participation and Farm-Level Impacts in 1995.* Agricultural Economics Research Institute, Helsinki.

Whitby, M. (ed.) (1994) *Incentives for Countryside Management.* CAB International, Wallingford.

Whitby, M. (ed.) (1996) *The European Environment and CAP Reform.* CAB International, Wallingford.

Whitby, P. and Lowe, P. (1994) The political and economic roots of environmental policy in agriculture. In: Whitby, M. (ed.) *Incentives for Countryside Management.* CAB International, Wallingford, pp. 1-24.

Zeddies, J. (1996) Analyse der laufende und geplanten Programme zur Förderung umweltgerechter Produktionsverfahern - Modifikationene une Perspektiven. In: Linckh, G. *et al.* (eds) *Nachhaltige Land - und Forstwirtschaft - Expertisen.* Springer, Berlin, pp. 655-700.

Organic Farming

13

Nicolas Lampkin, Susanne Padel and Carolyn Foster

INTRODUCTION

Organic farming is an approach to agriculture that emphasises environmental protection, animal welfare, sustainable resource use and social justice objectives, utilising the market to help support those objectives and compensate for the internalisation of externalities. Detailed production standards, protected by European Union (EU) legislation (Regulation 2092/91), severely restrict the use of artificial fertilisers, pesticides and growth promoters, and exclude the use of genetically modified organisms (GMOs). Instead, rotations, fertility-building legumes, mechanical weed control, biological pest control and agro-ecosystem management help to ensure acceptable yields of quality products.

Although organic farming as a concept has existed for over 70 years, only since the mid-1980s has it become the focus of significant attention from policy-makers, consumers, environmentalists and farmers in Europe. In 1991 the EU introduced legislation to define organic crop production (Regulation 2092/91). A more widespread application of policies to encourage conversion to and continued organic farming came into effect in 1992 when support to organic farming was included as one measure in the agri-environment programme (Regulation 2078/92), an accompanying measure of Common Agricultural Policy (CAP) reform. As a result, policy support for organic farming is now widely available across Europe, in recognition of its contribution to the policy objectives of surplus reduction, environmental management and rural development. Consumer demand for organic food has risen sharply, leading to the active involvement of multiple retailers and substantially higher prices at the farm gate than those received for non-organic produce.

However, the decision of producers to convert to organic farming in any country depends on the realisation of a need to explore alternatives to current farming practices and the confidence that organic farming represents an appropriate direction for the future. It is the intention of this chapter to document the development of the organic sector, review the support policies in the various EU countries in the three key areas of direct payments, market and regional support and information provision, prior to and after the reform of the CAP in 1992 and to discuss likely future directions under Agenda 2000.

METHODOLOGY AND DATA SOURCES

The chapter is based on work carried out as part of a wider research project on organic farming and the CAP. The overall objective of the project was to provide an assessment of the impact of the 1992 CAP Reform and possible future policy developments on organic farming, as well as the contribution that organic farming could make to EU agricultural and environmental policy goals (Foster and Lampkin, 1999; Lampkin *et al.*, 1999).

Data collection was based on standardised questionnaires and national experts in each EU country utilising various published and unpublished data sources and, where appropriate, consultations with key individuals in specific fields. Where possible the data were confirmed from other sources (Lampkin, 1996; Deblitz and Plankl, 1997; Willer, 1998, and various EU Commission documents).

THE GROWTH OF ORGANIC FARMING IN EUROPE

Recent years have seen very rapid growth in organic farming. In 1985, certified and policy-supported organic production accounted for just 100,000 ha in the EU, or less than 0.1% of the total agricultural area (Table 13.1). By the end of 1997, this figure had increased to 2.1 million ha, more than 1.5% of the total agricultural area. By the end of 1998, nearly 3.0 million ha was managed organically, representing a 30-fold increase in 13 years. These figures hide great variability within and between countries. Several countries have now achieved 5-10% of their agricultural area managed organically, and in some cases more than 30% on a regional basis. Countries like Austria, Italy, Sweden and Switzerland have seen the fastest rates of growth. But many others still languish below the 1% level.

Alongside the increase in the supply base, the market for organic produce has also grown, but statistics on the overall size of the market are still very limited. Recent estimates have suggested that the retail sales value of the European market for organic food was of the order of £3-5 billion (ECU 5-7 billion) in 1997 (Lampkin, 1997; Soil Association, 1998).

The majority of the growth of the sector (70% of the expansion in the land area) has taken place since the implementation in 1993 of EC Regulation 2092/91 defining organic crop production, and the widespread application of policies to support conversion to and continued organic farming as part of the agri-environment programme (Regulation 2078/92).

Policy support for organic production prior to CAP reform

Since the mid-1980s, policy makers have increasingly developed an interest in organic farming. Reasons for this include:

- The potential environmental benefits, including soil and habitat conservation, increased biodiversity, utilisation of local and renewable resources, and reduced soil and groundwater pollution.
- The existence of a distinct market for organically produced food (and fibre) as a means by which producers can be compensated for internalising external costs that would otherwise be carried by society.
- The increasing consumer demand for organic products, necessitating an increase in the supply base.
- The lower intensity of organic farming that can contribute to limiting surplus production.
- The opportunities for diversification of farms, and, because of increased labour requirements, the potential contribution to rural development.

Table 13.1   Area of certified organic and in-conversion land in the EU (ha) [a]

	1985	1990	1993	1995	1997	UAA (%) (1997)
Austria	5,880	21,546	135,982	335,865	345,375	10.12
Belgium	500	1,300	2,179	3,385	5,834 [b]	0.42
Germany [c]	24,940	90,021	246,458	309,487	389,693	2.25
Denmark	4,500	11,581	20,093	40,884	64,329	2.36
Spain	2,140	3,650	11,675	24,079	152,105	0.51
Finland	1,000	6,726	20,340	44,695	102,335	4.76
France	45,000	72,000	87,829	118,393	165,405	0.55
Great Britain	6,000	31,000	30,992	48,448	54,670	0.34
Greece		150	591	2,401	10,000	0.19
Ireland	1,000	3,800	5,460	12,634	23,591	0.52
Italy	5,000	13,281	88,437	204,494 [d]	641,149 [d]	4.08
Luxembourg	350	600	497	571	618	0.49
Netherlands	2,450	7,469	10,354	11,486	16,660	0.85
Portugal	50	1,000	3,060 [d]	10,719	12,193	0.31
Sweden [e]	1,500	28,500	36,627	83,326	117,669	3.70
EU-15	100,310	292,561	700,574	1,250,867	2,101,625	1.53

[a] Data at end of year. [b] Wallonia only. [c] Not including uncertified, policy-supported organic land (estimates): 1990:15,000; 1993:180,000; 1995:152,000; 1997:50,000. [d] Estimate. [e] Not including uncertified, policy-supported organic land (estimates): 1990:4,890; 1993:8,000; 1995:3,500; 1997:90,000.
Source: Own data from national agricultural administrations and certification bodies.

This has led to the introduction of support programmes in various countries. The pioneering Danish scheme, introduced in 1987, covered financial assistance to producers during the conversion period as well as the development of a market and of extension and information services (Dubgaard and Holst, 1994). Germany was the first country to introduce in 1989 support for conversion to organic farming in the context of the EU's extensification policy (Regulation 4115/88). France and Luxembourg introduced smaller programmes under the same regulation in 1992. Austria, Sweden and Finland had national conversion support programmes prior to their accession to the EU in 1995. The Swedish and Finnish programmes included support for a state advisory service for organic producers, and Sweden was unique at that time in providing support for continuation of organic production.

ORGANIC FARMING UNDER THE CAP REFORM

Support for organic farming under the agri-environment programme

Under EC Regulation 2078/92, subject to positive effects on the environment, aid is available for farmers who (among other options) introduce or continue with organic farming methods (EC, 1992). The majority of organic farming schemes under this regulation were introduced in 1994 (with some regional variations in Italy and Germany). Austria, Finland and Sweden followed in 1995 on accession to the EU. Greece and Spain did not start until 1996 and Luxembourg only introduced its organic farming scheme under 2078/92 in 1998. Most countries have a uniform national policy, but several (Germany, Spain, Finland, France, Italy, Sweden and the UK) have significant regional variations in rates of payment and requirements.

Nearly all countries (except France and the UK) support not just the conversion to, but also continuing organic production, often with lower payments, recognising the particular costs of conversion. However, Austria, Greece, Sweden and most regions of Italy do not offer higher payments for conversion. Austria has adopted this policy so as not to encourage entrants who are solely interested in the available subsidies (Posch, 1997).

Average rates of support for in-conversion and organic land in 1997 are presented in Table 13.2. Payment rates vary widely between countries, and within countries where regional variations exist. By October 1997, more than 65,000 holdings and nearly 1.3 million ha were covered by organic farming support measures at an annual cost of more than 260 million ECU. Organic farming's share of the total agri-environment programme amounted to 4% of agreements, 5% of land area and nearly 11% of expenditure, the differing shares reflecting in part the widespread uptake of baseline programmes in France, Austria, Germany and Finland (see Chapter 12 for an overview of agri-environmental measures in the EU).

There are reports from several countries that the types of farms converting are skewed towards moderate to low intensity livestock farms, particularly milk production in marginal areas, and farms with mixed cropping (Schulze Pals *et al.*, 1994; Schneeberger *et al.*, 1997). Specialist cropping farms (arable and horticulture) as well as intensive pig and poultry producers, seem to be less attracted by the available payment rates. To address this problem, Denmark introduced in 1997 a supplement of 230-266 ECU ha^{-1} year^{-1} for 3 years for arable farms without milk quota and for pig farms.

Requirements and eligibility conditions

Most schemes (except for Germany and Ireland) allow staged conversion during which experience can be gained and the risk of financially and environmentally damaging mistakes thus minimised. All schemes require organic management of crops to be maintained for at least 5 years. In nearly all cases (except Sweden and some regions in Germany and Italy) organic crop production must be controlled according to Regulation 2092/91 (EC, 1991). The intention in Sweden is to maintain a clear distinction between certified organic production for the market, and organic farming supported for agri-environmental policy reasons. Livestock production requirements are more complex because the Regulation 2092/91 was not extended to cover this aspect until 1999.

In a few countries (e.g. Spain, Greece, Portugal and parts of Italy), the payments are restricted to specific crops and, more commonly, permanent grassland and/or set-aside are excluded from the schemes. Some countries (Austria, Germany, Denmark, Finland, Ireland and Italy) introduced additional environmental requirements. In Ireland and Finland, participation in the main agri-environment programme is compulsory, for which additional payments are made (included in the payment levels shown in Table 13.2). In the UK, additional environmental restrictions have been incorporated into national organic production standards (UKROFS, 1997).

Other restrictions in the eligibility conditions are related to the principle of avoiding double payments for the achievement of the same objective under different agri-environment and mainstream measures, resulting in considerable variation between the schemes.

Regulation 2078/92 makes specific provision for training and demonstration in relation to good organic farming practice (see below).

Effects of the 1992 reforms of mainstream CAP measures

The impact of the mainstream CAP measures on organic farming is a topic that has received relatively little attention from policy makers, despite the potential for conflict between these measures and the agri-environmental measures. In many cases, the assumption is made that there is no difference between organic and conventional producers in terms of eligibility, and that therefore any impacts are likely to be negligible. Very few studies have attempted to quantify any possible impacts, so that the following analysis is unavoidably qualitative in nature.

In most countries, the mainstream measures of the CAP reform are seen as beneficial for the organic sector. Even though organic farmers do not contribute as much to surplus production, set-aside is seen to have potential to support the fertility-building phase of organic rotations during conversion and on arable farms with little or no livestock. This is confirmed by the higher use of set-aside on organic than on conventional farms in some countries. However, in most countries farm size is such that organic producers can qualify for the simplified scheme for arable area payments without the need to set land aside.

Only in a few cases have significant adverse impacts of other mainstream

Table 13.2 Uptake, public expenditure and average payments for organic farming schemes under Regulation 2078/92 compared to all agri-environment options (1997 data)

Country	Land area (ha) (% of total 2078/92 area)	No. of farms (x 1,000) (% of all 2078/92 agreements)	Public expenditure (MECU) (% of total 2078/92 payments)	Lowest conversion payment (ECU ha^{-1})	Highest conversion payment (ECU ha^{-1})	Average payment (conversion and continuing) (ECU ha^{-1})
Austria	246,000 (7.7)	18.5 (4.2)	65.03 (13.0)	217 (forage)	723 (hortic.)	264
Belgium	3,401 (17.9)	0.15 (8.0)	0.88 (23.7)	180 (cereals)	838 (fruit)	259
Germany	229,486 (4.17)	8.42 (1.5)	23.27 (6.0)	127 (cereals)	713 (fruit)	101 [a]
Denmark	50,281 (46.9)	1.45 (18.2)	9.44 (58.2)	87 (forage)	140 (high N)	188 [b]
Spain	50,000 (6.05) [c]	1.5 (5.0)	2.91 (3.9)	90 (forage)	362 (fruit)	58 [c]
Finland	89,403 (4.5)	4.16 (4.7)	21.07 (7.6)	280 (cereals)	1,056 (fruit)	236 [d]
France	41,976 (0.6)	1.55 (0.9)	4.02 (1.4)	106 (forage)	711 (fruit)	96 [a]
UK	29,127 (2.1)	0.3 (1.3)	0.82 (1)	20 (LFA)	101 (lowland)	28
Greece	42,600 (12.2)	0.89 (37.6)	4.25 (31.7)	182 (cereals)	1,217 (fruit)	100
Ireland	nd	nd	nd	337 (cereals)	398 (hortic.)	nd
Italy	308,367 (19.1)	17.12 (14.1)	102.9 (25.6)	185 (cereals)	1,235 (fruit)	334
Luxembourg	n/a	n/a	n/a	173 (all)	173 (all)	n/a [e]
Netherlands	4,640 (14.2)	0.27 (3.6)	0.34 (0.85)	226 (cereals)	837 (hortic.)	73 [a]
Portugal	9,938 (1.8)	0.23 (0.2)	1.18 (1.93)	217 (cereals)	723 (fruit)	119
Sweden	205,185 (11.7)	10.87 (14.5)	25.13 (17.1)	104 (crops)	254 (livestock)	123
EU-15	1,272,064 (5.1)	65.40 (3.9)	261.24 (10.7)	181 (cereals)	1,208 (fruit)	205

nd = no data, n/a = not applicable
[a] Lower payments for continuing organic farming. [b] Includes other forms of support. [c] Estimated. [d] Excludes payment for main agri-environment protection scheme. [e] Scheme was introduced in 1998.
Sources: European Commission and national agricultural administrations.

measures on organic farmers been identified and, in some cases, special provisions have been made to reduce these. The implications are different for existing organic producers as compared to producers in conversion and effects vary according to farm types.

Implications for existing organic producers

Existing organic crop producers have gained, because aid for crops is no longer linked to output, but to the areas of different types of crops grown. The previous system of price support and selling into intervention had been of little relevance to producers operating in an under-supplied premium market. Area aid calculated on the basis of regional average yields represents a bonus to many organic producers, particularly given that organic crop prices have not fallen as much as conventional prices as a consequence of the reforms. The higher level of support payments for protein crops such as beans and peas has also been of particular benefit to organic producers, given the contribution which these crops can make to the nitrogen and livestock feed requirements of the farm system.

However, in some cases the benefits gained may have been at the cost of setting land aside which might otherwise have been producing cash crops that were in demand, given that on most organic farms the fertility-building phase of the rotation is utilised by livestock. In addition, dairy and horticultural producers, who represent a relatively high proportion of organic production in most countries, saw few benefits from the 1992 CAP reform measures, as their crops, grassland and dairy cows were not eligible for support. To the extent that CAP support under the mainstream measures has been incorporated into land and rental values, the impacts may even have been negative.

For many producers operating rotational systems that included periods of fertility-building leys lasting longer than five years, the definition of eligible arable area according to land not in permanent grass (i.e. > 5 years old) at the end of 1991 meant that some of the rotational land would not qualify for support payments when it came back into production. This issue appears to have been resolved in some countries (e.g. UK and Ireland) by allowing producers to rotate eligible area around the farm or (as in the case of Belgium) through greater flexibility about the permanent/temporary nature of fodder area so that farmers could choose the optimum basis for the support regime.

Existing organic livestock producers, who had reduced livestock numbers before 1992, have in many cases received lower livestock quota allocations than would have been the case had they remained under more intensive, conventional management, with a potentially adverse impact on asset values. At the same time, they will have benefited (as other producers, but to a lesser extent given lower stocking rates) from the increases in headage support payments. The adverse impacts relating to lower stocking rates might have been less significant if support for livestock producers were also allocated on an area basis. However, organic producers would not have been as severely affected by the reductions in eligible stocking rates in the early years of the reforms. Indeed, many organic producers have benefited from the higher beef extensification payments for stocking rates less than 1.4 LU forage ha^{-1}.

Impacts on farmers converting to organic production

Negative effects might have occurred for farmers converting to organic farming because arable area payments differentiated by crop types and livestock aid eligibility quotas tend to freeze current production patterns and levels of intensity. This does not go well with the enterprise restructuring which conversion to organic farming entails.

To diversify their rotations arable farmers converting may therefore lose eligibility for some arable area payments, without compensation, but can only get access to some livestock premiums through quota purchase. In some areas, even quota purchase may not be possible because of the regional basis of quota allocations.

Livestock farmers converting are likely to receive livestock payments on fewer animals, yet will not be entitled to arable area payments for any new arable land introduced. There is therefore an active disincentive to producing cereals for livestock feed on the holding itself, in line with organic principles, when crops that have received support can be purchased relatively cheaply from elsewhere. On the other hand, the ability to trade quotas has facilitated the restructuring process during conversion and for many producers the ability to lease out or sell quotas during conversion has proved to be an important means of financing the conversion.

These blockages were seen as more of a problem in countries and regions with larger farm sizes, as the farms were too big to qualify for the simplified scheme, but in many cases creative use of the support measures could reduce the extent of the impacts significantly.

Special provision for organic producers

In order to mitigate negative impacts of CAP reform on organic producers several countries have made special provisions for organic producers or used investment aids and national/regional measures to provide additional assistance. Measures include:

- Less restrictive requirements compared with conventional producers, e.g. later cutting or cultivation dates (e.g. UK), exceeding of the maximum allowance of legume content for set-aside mixtures (UK and Sweden, although in most other EU countries there are no restrictions on the use of legumes in set-aside mixtures).
- Priority in allocation or free access to quota from the national reserve, e.g. suckler cow and sheep annual premium quota (UK), flexibility in choosing the reference time for milk quota (Sweden) and additional allocation of milk quota for organic and in-conversion producers (Denmark).
- Supplementary payments per LU or per ha to Less Favoured Area (LFA) payments under Regulation 950/97 for producers not receiving aid under the organic option under 2078/92 (one region in Italy since 1998).
- Rotation of eligible arable area land around the farm, provided the total area of eligible arable land on the farm remains the same (UK and Ireland) or

greater flexibility about the permanent/temporary nature of fodder area (Belgium).

- Priority status with respect to farm investment grants and loans (two regions in Italy).

In The Netherlands, special provisions exist with respect to the manure law which imply that, if organic farms have trouble meeting the standards for ammonia emissions, especially in poultry and pig-keeping, they will not have to farm within these norms. This exemption has to do with the fact that certain animal housing systems in organic farming (which do have advantages concerning animal health and well-being) may lead to higher ammonia emissions than certain housing systems in conventional farming.

SUPPORT FOR MARKET AND REGIONAL DEVELOPMENT

Legislation defining organic production (Regulation 2092/91)

Standards for organic agriculture are intended to promote consumer confidence and prevent the undermining of the market through fraudulent trading. In addition, because of their on-going revision and improvement, they 'reflect the progress and achievements' which are leading to an increasingly sustainable organic system (Geier, 1997). In some countries, several private sector bodies certify organic producers, each according to their own set of standards. The introduction of legislation defining organic agriculture is seen as a means to avoid confusion among consumers, protect the producer and hence assist the development of the market for organic food.

In 1991, the EC took up this challenge and introduced a common standard for organic crop production through the Regulation 2092/91 (EC, 1991). A Regulation to define organic livestock production was agreed in 1999. Prior to this, five countries (Austria, Denmark, Spain, Finland and France) had legal definitions of organic production. In most other countries (except Greece where no standards existed), standards were developed and operated by the private sector.

In some countries (Austria, Denmark, UK, some regions of Germany and Italy) producers receive some additional payments as part of the support schemes under Regulation 2078/92 to assist with the costs of certification. In other countries (Spain, France, UK, Ireland and Luxembourg) the costs are indirectly supported through grants to the inspection and certification bodies.

Whether or not the introduction of legislation has achieved the goal of assisting market development is a matter of debate. Regulations are seen by some to provide a more effective and objective control than a system that is essentially responsible for monitoring itself. In addition, the introduction of a single regulation, as opposed to a variety of standards, is likely to reduce confusion and increase consumer confidence. In Denmark, for example, the state regulation on organic farming is thought to be an important reason for the high consumer confidence in organically produced food in Denmark (Willer, 1998). However, the introduction of the EC legislation has had a differing impact on the number of approved certification bodies. In France the number has fallen to three, in Ireland it has risen to three and in Germany it has increased from six to over 50.

On the other hand, the private sector standards were largely developed with the involvement of the organic movement, taking regional variations of the conditions of the organic sector into account. Standards also can offer more guidance to the producer (and other operators) on how to fulfil their requirements. In the private sector, international standards have been developed by the International Federation of Organic Agricultural Movements (IFOAM, 1998).

The development of easily recognisable, common logos may play a more important role than legislation alone in improving consumer confidence and reducing confusion to a minimum. Denmark's single logo, for example, is recognised by half of Danish consumers, who seem to have high levels of confidence in the system. There has been a similar experience in Switzerland. This contrasts markedly with experiences from Germany, which, despite being one of the largest markets for organic produce, is only now introducing a common logo as a joint initiative between the organic producer umbrella organisation (AGÖL) and the national food marketing agency (CMA).

Support for improving the processing and marketing conditions of organic produce

Regulation (EEC) 2078/92 and other conversion support programmes have had a significant impact on the development of the supply base of organic food in most countries where they have been implemented. At the same time, developing the marketing structure and establishing new retail outlets is of key importance, if the sector is to be able to deal with this expansion and if premium prices are to be maintained (Hamm and Michelsen, 1996).

Policy support for marketing and processing in organic farming varies considerably. A number of countries, for example Austria, Germany and Denmark, have legislation, grants and/or support programmes available on a national level through which organic enterprises can and have received funding.

On an EU level, one of the established priorities for the application of Regulation 866/90 on improving the processing and marketing conditions for agricultural products (EC, 1990) is investment relating to organic farming products. In eight countries (Austria, Germany, Spain, Finland, UK, Italy, The Netherlands and Sweden), the organic sector has benefited under this regulation. Restrictive eligibility requirements have been identified as one of the barriers to greater uptake of EU support. In Germany, this resulted in the introduction of the national programme to support the marketing of products 'produced according to specific production rules' that takes the specific requirements of the organic sector, such as reduced turnover and smaller number of members for organic producer groups, into account.

Experiences in Denmark suggest that a more market-oriented approach to organic aid schemes can promote the development of a diverse marketing structure, provide help in entering into mainstream marketing, and help overcome problems such as discontinuity of supply and lack of widespread distribution.

Organic farming and regional development policy

Organic farming can help to meet many of the goals of regional development programmes, combining a sustainable model of agriculture with the encouragement of local production, processing and consumption patterns and local marketing networks, leading to an increase in the 'economic value' of a region (Vogtmann, 1996).

Particular organic farming projects have received support under Objectives 5b and 1 of the EU Structural Funds in ten countries (Austria, Belgium, Germany, Denmark, France, UK, Ireland, Italy, The Netherlands and Sweden). These projects cover a variety of activities, including direct marketing, promotion of regional products, research, technical advice and training. In nine countries (Germany, Spain, France, UK, Greece, Ireland, Italy, The Netherlands and Portugal), organic projects have received funding through the EU LEADER Community initiative. Table 13.3 presents examples of LEADER projects in which organic farming has featured and their main activities. Significant regional

Table 13.3 Examples of organic initiatives that have received LEADER funding in the EU

Country	Project	Main activity/objectives
Germany	Organic Milk Marketing, Saarland	Marketing of regionally produced organic milk in co-operation with a farm dairy, in order to encourage other enterprises to convert
	Wulkow, model of sustainable development, Brandenburg	This commune was set up by about ten villages as a model of 'global ecological development'. Its activities include diversification of agriculture to organic production
Spain	Organic crops and livestock rearing, Navarre	To adapt local farm produce to market trends, and protect the environment
France	European Ecological Centre, Terre Vivante, Rhône-Alpes	The aim of the centre is to present technical solutions, taking account of environmental protection. At the centre practical illustrations are offered, including organic vegetable gardening
	Du bleu bio à Lajoux, Franche-Comté	Organic production as a means to maintaining a cheese co-operative and its retail outlet
UK	Llanerchaeron Home Farm, Wales	A model farm which will integrate gardens, woodland projects and aims to convert to organic status. Biodynamic farming principles will be demonstrated

Table 13.3   Continued

Country	Project	Main activity/objectives
Greece	Ecological farm of Kia Vrissi, Central Macedonia	The farm houses a centre for research, experimentation and training in the area of organic farming
Ireland	Programme to develop organic agriculture in south and west Ireland (IOFGA)	Promotion of organic farming in conjunction with the Dutch organic group Agro Eco
Italy	Alce Nero co-operative, Marche	An integrated system for the harvesting, processing and marketing of organic cereals, as well as organised activities to promote organic methods
	Organic Farming and Rural Ecodevelopment, ARPA, Sardegna	Includes provision of advice to farmers, investments in small food processing plants, marketing promotion training of farmers and retailers
Netherlands	Marketing organic products in short channels: the EKO-Boerderijen Route, Drente	Development of an 'ecological cycling route' linking organic farms in order to develop the direct sale of organic products
Portugal	Support for organic agriculture in Beira Interior	Support fo ARAB (Associaçao Regional de Agricultores Biológicos) to promote organic agriculture

Source: Lampkin *et al.* (1999).

development initiatives for organic farming outside EU legislation have occurred in the Rhône-Alpes region of France.

Some regional development schemes include support for marketing and processing activities in the organic sector, mainly aimed at small-scale projects. Such schemes have been particularly successful in Germany in helping develop regional marketing networks, overcoming the problems of a small organic sector and encouraging the entry of new operators. The impact of grant aid on the organic sector and consequently the development of the region can be significant as evaluations of the Irish Objective 1 programme have shown (Fitzpatrick, 1997).

INFORMATION SUPPORT

To provide information and advice about organic farming is very important, because, as with other low input systems, inputs are replaced through management (Lockeretz, 1991). Only with access to suitable information can farmers who are considering conversion make an informed choice about the implications for their particular circumstances. In addition, the argument can be made that, as it was governmental advisory services that promoted intensification of agriculture, a shift towards more environmentally friendly farming systems should now receive similar support.

Everywhere, organic farmers and growers and their organisations are a very important source of information to those interested in organic production, and in seven EU countries the producer organisations receive public support in recognition of this role. Regional groups of producer organisations operating in ten countries facilitate the sharing of experience among organic farmers, act as a focal point for regional market development and give social support to the producers. The organisation of such groups is mostly voluntary, apart from four countries (Austria, Belgium, UK and Italy) where the co-ordination of regional groups received support under the Objective 5b of the structural funds.

Countries have given some direct support to information and advisory services and demonstration farm networks under national advisory support systems as well as under European Regulations (2078/92 and Objective 5b), mainly with the aim to increase the uptake of conversion support. In addition, indirect support to the provision of information has been given through training and research programmes including the second, third and fourth framework programmes from the EU.

Like the organisation of advisory services in general, organic advisory services vary considerably between the countries, ranging from widely available, subsidised advice for organic and in-conversion producers through the main agricultural extension service, partly supported under the agri-environment programme (e.g. Finland, Denmark and Germany) or private bodies (e.g. UK) to countries with an almost fully commercial basis for professional advice on organic production (Spain, Greece, Ireland and The Netherlands). Specific conversion information programmes in Sweden (under Regulation 2078/92) and the UK (national programme) have proved very popular.

The increasing involvement of the general agricultural extension services in organic agriculture is likely to increase the availability of advice for interested conventional producers. Organic farmers, on the other hand, are sceptical whether mainstream agricultural extension can provide information that is specifically tailored to organic systems. Personal role conflicts can arise if one person has to advise on organic as well as non-organic production methods (Gengenbach, 1996; Fersterer and Gruber, 1998).

The Netherlands set up a network of demonstration farms under a national regulation, whereas Belgium, Denmark, Portugal and Sweden used the framework of the agri-environment programme to support demonstration farms, for experienced and commercial organic producers willing to try new techniques and to show their farm to visiting groups. Through a network, the facilities are publicised, visits are co-ordinated and the farmers receive some form of compensation for their effort and support in developing information material about the farm.

Despite the traditional role of farmers in developing organic farming, more recently research has been recognised as very important for the development of the organic industry (Lindenthal *et al.*, 1996). In most countries there are increasing research activities and five countries (Germany, Denmark, Finland, UK and Sweden) have specific national research programmes in organic farming. The

EU funded a total of ten organic farming projects under the second, third and fourth framework programmes. However, there is a need for research to maintain links with the whole of the organic industry to ensure effective two-way communication of research needs and research results, and not to revert to the old R&D model, where researchers are seen as the only source of innovations.

PROSPECTS UNDER AGENDA 2000

Although growth trends in individual countries have varied considerably, with periods of rapid expansion followed by periods of consolidation and occasionally decline, overall growth in Europe has been consistently around 25% per year for the last ten years, i.e. exponential growth. There is no indication yet of this rate of growth declining. Similar growth rates are reported for organic farming in the US.

If these growth rates are projected forward to 2010, this gives some indication of the potential significance of organic farming within a relatively short period. Assuming a starting point of 1.6% of western European agriculture in 1997, continued 25% growth each year would imply a 10% share by 2005 and nearly 30% by 2010. Faster growth at 35% annually would lead to 18% by 2005 and 80% by 2010, but this seems highly unlikely on the basis of past performance. A slower rate of growth of 15% each year would still result in just under 5% by 2005 and 10% by 2010.

A total of 10% of western European agriculture, whether achieved by 2005 or 2010, may still sound like a small proportion, but it is very significant in absolute terms. It represents nearly 14 million ha and more than 800,000 farms, compared with the current total of 100,000 holdings, on which potential benefits of organic farming for the environment could be realised.

This level of growth has significant implications for the provision of training, advice and other information to farmers, as well as for the development of inspection and certification procedures. It also has major implications for the development of the market for organic food, as it progresses from niche to mainstream status, with a likely retail sales value in 2005 of EUR 30-50 billion.

Projections into the future based on past performance are not sufficient to realise the potential of organic farming. There is no guarantee that the rates of growth seen in the past will continue, and the normal expectation would be for rates of growth to decline eventually. A better understanding of the factors lying behind the growth of organic farming, and in particular the differences between countries, is needed.

In many respects, the development of organic farming has parallels to the traditional adoption-diffusion model for the adoption of innovations. Over time, the individualistic and socially-isolated innovators or pioneers are followed by the early adopters typified as community opinion leaders, to be followed in turn by the majority of farmers. In many countries, this shift can be clearly seen. However, the rate at which this change takes place depends on the complexity of the innovation, and the adoption of organic farming is clearly a complex innovation (Padel, 1994).

The adoption-diffusion model does not seem to explain why the development of organic farming may be characterised by periods of stagnation followed by very rapid growth. A possible explanation for this is that farmers must perceive the need to change before significant change will take place. A period of financial prosperity, as, for example, UK farmers experienced between 1992 and 1995 due to the CAP reform package combined with the low value of the pound, was clearly not conducive to change. The reversal of circumstances since 1996, with the BSE crisis, the high value of the pound, and falling prices and agricultural support levels, has changed this perception dramatically. Similarly unsettling circumstances have arisen in other countries, for example in eastern Germany following re-unification, and in Austria on accession to the EU, leading to large increases in the number of farms converting.

The perception of the need for change has to be accompanied by a conviction that organic farming is a suitable alternative. This requires a high degree of confidence building because of the perceived financial, social and psychological barriers to conversion. It is not simply a case of 'more profits = more farmers' as many might argue. Our preliminary assessment of this issue indicates that four key factors are involved:

- policy signals from government and other policy-related institutions;
- market signals from the food industry;
- access to information;
- the removal of institutional blockages or antagonisms.

Assuming that this analysis is correct, it provides a new basis for future policy development to encourage organic farming with a focus on integrated action plans rather than single measures like the direct organic support schemes under the agri-environment programme.

In this respect, potentially the most important Agenda 2000 reform is that to consolidate all existing agri-environment, rural development and structural policies into a single rural development regulation. By integrating all these measures into a single regulation, and requiring Member States to produce customised rural development plans, the reform has significant parallels to the 'action plans' for the development of organic farming. Action plans set a clear target for expansion of the producer base and integrate a variety of policy measures to achieve their goals, such as payments for production, harmonisation of certification procedures, market support, as well as support for advisory services, training and research and development. All the Nordic countries, The Netherlands, France and several *Bundesländer* in Germany have developed integrated policy programmes or action plans for the future development of the organic sector, setting targets of between 5 and 10% of area farmed organically by 2000 to 2005.

Of these action plans, the most detailed, and the one that has been most successfully implemented, is that of Denmark. In March 1995, the Minister of Agriculture and Fisheries produced the 'Action Plan for Promotion of Organic Food Products in Denmark'. A revised version was published in 1999 that includes recommendations to make conversion attractive, as well as securing the demand for organic produce and intensifying research, development and education and removing barriers for a sound organic development.

It is increasingly recognised that organic farming can meet many of the goals of rural development policies. For example, premium markets and opportunities for adding value through processing offer potential for improving incomes and employment in rural areas. So there is a strong case for the Agenda 2000 Rural Development Plans to include a specific focus on organic farming.

The commodity regime under the Agenda 2000 reforms is in general terms likely to be favourable for organic farming, as it represents a continuation of the process of change begun in 1992. However, it can be argued that organic farmers should be exempted from compulsory set-aside, but the option of voluntary set-aside should be retained. They are contributing already to surplus reduction by reducing crop yields per ha and by reducing the proportion of surplus crops in the rotation. At the same time, it makes no sense to reduce further the output of crops that are in very high demand from consumers. Organic farmers should be considered a high priority case for the allocation of any additional livestock quotas, and organic farming should be included as a management option wherever higher area payments in Environmentally Sensitive or Less Favoured Areas are introduced.

CONCLUSIONS

Organic farming has developed rapidly in Europe since 1993, against the background of significant policy support, mainly in the form of direct payments under agri-environmental support and indirectly through support for marketing and processing activities, certification, and information-related activities. The prospects are for continued growth, which may lead to 10% of EU agriculture managed organically in 2005, with organic farming moving from 'niche markets' to become a mainstream part of the agricultural sector. However, in order to achieve this, integrated policy support in all three key areas (production support, support for regional and market development and support for knowledge networks) is essential. Agricultural support programmes need to be designed taking into account the special characteristics of the sector, such as production standards, farming and market structure, certification requirements and information needs.

Acknowledgements

The research reported in this chapter was carried out with financial support from the Commission of the European Communities' Agriculture and Fisheries (FAIR) specific RTD programme, FAIR3-CT96-1794, 'Effects of the CAP Reform and possible further development on organic farming in the EU'. It does not necessarily reflect the Commission's views and in no way anticipates the Commission's future policy in this area.

REFERENCES

Deblitz, C. and Plankl, R. (1997) *EU-wide Synopsis of Measures According to Regulation (EEC) 2078/92 in the EU.* Federal Agricultural Research Centre, Braunschweig.

Dubgaard, A. and Holst, H. (1994) Policy issues and impacts of government assistance for conversion to organic farming: the Danish experience. In: Lampkin, N.H. and Padel, S. (eds) *The Economics of Organic Farming.* CAB International, Wallingford, pp. 383-391.

EC (1990) Council Regulation (EEC) No. 866/90 on improving the processing and marketing conditions of agricultural products. *Official Journal of the European Communities* L91 (06.04.90), 1-6.

EC (1991) Council Regulation (EEC) No. 2092/91 of 24 June 1991 on organic production of agricultural products and indications referring thereto on agricultural products and foodstuffs. *Official Journal of the European Communities* L198 (22.7.91), 1-15.

EC (1992) Council Regulation 2078/92 on the Introduction and Maintenance of Agricultural Production Methods Compatible with the Requirements of the Preservation of the Environment and the Management of the Countryside. *Office for Official Publications of the European Communities,* Luxembourg.

Fersterer, S. and Gruber, A. (1998) *Beratungsstrukturen für die Biologische Landwirtschaft in Österreich im Vergleich mit Ausgewählten Europäischen Ländern.* MECCA-Environmental Consulting, Vienna.

Fitzpatrick (1997) *Mid-term evaluation: Development of Organic Farming (Measure 1.3 (e)).* Fitzpatrick Associates, Dublin.

Foster, C. and Lampkin, N. (1999) *European Organic Production Statistics, 1993-1996.* Organic farming in Europe: Economics and Policy, 3. University of Hohenheim, Hohenheim.

Geier, B. (1997) Reflections on standards for organic agriculture. *IFOAM-Ecology and Farming* 15, 10-11.

Gengenbach, H. (1996) Fachberatung biologisch-dynamischer Landbau in Hessen. *Lebendige Erde* 1996(3), 237-243.

Hamm, U. and Michelsen, J. (1996) Organic agriculture in a market economy: perspectives from Germany and Denmark. *Fundamentals of Organic Farming, Proceedings 11th IFOAM Conference,* IFOAM. Tholey-Theley, Germany, pp. 208-222.

IFOAM (1998) *Basic Standards of Organic Agriculture.* International Federation of Organic Agriculture Movements, Tholey-Theley, Germany.

Lampkin, N. (1996) *Impact of EC Regulation 2078/92 on the Development of Organic Farming in the European Union.* Working Paper, No. 7, Welsh Institute of Rural Studies, Aberystwyth.

Lampkin, N. (1997) Opportunities for profit from organic farming. Paper presented at the *RASE-Conference: Organic Farming - Science into Practice,* RASE, Stoneleigh.

Lampkin, N., Foster, C., Padel, S. and Midmore, P. (1999) *The Policy and Regulatory Environment for Organic Farming in Europe.* Organic Farming in Europe: Economics and Policy, Vol. 1 and 2, University of Hohenheim, Hohenheim.

Lindenthal, T., Vogl, C.R. and Hess, J. (1996) *Forschung im Ökologischen Landbau.* Bundesministerium für Land- und Forstwirtschaft, Vienna.

Lockeretz, W. (1991) Information requirements of reduced chemical production methods. *Amercian Journal of Alternative Agriculture* 6(2), 97-103.

Padel, S. (1994) *Adoption of Organic Farming as an Example of the Diffusion of an Innovation - A Literature Review on the Conversion to Organic Farming.* Discussion Paper, 94/1, Centre for Organic Husbandry and Agroecology, Aberystwyth.

Posch, A. (1997) Making growth in organic trade a priority. In: *The Future Agenda for Organic Trade, Proceedings of the 5th IFOAM Conference for Trade in Organic*

Products. (T. Maxted-Frost). IFOAM, Tholey-Theley and Soil Association, Bristol, pp. 9-12.

Soil Association (1998) *The Organic Farming Report 1998.* Bristol.

Schneeberger, W., Eder, M. and Posch, A. (1997) Strukturanalyse der Biobetriebe in Österreich. *Der Förderungsdienst - Spezial* 45, 1-12.

Schulze Pals, L., Braun, J. and Dabbert, S. (1994) Financial assistance to organic farming in Germany as part of the EC extensification programme. In: Lampkin, N.H. and Padel, S. (eds) *Economics of Organic Farming: An International Perspective.* CAB International, Wallingford, pp. 411-436.

UKROFS (1997) *Standards for Organic Food Production.* United Kingdom Register of Organic Food Standards, London.

Vogtmann, H. (1996) Regionale Wirtschaftskreisläufe - Perspektiven und Programme für die Landwirtschaft in Hessen. Paper presented at conference *Für den ländlichen Raum und seine Menschen.* Landessynode der Evangelischen Kirchen Kurhessen Waldeck, Hofgeismar, pp. C10-C18.

Willer, H. (ed.) (1998) *Ökologischer Landbau in Europa - Perspektiven und Berichte aus den Ländern der EU und den EFTA Staaten.* Ökologische Konzepte, Bd 98. Deukalion Verlag, Holm.

Part III

Institutional Factors in Reorienting Agriculture

Do 'Soft' Regulations Matter? 14

Flemming Just and Ingo Heinz

INTRODUCTION

The growing number of statutory rules and plans currently requires increasing administrative efforts regarding implementation, control and monitoring. New policy instruments are formulated, which are characterised by being voluntary in nature, compared to traditional compulsory regulations like prohibitions, orders and taxes. The European Commission now calls for increased dialogue and intensified co-operation between European Union (EU) institutions, Member States, regional and local authorities, industries and citizens. To a certain extent this may be seen as a shift from a top-down to a bottom-up approach with more emphasis on integrating industrial and business sectors, including the farming community, in problem-solving (Wurzel, 1993). Article 19 of Council Regulation 797/85 on 'Improving the Efficiency of Agricultural Structures', for example, introduced special national schemes in environmentally sensitive areas and authorised Member States to subsidise farming practices favourable to the environment. Much emphasis has been laid on education of farmers and influencing attitudes assuming that greener attitudes will result in a greener behaviour. Outstanding is the EU Regulation 2078/92 with its many different programmes for environmentally sound production methods in agriculture and measures for securing landscape and biodiversity.

The cost-effectiveness of policy measures increasingly also becomes an important decision criterion for assessing agricultural and environmental policies. Voluntary measures generally seem to meet these requirements. One important reason is that 'hard' regulations like prohibitions, orders and taxes are often ineffective in achieving environmental goals in agriculture This is because they are usually applied to inputs into farming, and the consequences for the environment are highly contingent (affected, for example, by the behaviour of farmers, animals, machinery and the weather) and dependent on complex processes (such as the transport of leachates through the soil). The impacts on the environment also tend to be diffuse. Consequently, the polluter-pays principle can be applied only to a limited extent in the context of agriculture.

Below, the different policy instruments available in agri-environmental regulation will be considered. A first claim in this chapter is that there are strong indications that 'soft' regulations - under certain circumstances and in combination with other policy instruments - can contribute to an improvement of nature (landscape, biodiversity) and environment (soil, water, air).

A second contention is that voluntary instruments are very dependent on national/regional political and institutional conditions. In Germany, for example, voluntary co-operative agreements between farmers and water supply companies and/or nature conservation groups are very widespread, while in Denmark they are almost non-existent. Water supply problems - like other environmental problems - are here predominantly solved through direct state intervention, while nature and landscape regulations are voluntary and initiated by farmers, the farmer-governed advisory system, and state and county authorities (Heinz *et al.,* 1995; Just, 1996).

AGRI-ENVIRONMENTAL INSTRUMENTS

Intervention by society on agri-environmental issues may take place through different types of regulatory instrument. Some are initiated by the state, others by the state in close collaboration with organisations, and others are totally an expression of self-regulation. The literature on instruments is growing, and the typologies likewise. Bruckmeier and Teherani-Krönner (1992), for example, distinguish between statutory regulation, financial incentives, moral persuasion and self-regulation. One of the latest examples is Vedung's typology which boils instruments down to three basic combinations: regulative, economic and communicative (Vedung, 1998).

As a rough distinction the latter typology is reasonable. If it is applied empirically, however, it will soon be realised that it does not catch the nuances of different political problem-solving mechanisms. For instance, public authorities do not regulate solely through Acts, prohibitions, taxes and the like. Different authorities play independent roles in the interpretation and implementation of rules. To reflect this complexity, therefore, the number of instrument types used below will be extended to six. The schema has primarily been developed to understand modes of environmental regulation of agriculture, but the classification may also apply to other sectors.

Moving from the top to the bottom in Table 14.1, the six types represent decreasing state intervention and increasing voluntary self-regulation. The first four instruments are characterised by being compulsory and implemented by state/public agencies, whereas the last two are voluntary and implemented through organisations, advisory services, networking, etc.

Many regulations are a mixture of more forms. Measures under Regulation 2078/92, for example, can be seen as an economic instrument as farmers are compensated for applying environmentally sound practices. However, for the individual farmer and for the success of the measure it is more important that the scheme be voluntary. The payment provided is an incentive for making this normative scheme with contract farming attractive. Information and influence from farm unions and advisory service may also be incentives. For the end-users the main difference between the different forms of regulation is whether they are compulsory or voluntary, as the former has an (almost) immediate and unavoidable effect.

Hierarchies also play an important and partly independent role in interpreting statutory norms. Hierarchies are public authorities either at state level or at county and municipal level with the task of controlling and monitoring statutory norms and plans. Furthermore, hierarchies are involved in administrative decisions which as a consequence will regulate agri-environmental conditions. An example of the latter is the Danish Environment Complaints Board. As the final court of appeal it has an important role in the interpretation of the Environment Protection Act.

Table 14.1   Modes of environmental regulation of agriculture

Instrument	Actor	Medium	Implementation form
Statutory	State	*Compulsory:* power	Statutory norms, acts, prohibitions, permits
Hierarchical	State, local authorities	*Compulsory:* power	Control monitoring administration
Planning	State, local authorities	*Compulsory:* power	Plans
Economic	State	*Compulsory:* money *Voluntary:* money	Taxes Subsidies
Communicative	State, EU, private	*Voluntary:* communication value commitment, voluntary institutions	Information, education, consulting, voluntary associations, contract farming
Self-regulatory	Private	*Voluntary:* value commitments, contract	Voluntary associations, voluntary contracts

Another means of public regulation is physical planning which is increasingly setting limits on the use of open land in order to secure the environment, nature, landscape and biodiversity. Regional plans and local plans are the most direct way of regulating farming activities, but designation of areas is also contained in many nature and environment Acts and through EU directives and international conventions.

Economic instruments may be taxes or subsidies. Until now subsidies have been much used, for instance as support to investments in storing facilities for slurry or as an incentive to make nature management contracts. Taxes and fees have been applied to a limited extent only, and even then are often used to fund other environmental measures or the proceeds are reverted to agriculture. Other economic instruments discussed in the literature are tradable permits, insurance against yield risks and special compliance incentives (Oskam *et al.*, 1998).

Subsidies fit at best into the 'voluntary' approach, because farmers can decide independently whether they act or not. The impacts of taxes on pesticides or on chemical fertilisers are uncertain due to the farmers' attempts to avoid the payment or to find loopholes. The acceptability of taxes as a compulsory regulation is generally low. Furthermore, taxes are insensitive to differences in local circumstances with respect to farming practices and natural environment, so that the cost-efficiency of this economic instrument is uncertain. However, there are several possible variants on the tax instrument, including levies on the nitrogen content of chemical fertilisers or on the surplus resulting from the nitrogen balance. The primary purpose of a tax can be to reduce the use of agro-chemicals (a regulatory levy) or to generate funds for environmental measures (a financial levy). The efficiency of charging financial levies depends particularly on the kind of reimbursement to the farmers and how this actually changes their behaviour. Consequently, further research is needed to investigate the impact of economic instruments, and to explore under which circumstances they might be effective and economically efficient (Centre for Agriculture and Environment, 1998).

Communicative instruments have been in the forefront in many countries as an effort to stimulate collaboration between the sector agencies and the private interest organisations. The underlying hypothesis seems to be, on the one hand, that the effects of using compulsory regulations are not always commensurate with the costs involved (for example, because of political resistance from the farm lobby). On the other hand, voluntary measures offer themselves as a relatively cheap instrument based on the assumption that farmers will be more collaborative and positive if the measures are taken on a voluntary basis (Cox *et al.*, 1990). Besides it is assumed that a change towards greener attitudes will lead to greener behaviour. There is an emphasis on changing the outlook of, and working through, intermediary organisations, such as farmers' unions, extension services and farmers' networks.

Regarding pesticides, policies in most countries are primarily focusing on awareness building and the participation of farmers in spraying courses (OECD, 1996). Besides that, many countries attach much emphasis to the role of the extension service in turning farmers from a productivist orientation to more sustainable methods (Bruckmeier and Teherani-Krönner, 1992).

Self-regulation takes place when no public authority at all is involved. Organic farming is an example: in the first phases of the organic movement self-regulation was a sufficient instrument to make farmers adhere to the organic rules (Hamm and Michelsen, 1996). Market-based contract farming is another type of self-regulation, for instance if a food processor sets strong limitations on producers' use of pesticides. Co-operative agreements between farmers and water supply companies and/or nature conservation organisations, as important examples of voluntary contracts, are discussed in detail below.

These various regulatory forms are not mutually exclusive. Actual regulatory measures often include combinations of several regulatory forms and different means of implementation. An example is the regulation of pesticide use in Denmark. It combines statutory norms, an economic instrument (fee), and communicative measures in the form of information, training and the involvement of the advisory service. In the 1980s the regulation of pesticides laid emphasis on education and the role of advisers. There was also a modest levy which had an insignificant financial impact on farm businesses. From the mid-1990s pesticides policy has been much more interventionist due to great public debate about water quality (Just and Nielsen, 1996). Communicative forms are still involved, but education and advising are so much integrated in the education of young farmers and in the normal behaviour of production planning that they are not recognised as specific instruments. Instead there is major awareness about the intensified role of compulsory methods like an increased levy and a restrictive policy for the registration and prohibition of pesticides (Svendsen *et al.*, 1997).

In theory, there should be a close link between policy objectives and the instruments chosen, and thus between implementation and effectiveness. However, the implementation conditions vary from policy area to policy area which makes it very difficult to generalise the relationship between goals and instruments. The theory of bargaining has shown that negotiations between polluters and victims tend to solutions based on what is best for the economy (Kuckshinrichs, 1990). This particularly applies to water supply companies and farmers. While water supply companies are interested in keeping compensation payments as low as possible, the farmers try to maximise the difference between the costs of changing to more environmentally friendly production methods and the compensation payments they receive. However, it is not given that the most (cost) efficient instrument will be chosen. Contrary to this, in many cases efficiency only plays a limited role. Political actors' self-interest, distributional consequences, the strength of sectoral interests, the regulatory traditions, the political institutions, and organisational arrangements often seem to be more important factors in explaining the choice of instrument (Daugbjerg, 1998).

With special regard to voluntary contracts, co-operative agreements do not fit into the pattern of the 'polluter-pays' principle: the external effects caused by agriculture are not internalised into the cost functions of the farmers. But the victims, namely water supply companies and/or nature preservation groups, are prepared to compensate the polluters, because they have something to gain from it. The farmers actually control the property rights, as far as the use of nature (especially water resources) is concerned, even though the law may suggest differently. It is particularly significant that the actual establishing of rights to use nature is determined not solely by law, but also by market forces which aim to reach the best economic results. The reason for this surprising phenomenon is the high costs for the victims of successfully claiming for damages. The problem lies in the fact that proving environmental damage against individual farmers is extremely difficult. Consequently, the pursuit of the 'polluter-pays' principle would, in most cases, lead to welfare losses, because the follow-up costs would exceed the benefits of doing so (Heinz, 1998).

The assumptions to be tested are:

- Attitudes influence behaviour, hence information and education are relatively cheap instruments and will stimulate farmers' commitment to use more environmentally friendly production methods.

- Farmers' institutions are positive towards voluntary regulations as they see them as a challenge to farmers' craftsmanship and a way to avoid tougher regulations. Hence these institutions will play an active role in persuading farmers to engage in voluntary regulations.
- Co-operative agreements between farmers and water supply companies and/or nature conservation organisations can improve the implementation of environmental goals, because they are based on the self-interests of the participants.

ATTITUDES AND BEHAVIOUR

According to Søgaard (1997) three layers of conditions need to be met if a farmer is to act in an environmentally friendly way:

- The farmer should be able to recognise adverse environmental effects of conventional farming (the cognitive dimension).
- The farmer should be emotionally affected by this understanding (the affective dimension).
- The farmer should accept it as a personal responsibility to do something about the environment (the Kantian ethics dimension).

These conditions stress the vital role of individual farmers' attitudes as one of the primary conditions for a change of environmental behaviour. A major Danish survey (Bager and Søgaard, 1994) and other surveys based on questionnaires sent to farmers (Beus and Dunlap, 1990; Vanclay and Lawrence, 1995) do suggest a link between attitudes and behaviour. However, methodologically it is a very difficult and sensitive task to verify the connection. Organic farming is an example of a clear correspondence between attitudes and behaviour. But in general it is difficult to detect a clear correlation. In most countries authorities try to change farming practices through seeking to inculcate appropriate attitudes, but surveys confirm the Danish findings: that attitudes only slightly motivate behaviour. An Australian survey found no differences in attitudes between members and non-members of land-care groups. Moreover, it appeared that farmers using new and more environmentally and nature friendly production methods were not more green than non-users. The conclusion was that farmers in general had a strong ethical attitude towards landscape and nature, and that other factors like resources and the farmer's risk perception were more salient barriers to the adoption of new methods (Vanclay, 1992).

There has been some change in farmers' attitudes in recent years. Measured on the so-called Likert-scale, where 1 is very alternative and 5 is very conventional, the environmental attitudes of Danish farmers typically ranged between 3 and 4 in 1993, whereas in 1988 the range had been between 4 and 5. The general greening of attitudes in society most likely plays a significant role. Very contentious topics of 15 years ago are now characterised by broad consensus.

However, the change in behaviour towards more environmentally friendly production methods has been deeper than the change of attitudes, because farmers have had to adjust to orders, prohibitions and new techniques.

Although it is first and foremost a change in attitudes and behaviour among conventional farmers, it should be noted that attitudes amongst organic farmers are also changing somewhat. An investigation in 1995 of converters showed that, although environmental concerns still play the same decisive role in the motivation for conversion, growing importance is also attached to aspects of self-realisation and professional challenge. With many new converters in both 1996 and 1997, most recruited from the ranks of conventional farmers, a convergence in outlook and background is expected between conventional and organic farmers (Michelsen, 1997b).

Farmers' attitudes may also gradually become greener as they come to terms with concrete changes in behaviour. By nature most farmers do not ride the first fashion wave, but require demonstrable proof. If a new method appears to be successful, the attitude towards nature keeping and more environmentally friendly farming practices should gradually change.

INSTITUTIONS

Even though the three criteria developed by Søgaard correctly point to the vital role of the individual farmer, institutional factors (study groups, extension service and agricultural organisations) also affect the attitudes of farmers. With the growing interest in the new institutionalism the role of meso-level and micro-level institutions in agriculture has also attracted attention. Recent studies emphasise the very important role of institutions - such as research, extension service, organisations, networking (e.g. study groups), pioneers, commercialization - in farmers' conversion to organic farming or other kinds of more environmentally friendly production methods. Michelsen (1997a) argues that the institutional set-up is a key vehicle for furthering policy objectives:

> The institutional design reflects the discourse upon which the political regulation is built and thus the openness of the institutions towards different ways of defining problems and hence towards helping different social groupings in promoting their ways of thinking and acting (Michelsen, 1997a, p. 11).

Conversion to organic farming or other more sustainable farming practices represents such a shift in perception. If changes of this kind take place to a significant extent, it will inevitably result in a different institutional design reflecting the new discourse. Evidence for this comes from Assouline *et al.*'s (1996) analysis of why organic farming has been successful in some areas. Based on a comparison between successful areas in Denmark and France, they conclude that institutions matter: the conversion process to organic agriculture needs to be supported by the involvement of technical advice; supporting measures must be in keeping with local farming circumstances; and specific organic organisations need to be built.

Study groups

In the deliberate use of farmers' networking as an important teaching instrument, study groups/experience groups are used as an element in fulfilling nature and environment goals. In Denmark there were 621 crop protection groups in 1995 with more than 4,300 participants, or one in seven full-time farmers (Svendsen *et al.*, 1997). In The Netherlands study groups with an attached extensionist are more widespread, especially in horticulture where the study groups even have formed an influential umbrella organisation. In both countries study groups seem to have led to a reduced use of pesticides and nitrogen (Bager and Proost, 1997).

In The Netherlands there is a strong belief in study groups as both a self-regulatory instrument, enabling farmers to cope with environmental requirements without losing competitiveness, and as an instrument to advance 'development, adaptation and improvement, exchange and application of technologies and techniques' (Oerlemans *et al.*, 1997, p. 273). Farmer-led research and development activities, named Participatory Technology Development, are seen by both farming organisations and the Ministry of Agriculture as a prerequisite for a successful adaptation to competitive markets, and also as a means to make environmental regulations work better. Thus another Dutch innovation is the establishing of around 25 environmental co-operatives. The idea is that farmers take a joint responsibility for fulfilling certain nature and environment objectives, but they decide themselves the means to reach these objectives (De Bruin, 1995).

Agricultural extension service

Agricultural advisers have played a major role in agriculture's productivistic development since World War II. As intermediaries between science and farming, the advisers have translated the latest research into new farming practices very quickly and very extensively, to the majority of farmers. In many countries they have been regarded as representing impartial expertise. Danish surveys on different aspects of pesticides show that 85% of farmers consider plant breeder advisers as being well or extremely well qualified. At the same time, the advice is considered to be unbiased (Bager and Søgaard, 1994, p. 94). Together with farmers' own experience, advisers are the most important source of information to farmers in Denmark regarding the selection of pesticides and of the dosages to be applied. Sixty per cent of the farmers attached great importance to this kind of advice. However, the situation may vary between countries: the corresponding figure was only 13% in a 1992 survey of Dutch farmers. Instead, 70% of them said that the label was the most important source of information, and advisers from commercial companies were the most important for 23% of them (Ley and Proost, 1992).

A special investigation of all plant breeder advisers in Denmark shows that they are highly aware of the environmental consequences of farming practices

(Svendsen *et al.*, 1997).[1] On the Likert-scale agricultural advisers seem to be a little more green than farmers. However, there is only a very weak correspondence between their attitudes and their advice to farmers regarding the use of pesticides. The advising ideal seems to be one of impartiality based on secure economic and environmental knowledge. Hence, many advisers are asking for more experiments and research highlighting precisely the economic and environmental consequences of reduced pesticide consumption. The survey demonstrated that the advisers as a whole had influenced farmers:

- to reduce the number of treatments and dosages applied;
- to estimate better the need for sprayings;
- to spray at optimal times of the day;
- to use pesticides with lower risks to the environment.

Another finding was a clear correspondence between farmers' environmental behaviour and the frequency of the use of the advising system.

Role of agricultural organisations

The type and degree of national regulation will depend on political circumstances, traditions, and the power of agriculture. Industrialised countries with a pronounced environmental policy and relatively strong agricultural organisations (farmers' unions, co-operatives, etc.) are able to influence legislation towards very technical solutions, in which the individual farmer and the many experts tied to agriculture can influence the practical implementation and the details of the regulations. This is the case, for example, in The Netherlands and in Denmark. In Sweden, in contrast, where agriculture always has been weak in relation to consumer and labour interests, the farming community has to a greater degree experienced the general legislation applied to other sectors. One consequence is that taxes have been more used in Sweden to regulate agriculture.

In countries with a less developed environmental policy and weak agricultural organisations, there will be a tendency to find specific agri-environmental regulations, i.e. measures directed towards solving specific local problems and with agricultural organisations participating to varying degrees reflecting the weakly institutionalised position of environmental policy. A case in point is Spain where both the general public awareness and the agricultural organisations are weak, the latter with only 15-20% of farmers organised (Garrido and Moyano, 1996).

Is it beneficial to the environment then to have weak agricultural organisations? Before answering that question it is necessary to distinguish between weak organisations in absolute terms as in Spain and weak in relative terms as in Sweden. From an environmental perspective, the Swedish situation of highly developed, but politically weak, agricultural organisations seems the most advantageous. In contrast, on the one hand, politically powerful farmers' organisations in other counties are able to

---

[1] The survey was financed by the Danish Environment Protection Agency in 1997 as part of an evaluation of the 10-year-old Pesticide Action Plan in which agricultural advisers were allocated a central role.

avoid some of the regulations and to postpone the fulfilment of environmental targets for a longer period than economists or environmentalists find desirable. On the other hand, well developed farmers' organisations are able to take responsibility and contribute to widespread efforts to comply with the rules. It is a long-term project to turn entrenched productivist practices and attitudes into more sustainable ones, which requires the active collaboration of farmers' organisations.

The local level is here of special importance. Assouline *et al.* (1996) show in their comparison between regions in Denmark and France that one of the success factors for conversion to organic farming is active collaboration between converters and local farmers' unions. In the Lemvig region of Denmark it was important to have respected pioneers where one of them was also a member of the steering committee of the old local farmers' union. This meant that conversion became part of the union's local strategy, which made it socially acceptable to convert. As the advisory system in Denmark is attached to the unions it also meant that extensionists could devote some of their time to advising (potential) converters.

CO-OPERATIVE AGREEMENTS

Germany is also a case of a strong environmental policy framework and strong agricultural organisations, especially at the local level, and there co-operative agreements between nature conservation or water supply organisations and farmers are a well established form of voluntary measure. In Germany, groundwater makes up a high proportion of drinking water sources and drinking water production is organised in a very distributed fashion (Heinz *et al.*, 1995). With the water supply industry so decentralised, negotiations between water supply companies and farmers usually occur at the local levels. In contrast, such agreements are rarely found in countries where river water forms a high proportion of water used in drinking water production and large supraregional water supply companies are predominant, as is the case especially in the UK and France. In these cases, the transaction costs of negotiating agreements with farmers across vast water catchment areas could be prohibitive.

Local conditions in the individual countries influence behaviour of both the farmers and the water supply companies and largely affect the costs involved in complying with the EC Maximum Admissible Concentrations (MACs) for pesticides and other agro-chemicals in drinking water. As expected, surface water catchments tend to be more expensive than catchments from groundwater or spring water, especially in the case of small drinking water works. Furthermore, the costs of complying with the pesticide standards are usually lower in water catchment areas where co-operative agreements are in place than in other regions where measures to purify the drinking water predominate (Heinz *et al.*, 1995). In other words, prevention of water pollution tends to be less costly than activities aimed at repairing environmental damage. However, in certain circumstances, the costs of compensation payments to farmers or of land purchases by water supply companies can exceed the costs of water purification or closing down of wells.

In Germany, there is a growing number of reports describing co-operative agreements between farmers and water supply companies (Bund für Umwelt,

1997; Rheinisch-Westfälisches Institut, 1997). Since regulations (e.g. §19 Federal Water Act - WHG) have not led to an effective change in the behaviour of the farmers in many regions, more and more water supply companies have decided to enter into direct negotiations with them. Because of the lack of consequent action on the part of the authorities these negotiations are more effective in persuading the farmers to use environmentally friendly production methods. Furthermore, based on confidence generated in voluntary agreements, it is far easier to check the continued enforcement of the contracts than in the case of regulatory conditions. But what is especially significant is the fact that bargains between water supply companies and farmers have tended to lead to the most economically and ecologically efficient solutions complying with MACs at the local level. In contrast, particularly the definition of the ecological requirements set by authorities for individual water catchment areas, which differ with respect to hydrogeological and agricultural conditions, has proved to be far more complicated than expected. It is remarkable that many water supply companies are carrying out costly research programmes to analyse the local soil permeability and susceptibility of water resources for different rotations of crops.

In most of the cases where water supply companies have decided to negotiate with farmers, the expenditure for preventive measures was less than the costs which would have been incurred for drinking water treatment or prospecting new water resources. Additionally, upholding the image of the customers, of drinking water being pumped from unpolluted water bodies, is a very important motivation for the water supply companies. Therefore they are keen to change the farmers' behaviour. Indeed, the Germans are particularly willing to pay large sums for natural, nearly untreated drinking water. Consequently, many water supply companies compensate farmers who are ready to extensify their use of grassland and to change arable land to meadows. Furthermore, they co-operate with the farmers in marketing products from ecologically-orientated agriculture.

In some German Bundesländer, as for instance in Hesse and Lower Saxony, in order to promote local co-operative agreements the Länder are charging water supply companies and private firms a fee for the water intake. The revenue is being used to compensate the farmers and to support the marketing of their products. Furthermore, water supply companies are granted subsidies to purchase and to lease agricultural land. Last but not least, the funds can be used for preserving water resources which are presently not being used for water supply. This option is seen as a first step to protect *all* areas of groundwater resources.

In contrast, the government of the Bundesland North Rhine-Westphalia refuses in principle to charge for water intake and instead promotes co-operative agreements by especially assisting arrangements at the Bundesland level. In 1989 it established the programme '12-Punkte Vereinbarung' which is aimed at implementing the EC Drinking Water Directive as far as agro-chemicals are concerned. It represents the organisational framework for almost 100 co-operative agreements covering a water supply of 1,300 million m³ year[-1] and an area of approximately 1.2 million ha (Rheinisch-Westfälisches Institut, 1997). The organisation of this programme consists of three levels: a) three Ministries (Environment, Spatial Planning and Agriculture); b) six regional associations of representatives from water authorities, water supply companies and agriculture;

and c) working groups and co-operation between water supply companies and farmers at the local level. Thereby each of the local co-operative agreements can be linked both with the political objectives of the government and the regional water management of river basins or subterranean catchment areas. The tasks of the associations at level b) include investigations into the developments in agriculture (e.g. changes of the farmers' economic situation with regard to the European market), the demands on water supply companies (e.g. new drinking water requirements) and the existing co-operative agreements (e.g. plans for the quality-restoration of waters). Various action programmes have been developed regarding water monitoring, water pollution by agro-chemicals, application of environmentally friendly production methods in agriculture and reactions of the water supply companies (e.g. prospecting of water resources). Furthermore the costs of these action programmes are estimated. More than 30 water advisers who are paid by the water supply companies play a decisive role in the implementation of the action programmes. These advisers consult the farmers and co-ordinate the information transfer between levels b) and c). Local co-operation between water supply companies and farmers can obtain subsidies from the Land government.

Subsidies can particularly promote the foundation and running of co-operative schemes. As mentioned already, under certain circumstances extremely high transaction costs prohibit co-operative agreements and might thus cause the loss of economic and ecological benefits. However, if the transaction costs exceed the potential economic benefits, the ecological benefits could outweigh the excessive costs.

The ecological benefits from co-operative agreements can include favourable side-effects for the preservation and restoration of aquatic ecosystems (particularly due to reduced use of pesticides). Economic assessments into the side-effects of co-operative agreements, however, remain scarce. There are, though, several cases in Germany where the economical efficiency of co-operative agreements between water supply companies and farmers can be demonstrated. The financial participation of a water supply company in Rheine near Münster has been around 80,000 to 134,000 ECU year^{-1}. Pesticide concentrations have been practically undetectable in the replenishment water for 4 years now. Without these successful co-operative negotiations, the water authorities would have ordered the installation of an activated carbon plant, incurring investment and operating costs of around 321,000 ECU annually. The economic net benefit of this co-operative agreement of more than 187,000 ECU year^{-1} does not include the economic value of the preserved natural habitats.

CONCLUSIONS AND PERSPECTIVES

Efforts to reform the Common Agricultural Policy (CAP) have partly been motivated by acknowledgement of environmental issues. There is presently intense debate on whether the current activities should be assisted by further environmentally oriented policies to internalise the external costs caused by agriculture, e.g. taxes per unit of pollution

However, the broad concept of voluntary measures are deeply embedded in national and now also EU legislation, both because that requires less control, and because the environmental consequences of the use of other instruments (e.g. taxes) are very insecure, and perhaps not least because soft regulations meet less resistance from the farming community. A strengthening of education and advice combined with voluntariness and premiums are, of course, much more attractive for farmers and their unions than fees and prohibitions.

Voluntary institutions do matter. Study groups/experience groups with extensionists attached seem to contribute to reduced use of pesticides. Agricultural organisations may also be a positive element, because they can be decisive for a broad implementation of environmental measures.

The disappointments which exist partly with regard to the environmental benefits of the agri-environmental measures under Regulation 2078/92 should not blur the fact that different 'soft' measures contribute to make regulations work. Especially the use of agricultural advisers seems to be effectual, when used in individual advising, study groups and organisations.

Advisory services also play an important role in co-operative agreements between farmers, and water supply companies or nature conservation bodies. In particular, there are some water supply companies in Germany which are paying agricultural advisers. The promotion of co-operative agreements as a special type of self-regulation does not focus primarily on the introduction of new prices for the use of the environmental resources, but on compensating the efforts of farmers who promote the environment. This instrument appears to have some advantages over taxes whose implementation imposes some practical difficulties. The main problem for a cost internalising strategy seems to be the lack of an effective check on whether the farmers' behaviour has changed or not: the farmers are not necessarily convinced of the imperative to change. Co-operative agreements, however, are based primarily on confidence in the agreement and the voluntary decisions made by the farmers. Consequently the cost of controlling this are relatively low and the motivation to find out the most cost-effective measures to fulfil the agreement is high. But it is also conceded that such 'soft' instruments will have to be set within a regulatory framework which establishes quantitative goals, monitors progress and provides enforcement mechanisms, if they are to guarantee pollution reduction.

The Dutch and German experiences should be used in particular to apply further voluntary agri-environmental regulations in the EU, so that they contribute to the CAP reform in a highly efficient way. Cost-benefit analysis can show where and how voluntary measures and co-operative agreements should be assisted by the authorities.

REFERENCES

Assouline, G. *et al.* (1996) *Stimulating the Implementation of Environmentally Sound Agricultural Practices in Europe. The Role of Farmers' Organisational Dynamics.* Report for European Commission, DGXI. QAP Decision, Theys.

Bager, T. and Proost, J. (1997) Voluntary regulation and farmers' environmental behaviour in Denmark and the Netherlands. *Sociologia Ruralis* 37(1), 77-96.

Bager, T. and Søgaard, V. (1994) *Landmanden og Miljøet - Holdninger og Adfærd belyst ved en Spørgeskemaundersøgelse.* [The Farmer and the Environment. Attitudes and Behaviour Investigated through a Questionnaire]. South Jutland University Press, Esbjerg.

Beus, C.E. and Dunlap, R.E. (1990) Measuring adherence to alternative vs. conventional agricultural paradigms: a proposed scale. *Rural Sociology* 56(3), 432-460.

Bruckmeier, K. and Teherani-Krönner, P. (1992) Farmers and environmental regulation: experiences in the Federal Republic of Germany. *Sociologia Ruralis*, 32(1), 98-106.

Bruin, R. de (1995) Local cooperatives as carriers of endogeneous development. In: Ploeg, J.D. van der and Van Dijk, G. (eds) *Beyond Modernization. The Impact of Endogeneous Rural Development.* Van Gorcum, Assen, pp. 256-273.

Bund für Umwelt und Naturschutz Deutschland e.V. Landesverband Hessen (1997) *Grundwasserschutz und Landwirtschaft in Hessen - Konflikt oder Kooperation, Fachtagung am 15./16. Mai 1997.* Mörfelden-Walldorf.

Centre for Agriculture and Environment (1998) *Economic Instruments for Nitrogen Control in European Agriculture.* Workshop document. Utrecht.

Cox, G., Lowe, P. and Winter, M. (1990) *The Voluntary Principle in Conservation.* Packard, Chichester.

Daugbjerg, C. (1998) Power and policy design: a comparison of green taxation in Scandinavian agriculture. *Scandinavian Political Studies*, 21(3), 253-284.

Garrido, F. and Moyano, E. (1996) The implementation of agri-environmental policy in Spain. In: Just, F. (ed.) *Incentives and Obstacles Towards the Implementation of more Sustainable Methods in Agriculture in Denmark, the Netherlands and Spain. Final Report.* Esbjerg, pp. 261-306.

Hamm, U. and Michelsen, J. (1996) Organic agriculture in a market economy. Perspectives from Germany and Denmark. In: Østergaard, T. (ed.) *Fundamentals of Organic Agriculture.* Copenhagen, pp. 208-222.

Heinz, I. (1998) Costs and benefits of pesticide reduction in agriculture: the best solutions. In: Wossink, G.A., Van Kooten, G.C. and Peters, G.H. (eds) *Economics of Agrochemicals.* Ashgate Publishing limited, Aldershot, pp. 333-344.

Heinz, I., Fleischer, G., Michels, M. and Zullei-Seibert, N. (1995) *Economic Efficiency Calculations in Conjunction with the Drinking Water Directive (80/778/EEC), Part III: The Parameter for Pesticides and Related Products*, Contract No. B4 - 3040 /94 / 000223 / MAR / B1. Brussels.

Just, F. (1996) *Incentives and Obstacles towards the Implementation of more Sustainable Methods in Agriculture in Denmark, The Netherlands and Spain. Final Report.* Esbjerg.

Just, F. and Nielsen, D.E. (1996) Intensive farming and water pollution in western Jutland. In: Mydske, P.K. (ed.) *Implementing Environmental Policy Instruments in Nordic and Post-Communist Countries.* University of Oslo, Department of Political Science, Research Report 03/96, pp. 5-25.

Kuckshinrichs, W. (1990) *Zur ökonomischen Theorie der Grundwassernutzung.* Lit-Verlag, Münster.

Ley, H.A. van der and Proost, M.D.C. (1992) *Gewasbescherming met een Toekomst: de Visie van Agrarische Ondernemers. Tabellen.* Landbouwuniversiteit, Wageningen.

Michelsen, J. (1997a) *Institutional Preconditions for Promoting Conversion to Organic Farming.* Paper prepared for the XVII Congress of the European Society for Rural Sociology, 25-29 August, Chania, Crete, 20 pp.

Michelsen, J. (1997b) *Danske Økologiske Jordbrugere 1995.* [Danish organic farmers 1995]. Esbjerg.

OECD (1996) *Activities to Reduce Pesticide Risks in OECD and Selected FAO Countries. Part I: Summary Report.* Series on Pesticides No. 4. Organisation for Economic Co-operation and Development, Paris.

Oerlemans, N., Proost, J. and Rauwhorst, J. (1997) Farmers' study groups in the Nether-lands. In: Veldhuizen *et al.* (eds) *Farmers' Research in Practice. Lessons from the Field.* ILEIA, Leusden.

Oskam, A.J., Vijftigschild, R.A.N. and Graveland, C. (1998) *Additional EU Policy Instru-ments for Plant Protection Products.* Wageningen Pers, Wageningen.

Rheinisch-Westfälisches Institut für Wasserchemie und Wasserforschung GmbH (1997) *Kooperativer Gewässerschutz in Nordrhein-Westfalen - Sachstand, Trends, Strate-gien.* Mülheim an der Ruhr.

Søgaard, V. (1997) *Attitudes and Behaviour - an Exercise in Hindsight.* Paper presented at the XVII Congress of the European Society for Rural Sociology, 25-29 August 1997, Chania, Crete. 11 pp.

Svendsen, S.V., Søgaard, V. and Just, F. (1997) *Farmers, Advisers and the Use of Pesti-cides. Survey of Attitudes and Behaviour in Connection with Reduction of Pesticide Use.* South Jutland University Press, Esbjerg.

Vanclay, F. (1992) The social context of farmers' adoption of environmentally-sound farming practices. In: Lawrence, G., Vanclay, F. and Furze, B. (eds) *Agriculture, En-vironment and Society.* Macmillan, Melbourne.

Vanclay, F. and Lawrence, G.L. (1995) *The Environmental Imperative - Eco-social Con-cerns for Australian Agriculture.* Central Queensland University Press, Rockhampton.

Vedung, E. (1998) Policy instruments: typologies and theories. In: Bemelmans-Videc, M.L., Rist, R.C. and Vedung, E. (eds) *Carrots, Sticks, and Sermons: Policy Instru-ments and their Evaluation.* Transaction Books, New Brunswick.

Wurzel, R. (1993) Environmental policy. In: Lodge, J. (ed.) *The European Community and the Challenge of the Future.* Pinter Publishers, London, 2nd edition, pp. 178-199.

National Cultural and Institutional Factors in CAP and Environment

15

Philip Lowe, Brendan Flynn, Flemming Just, Aida Valadas de Lima, Teresa Patrício and Andrea Povellato

---

INTRODUCTION

This chapter reports the results of a comparative study of national perspectives on Common Agricultural Policy (CAP) reform and the environment. The study surveyed opinions in national policy communities of selected European Union (EU) Member States. The work was conducted in 1996 and 1997, and looked back at the 1992 reform and forwards to the prospect of further rounds of reform. The intention was to identify the place of the environment in shaping national positions towards reform of the CAP.

The countries chosen for the study were Denmark, Ireland, Italy and Portugal. In all of them agriculture is of economic significance although its economic role differs greatly between them. Likewise, agriculture is a very important land use but the environmental and geographical circumstances are considerably different. Between them the countries well represent the range of production systems, from intensive to extensive ones; the types of environmental problem associated with different farming systems; and varying degrees of politicisation and institutionalisation of environmental concerns.

In the research, an institutional focus was adopted which involved identifying the leading groups and organisations in CAP reform in the separate countries; followed by interviews with senior representatives to clarify their outlook, concerns and strategies. Our approach thus included the following elements:

- Preparation for each country of an initial profile identifying the key issues and actors in CAP reform and the environment.
- An interview strategy covering agricultural and environmental departments, environment and rural development agencies, NGOs, farming and agricultural co-operative groups and consumer organisations. Between 10 and 15 organisations were covered in each country.

The chapter presents a comparative analysis of the results. It is structured around the following sections: agricultural characteristics; the rural environment; attitudes towards CAP reform; perceptions of the relationship between agriculture and the environment; institutional structures for agriculture and the environment;

and property relations surrounding rural landownership and the public interest. The concluding section reflects on different national strategies towards the development of EU policies.

AGRICULTURAL CHARACTERISTICS

The economic significance of agriculture

Between the four countries, the economic significance of agriculture differs considerably (see Table 15.1). Italy has the largest number of people employed in agriculture of any EU Member State, although Portugal and Ireland have larger proportions of their working populations in farming. For Ireland, but not Italy or Portugal, this employment importance extends through into a significant proportion of GDP and a major contribution to exports. For Denmark, food products make up an even greater share of national exports and likewise contribute to a large positive trade balance in food; however, Danish agriculture makes only modest contributions to employment and GDP. In Italy and Portugal, with higher ratios of population to agricultural area, the national agriculture is oriented largely to domestic consumption, with negative external trade balances in food products. Inevitably, the overall orientation of each country towards its agricultural sector is strongly shaped by the sector's distinctive economic significance:

- Denmark: a highly productive, export-oriented agriculture.
- Ireland: a productive agriculture of significance for exports, GDP and employment.
- Italy: an agriculture oriented largely towards domestic consumption employing a great number of people.

Table 15.1   Agriculture in national economies

	Denmark	Ireland	Italy	Portugal	EU-15
Number employed in agriculture (x 1,000)	102.0	146.0	1,332.0	541.0	7,514.0
Proportion of civilian employment in agriculture (%)	3.9	11.2	6.7	12.2	5.1
Share of agriculture in the GDP (GVA/GDP, %)	2.4	4.1	2.7	3.3	1.7
Share of food and agricultural products in total exports (%)	25.7	15.9	6.9	7.4	7.4
External trade balance in food and agricultural products by population (ECU per capita)	1,039.0	983.0	-105.0	-232.0	-27.0
Ratio of total population to agricultural area (persons ha^{-1})	1.9	0.8	3.9	2.5	2.8

Source: European Commission (1998). The statistics are for 1996.

- Portugal: a mainly traditional, domestic-oriented agriculture supporting a significant proportion of the population.

The different national agricultures can be summarised graphically, if in an oversimplified way, as follows (Figure 15.1).

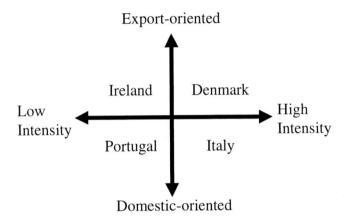

Figure 15.1   Characteristics of the different national agricultures

The structure of agriculture

Portugal and Italy have farm structures that are dominated numerically by large numbers of very small farms (see Table 15.2). Ireland has mainly small and medium-sized holdings. Denmark has mainly medium-sized holdings. All have experienced relentless declines in their farming populations: the greatest rates of decline have been in Italy and Portugal. In Portugal there has been an associated increase in part-time farming as farm families have sought means to supplement their incomes. In Italy, the continued economic viability of the small-farm structure depends largely on the widespread use of contractors who, in practice, manage many of the farms. In Ireland, Italy and Portugal there have been marked declines in the regularly employed non-family workforce, offset somewhat by a growth of casual employment, but with a general tendency towards greater reliance on family labour. Irish agriculture comes closest to the ideal model of the full-time family farm. Danish agriculture, in contrast, shows increasing reliance on regularly employed, non-family members. The farming population is an ageing one in all the countries except Denmark. The problem is most extreme in the Portuguese case where small or part-time family holdings offer little opportunity for the young and are occupied mainly by the middle aged and elderly. Italy in its mountainous areas and Portugal in its interior regions have suffered extensive land abandonment as once largely self-sufficient rural societies have collapsed and have abandoned subsistence cultivation.

Table 15.2   The structure of agriculture

	Denmark	Ireland	Italy	Portugal	EU-15
Average size of holdings (ha)	37	27	6	8	16 [a]
Proportion of holdings 1-5 ha (%)	2	10	77	78	58 [a]
Proportion of holdings of 50 ha or more (%)	22	12	2	2	7 [a]
Agricultural employment change 1990/1980 (% per annum)	-3.0	-1.9	-4.1	-2.9	-2.8
Agricultural employment change 1996/1990 (% per annum)	-4.6	-2.2	-4.6	-5.5	-3.1
Proportion of part-time employment in agriculture (%)	19	6	12	26	15
Proportion of self-employed in agriculture (%)	50	81	63	84	69
Proportion of those in agriculture aged under 25 (%)	23	9	11	5	9
Proportion of those in agriculture aged 55 and over (%)	21	31	24	53	28

[a] EU-12.
Source: European Commission (1998). The statistics are for 1996, except for those on the size distribution of holdings which are for 1993.

THE RURAL ENVIRONMENT

Rural land use

The four countries have markedly different patterns of land use (see Table 15.3). Denmark is dominated by arable land for cereal production. Ireland is dominated by grassland for cattle and sheep production. Both have only small areas of woodland. Portugal has roughly equal proportions of arable land (mainly cereals), of permanent crops (olive trees, vines, fruit and nut trees) and of pastures. A third of its land area is also woodland; about half coniferous and half broadleaved. Italy shows a broadly similar mixture of land uses to that of Portugal, but its woodlands are predominantly broadleaved and traditionally most were coppiced although a large proportion are now almost abandoned. These broad patterns of land use largely reflect differences in topography and climate (in both of which respects, Portugal and Italy exhibit much greater variety than Denmark or Ireland). Ireland has proportionately the most extensive agricultural area which embraces a great deal of low productivity grazing land, which in the other countries would not be farmed but might be wooded. In Italy only slightly more than half of the land is farmed, and in Portugal less than half: that leaves (apart from the urban areas) extensive tracts that are practically uncultivated including woodland, mountains, wetlands, coastal areas and barren land.

Table 15.3 Land use (% of total area) in 1990

| Country | Agricultural land | | Wooded |
	Arable	Pastures	
Denmark	59.7	5.0	11.4
Ireland	13.4	66.8	4.9
Italy	39.7	16.2	22.4
Portugal	34.3	9.1	32.1

Source: Stanners and Bourdeau (1995).

The environment and agriculture

The general national statistics on land use tell us little about the environments associated with agriculture. Much depends upon the intensity of production (see Table 15.4), which, above certain levels, is generally inversely related to ecological diversity. At too high levels of inputs, the absorptive capacities of soils, air and water may be surpassed, leading to pollution. The key issues in the countries studied are as follows:

- Danish farming is highly intensive. A seventh of its agricultural area is irrigated. It has a very high level of use of nitrogen fertilisers (second only to The Netherlands) and also a very large pig herd (the fifth largest in the EU), almost all intensively housed. The consequence is a large nitrogen surplus.
- Locally, in all countries, the spread of intensive systems of livestock production to include most pig and poultry production, as well as some dairy (e.g. in

Table 15.4 Indicators of agricultural intensity

	Denmark	Ireland	Italy	Portugal
Nitrogen surplus by agricultural area (kg N ha^{-1}) 1995	138	55	48	59
Sales of pesticides by cropped area (kg ha^{-1}) (various dates)	2.2	2.2	7.6	1.9
Average number of pigs per holding 1995	548	514	29	17
Proportion of total cattle on holdings with more than 300 stock (%) 1995	6.7	6.1	22.3	9.8
Proportion of dairy herds above 50 cows (%) 1995	37.6	17.0	9.5	0.9
Irrigated area as proportion of agricultural area (%) 1990	15.4	-	18.5	15.7

Source: Brouwer (1995); Stanners and Bourdeau (1995); OECD (1997); European Commission (1998).

Ireland) and beef production (e.g. in Italy) is leading to localised problems of point source pollution.

- Levels of pesticide use, mainly fungicides, are particularly high in Italy, which accounts for a fifth of the total sales of plant protection products in the EU.

National agri-environmental indicators may gloss over local and regional problems. For Italy, in particular, the national data hide a great geographical variability with areas of very intensive, high-input farming (largely in the irrigated lowlands, especially in the Po valley) and areas of very extensive agriculture (in the uplands and mountains) (Bordin *et al.*, 1998).

Low intensity farming and fragile landscapes and habitats

Much of the wildlife and landscape value of the European countryside depends on low intensity farming and might be threatened by both increased and decreased production pressures. According to Bignall and McCracken (1996), low intensity farming systems 'have tended to adapt their management techniques to the natural environmental constraints of the region rather than adapting the environment (and in many cases the livestock breeds and crop varieties) to meet a standardised, often industrialised, production system'. In Italy and Ireland low intensity systems make up about one-third of the agricultural area, and in Portugal as much as 60%, including both livestock and crops (Table 15.5). Extensive agricultural systems have lasted longer and up to more recent times in Portugal than in the rest of Europe. This compiled with the country's diversity of topography and climate has resulted in a rich variety of semi-natural habitats and wildlife, whose conservation to a large extent depends on the continuation of traditional farming practices. Examples include the 'montado' (oak and cork oak groves) agro-pastoral systems of the Alentejo region and the 'lameiros' (permanent pasture and meadows) of the Northern Interior (Caldas, 1998).

The need for some support for extensive farming systems operating under different natural conditions has been recognised for some time in the Less Favoured Areas Directive (75/268). The support under the Directive has been more or less effective in helping maintain farming in different regions of Europe (see Chapter 11). However, it has not been able to counter widespread processes of land aban-

Table 15.5 Agricultural area in low intensity systems and LFAs (share of total agricultural area, in %)

Country	Low intensity systems	LFAs
Denmark	5	0
Ireland	35	70.9
Italy	31	53.6
Portugal	60	86.9

Source: Danish Ministry of Environment (1995); Bignall and McCracken (1996). The statistics on low intensity systems are estimates.

donment, particularly in southern Europe. Conversely, in some cases, the payments available may encourage overstocking and overgrazing. Surveys conducted in Ireland have shown that high stocking densities (encouraged also by the EU sheepmeat regime) in upland heather, moor and blanket bog have led to the disappearance of heather, sedges, moorgrasses and mosses, depriving wildlife of food and cover (Sheehy *et al.,* 1991). In the Apennines in Italy, stocking levels as low as 0.3 sheep ha^{-1} are causing widespread overgrazing of arid summer grazing lands, affecting about one and a half million ha, including alpine pasture, wooded pasture, scrub and forest (Baldock *et al.,* 1994).

Soil erosion is one of the most serious rural environmental problems in the Mediterranean countries. It is particularly acute in Portugal, where 30% of the land is judged to suffer a high erosion risk and 57% an intermediate risk of erosion (Caldas, 1998). The problem may be exacerbated by a number of factors, including land abandonment, forest fires, overgrazing and the cropping of unsuitable land.

ATTITUDES TOWARDS THE CAP

National outlooks towards reform

The outlook towards the reform of agricultural policy varied between the four national capitals: from consensus in Copenhagen that CAP reform was unfinished, and that liberalisation was both desirable and inevitable; to the implicit acceptance of the *status quo* and reluctance to reopen the issue, in Lisbon and Rome. In the latter two capitals the opinion was also expressed (during the interviews, which were conducted in 1996) that the 1992 reform was still being worked through and it was much too soon to be considering further changes. In Dublin, attitudes were also mainly complacent, although some people privately conceded the need for change. It was beginning to be recognised that all of Ireland was unlikely to qualify for the next round of the structural funds. There was also growing disgruntlement amongst younger farmers with the restrictions of milk quotas. An undercurrent of opinion thus saw further liberalisation of the CAP as unavoidable and probably advantageous to the more commercial, export-oriented sectors of Irish agriculture.

We can relate these positions back to the national economic significance of agriculture (see Table 15.1) as well as to the more general outlook that states have towards free-trade and protectionism. Although Danish agriculture has benefited considerably from CAP supports, its future prosperity is seen to depend upon its unfettered access to world markets. On the other hand, in Portugal and Italy, the CAP is seen largely as a key means of income transfer to rural areas which it would be politically unwise to disrupt. Ireland, with a significant proportion of both employment and exports due to agriculture, shows some ambivalence towards CAP reform. It should be borne in mind that respondents were asked general questions concerning their attitudes towards medium and long-term policy developments. Such attitudes may not be that useful a guide as to how governments will negotiate on specific policy issues when narrow considerations of the costs and benefits to the agricultural sector may be to the fore. A country such as

Denmark, for example, which is a major net beneficiary of the CAP, may thus adopt a fairly conservative stance in the EU Agricultural Council over changes to specific commodity regimes, even though there is a strong national consensus behind the need for the liberalisation of the CAP - an outlook that is more likely to be to the fore at the European Council level.

Changing attitudes toward the CAP

The process of CAP reform, initiated by MacSharry, has itself changed the nature of the views expressed. For a start, it has given prominence to the instruments of the CAP and the role of the EU in national legislation. Where the old system of price support disguised the effects of the CAP, the reform has made policy instruments more transparent and opened a debate on the consequences of the policy pursued. In all the four countries, indeed, there was little if any public debate on the CAP previously. The opening up of debate has brought additional policy actors into play beyond the traditional agricultural community.

The accompanying measures have specifically drawn environmental groups into discussions over the environmental merits of the CAP. But other groups have also taken an interest: for example, consumer, trade union and industrial groups, all beginning to question some of the costs of the CAP. More generally, public opinion has become more informed of the CAP and as a result has become more unhappy with its inefficiencies. The MacSharry reforms have also affected opinions amongst farmers. Among Danish farmers there has been a·strong reaction to the bureaucracy involved, amidst considerable apprehension that the price of continued subsidies may be many regulations. In the other countries the availability of agri-environmental payments for small and traditional farmers has given a new voice to minority farming groups, such as those representing particular regions, or smaller farmers, or organic producers.

PERCEPTIONS OF THE RELATIONSHIP BETWEEN AGRICULTURE AND THE RURAL ENVIRONMENT

The framing of the relationship

It is evident that not only are there marked national variations in the degree of concern for rural environmental problems, but the way in which these problems are perceived also differs markedly. Much depends upon the different perspectives from which the problems are framed, whether predominantly in terms of pollution or countryside management or rural desertification. This framing determines not only how the problem is evaluated but also how it intersects with other policy fields (e.g. pollution regulation, resource management, protected areas, land use planning, regional policy, rural development). In some cases, indeed, agri-environment problems are actually subsumed to these other dimensions. The framing of the problem partly reflects its history (what was the agri-environment problem that first broke through to public consciousness and debate). It also obviously reflects the fundamentals of the rural environment and the nature of the farming systems that maintain and exploit it. But there are also evident cultural factors too, particularly regarding the place of farming and rurality in national imagery. By and large one can express the different approaches to the national framing of the problem in the following terms:

- Denmark: a pollution discourse;
- Ireland: an agrarian discourse;
- Italy: a consumer and regionalist discourse;
- Portugal: a rural development discourse.

Denmark

The Danes are strongly concerned with the relationship between agriculture and the environment, primarily because of pollution from intensive agriculture. In an affluent society but one that industrialised relatively late, environmental and green consumer concerns are strong and widespread, but Denmark has fewer of the urban-industrial problems of societies that industrialised earlier. Instead, a rather intensive agriculture has come to be seen to pose unacceptable burdens on the environment. Thus, whereas other national environmental lobbies have been more preoccupied with urban-industrial sources of pollution and hazards, the Danish lobby has had a primary focus on farm pollution. These concerns emerged politically in the mid-1980s around the eutrophication of surface and coastal waters which led to the passing of several remedial action plans (Linddal, 1998). More recently, in the wake of the MacSharry reforms, the Danes have shown increased interest in some of the wider consequences of agricultural intensification, for example, on landscape, animal welfare and food quality. Ecological concerns, though, have tended to lead the way. For example, problems of pesticides in the environment were politicised first before the possible health risks of pesticide residues; the broadening of concern being catalysed by the fact that the more researchers looked for traces of different pesticides in water the more they found.

In the 1980s popular environmental groups, with the Nature Conservation Association in the lead, played a crucial role in forming and promoting an agri-environmental agenda. At the same time the Social Democratic Party shifted from a productivist to a more environmental and consumerist orientation (Andersen and Hansen, 1991). A coalition between this and other centre and left-wing parties - assisted by expertise in the national Environmental Protection Agency and the environmental lobby - succeeded in penetrating the hitherto very strong and closed agricultural policy community. Initially the agricultural organisations and their ministry repelled these challenges, but gradually they were obliged to accept and institutionalise the new issues on the political agenda (Daugbjerg, 1998). This brought environmental groups into the policy process and the issue is no longer so salient politically even though there have been difficulties in achieving the intended reductions in agricultural pollution.

In the 1990s the main political focus has turned more to consumer questions, both in terms of the consumption of healthy and tasty food produced in an acceptable way and the consumption of nature and landscape. Almost all political parties have pursued such middle class, consumerist values, including parties that hitherto championed agrarian interests, as it has been generally realised that the road to power requires support from a broad urban public. A symbol of the shifting public priorities was the change in name of the Ministry of Agriculture to the Ministry of Food in 1996. This re-orientation has gained momentum, especially through the success of organic farming which by most town people is considered to be a solution that saves equally the environment, nature, animal welfare, human health, working conditions and rural life. As of January 1999, 5.4% of Danish agricultural land (143,000 ha) was organic or under conversion, and almost 20% of milk for drinking was organic (Oldrup and Just, 1999).

All in all, the development in agricultural policy has been characterised by the erosion of any vestigial sense of agricultural exceptionalism. Politicians and consumers demand much the same from agriculture as from industry and the service sector. Although the term is not used, there is a broad consensus in society, including within agriculture, over ecological modernisation. Translated into the traditionally strong Danish orientation to world markets, this has meant that rather strict regulations about the environment and animal welfare are regarded as a means to maintain international competitiveness. Both the government and lobbying groups are therefore keen to extend high domestic standards to the EU as well (Andersen and Liefferink, 1997).

Rural social and economic problems are very low on the agenda, although there is some ambivalence towards them. Both inside the agricultural community and perhaps even more in the towns and cities it is regarded as a shame that the actual number of farms has decreased to one-third of its 1960 level. On the other hand, it is widely accepted that farms need to be specialised and not too small if they are to maintain their international competitiveness. This ambivalence is also reflected in political actions where a majority of parties pay lip-service to the values of family farms, country life and stable rural communities, but on the other hand pass land laws that fail to curb the development of big holdings. A major reason is that rural problems are relatively small in their scale and intensity. Economic development is widely diffused and alternative job possibilities in rural areas are quite good. This also means that EU subsidies for rural development are insignificant and generate very little public awareness.

Ireland

A traditional agrarian discourse prevails in Ireland which projects the family farm as the focus of country life. In recent years this discourse has not gone entirely unchallenged by public figures associated with urban interests and the public sector trade unions, most especially during the linked CAP reform and GATT negotiations of 1992 (Flynn, 1998). A brief but intense challenge was also raised in 1994 over the issue of animal welfare and the live export trade, but since then this issue has subsided. In any event, the broad ideology prevails that the rural economy and the countryside are best secured by safeguarding the position of family farmers (McCullagh, 1991). Consequently, efforts to improve the level or extent of countryside or woodland management can fall foul of a populist rhetoric which stresses that people come first in the countryside. Indeed, the worldview held by many farmers is one which argues that, first and foremost, land must be occupied and used for production purposes, to support 'proper family farming' (Allen and Jones, 1990; Hickie *et al.*, 1995; Varley, 1996). Efforts in particular under the EU Habitats and Birds Directives to secure workable conservation designations on ru-

ral land have foundered against local and populist politics which respond with te-
nacious hostility to any restrictions being placed on private property (Hickie,
1996). At the same time, drawing on the traditional pastoral and small-scale nature of Irish farming, the notion of 'Green Ireland' is being strongly promoted as a
marketing image that emphasises the wholesome qualities of Irish agricultural
produce.

While it is difficult from such an ideological perspective to perceive farmers
as presenting a threat to the rural environment, farming is implicated in the decrease in ambient water quality, some very public gross pollution incidents and
losses of biodiversity. Intriguingly however, Ireland's well run and funded farming
lobby, led by the Irish Farmers' Association (IFA), has responded in a deliberate
way to meet its relatively disparate and small number of critics head on and rebuff
their claims. For example, prominent advertisements have been placed in daily
papers to emphasise that farmers do care for their livestock. Poor state provision
of sewage treatment is blamed as the real cause of water enrichment, rather than
farmers' actions. Conversely, threats to the family farm, including various economic and regulatory measures associated with CAP reform, are seen and
portrayed by the farmers' organisations as threatening the vitality of the countryside.

The main picture, then, from Ireland, is of an agri-environment debate that is
comparatively in its infancy. Unlike Denmark, Ireland has few if any environmental NGOs that specifically target and lobby on agriculture or related matters.
The principal organisations, Birdwatch Ireland and the Irish National Trust (An
Taisce), which do have an impact are not dedicated to agri-environment issues *per
se* but do make informed contributions in their lobbying of the EU and Irish state.
But these issues are not politicised as environmental concerns among the public at
large; indeed fears about litter, traffic congestion and nuclear energy would rank
higher in popular perception. Equally there can be no doubt that policy development is influenced by a strong neo-corporatist relationship between the Irish state
and farmers, largely to the exclusion of environmentalists and other dissenting
voices. Debate, such that it exists, remains centred around an opaque and technical discourse between farmers' representatives and the state's Department of
Agriculture and its agricultural field advisory service (Teagasc). In this case the
focus of policy and perception is the Rural Environmental Protection Scheme,
through which well over £300 million in financial support from the EU has been
granted under 2078/92, the agri-environment Regulation (McLoughlin, 1998, p.
48). This scheme, which involves a farm management contract, and an extensive
list of compulsory and additional voluntary management measures is portrayed by
the farmers' organisations and state agents alike as a 'flagship' policy which is addressing the issues of agricultural pollution and farm conservation. Yet it is not at
all clear whether the scheme is being successful in curbing farm pollution incidents or conserving wildlife, or even if it enjoys the continued confidence of
Ireland's small environmental NGO sector (Murphy, 1996). Moreover, most
farmers joining the scheme appear to do so less out of any environmental motiva-
tions, and more out of a sense of maximising yet another source of income
transfers from Brussels. In conclusion then, Ireland appears to present a picture of
a somewhat reactive policy debate, led and dominated by farmers and the state,

with a relatively low level of public environmental concern, affecting policy outcomes.

Italy

In a country where there is little public interest in agriculture, there is nevertheless
general recognition that the very different agricultural systems and conditions
found across Italy have decisively shaped the country's great regional diversity,
and continue to do so. That diversity is particularly appreciated for its contribution to regional cultural identities and, increasingly in many areas, to the
competitive position of local places and products. Agricultural change has
aroused some concern in recent years, although the degree and focus of concern
have been regionally differentiated. Typically, though, the issues have been consumption oriented, for example pesticide residues in food and agricultural
chemicals in water supplies. This reflects not only the national cultural preoccupation with food but also the high status and attention accorded to notions of
quality and authenticity in consumption.

Since the 1960s there has been a progressive dwindling of public interest in
agriculture in Italy. Nowadays public opinion is little informed about agricultural
issues and even less so about agricultural support policies. In consequence there is
very limited awareness of the cause-effect relationship between agricultural policies and the state of the rural environment.

The demise of the once-prominent agrarian ideology seems to be due more to
the atrophying of the 'social contract' established between farmers and society after World War II, than to the decreasing weight of the sector in the economy. The
family farm has remained the basis of that contract since the Constitution, approved in 1948 and still in force, gave significant protection to the *small* family
farm. Between the 1950s and the 1980s the Christian Democrats, who were in
power throughout this period, centred their political base on the rural vote, pursuing clientelistic agricultural policies intended to secure the continued allegiance
of farmers and rural landowners. This did not prevent market-driven technical
changes which saw a huge manpower exodus from agriculture - some two million
jobs being lost from the sector between 1970 and 1990 - but it did help keep a
substantial number of small farms in being, such that Italy has retained by far the
largest number of farmers of any EU country. Many farms though are not actually
managed by the owners but by contractors (Povellato, 1993). The owners still oc-
cupy the farm house and may engage in some part-time farming (Fabiani, 1986;
Barberis and Siesto, 1993). There are tax advantages for these arrangements, but
the owners also tend to regard their land as an inalienable asset, a source of family
security and an essential connection to their roots (Mantino, 1995).

The great number of small farms has made land consolidation programmes
unfeasible (Povellato, 1996), which has meant that most Italian regions have
largely avoided the drastic simplification of the farmed landscape that has occurred across north-west Europe (Sillani, 1987). On the other hand, the
widespread diffusion of dwellings has brought about a progressive urban degradation of rural areas particularly in the economically dynamic north-east and
central regions of Italy (Bagnasco, 1977; Merlo and Manente, 1994; Saraceno,
1994). This has involved a loss of well over 2 million ha of fertile agricultural

land to urban growth since the 1960s. Over the same period, a similar amount of marginal agricultural land has been abandoned in mountainous and hilly areas. The environmental consequences have been complex but the main issue of public concern has been the increased incidence of avalanches and flash floods due to the abandonment of the traditional grazing and watershed management of mountain pastures and meadows.

The clientelistic relationship between farmers and the main governing party throughout the post-war period overshadowed any attempt to establish strategic directions for agricultural policy or to maintain the public image of farming. The relationship remained unchallenged until the 1980s when the spread of economic development in many rural areas and new consumer demands demonstrated the need to reconsider the role of agriculture in the society. In the late 1980s, some sensational events - particularly the revelation of widespread atrazine contamination and of methanol in wine - were catalytic in arousing public concern over harmful agricultural practices that threatened the quality of food and drinking water.

Green campaigners took up the issue of pollution from agriculture and managed to achieve a national referendum in 1990 that would have placed severe limitations on pesticide usage, but the voter turn-out was insufficient for the result to be binding. Nevertheless, there are widespread public anxieties over the risks posed by intensive agricultural practices. In contrast, the role of agriculture in the environmental management or maintenance of the countryside has not become a significant public issue. One reason for this is the public perception that the rural environment comprises the woodlands, forests, mountains, hills and coastal areas that cover almost half the country, beyond the farmed landscape.

Farmers and their unions, having lost their traditional political role (following the collapse of the old Christian Democrat power structure), find it difficult to choose a new strategy confronted by the demands of increasing international competitiveness and a sceptical and sensitised public opinion (De Benedictis and De Filippis, 1998). Increasingly they are looking to the opportunities in green consumer trends that put more and more importance on the 'environmental value' of goods and services. In response, the Italian regions have sought to create regional trademarks to enhance local production. Farmers show a great interest in this kind of 'competitive advantage', and the most advanced operators in both farming and food processing are aware that the environmental quality of the product will be more and more decisive in ensuring market outlets. The positive relationship between agricultural development and the environment is also a key feature of the promotion of agriculture in certain marginal areas through product differentiation on the basis of quality and product origin, to improve the returns to traditional agriculture.

Portugal

The problems that necessitated CAP reform - the over-production and over-intensification of agriculture - seem irrelevant to the Portuguese context (Patrício and Lima, 1996). It is still seen that the agricultural sector needs to modernise so as to reduce food imports. The major priority for rural areas is seen to lie with new markets for rural products and forms of economic development that will help to stabilise rural populations (Ministério da Agricultura, 1993).

Portuguese agricultural discourse, as with the Irish case, has traditionally portrayed the family farm as a social ideal. However, the economic difficulties associated with farming have led to a massive reduction in the number of farms without a corresponding consolidation of land holdings (Lima, 1991; Baptista, 1995). The result is land abandonment and an alarming rural exodus. Agriculture is thus widely seen to be in crisis, with concern over not only its fundamental viability as an economic sector but also the associated undermining of rural society that is leading to widespread desertification of the countryside. The growing perception is that agriculture is systematically being destroyed and that concerted political action is required in favour of some type of agricultural modernisation.

While the dominant agricultural discourse expresses a pessimistic outlook, a new discourse centered on rurality seems to be gaining support, with novel perspectives emerging regarding the use of rural space (Portela, 1994). The economic appropriation of rural space suggests new forms of production such as rural tourism and regional produce as viable alternatives to primary commodities. These forms are also more conducive to stabilising rural populations, through the encouragement of pluriactivity, and are in keeping with the new emphasis on the protection and active management of the rural environment.

The debate on the relationship between the environment and agriculture is fairly recent, having been given its impetus with Portugal's entry into the EU and the need to address specific agri-environmental problems (Patrício and Lima, 1996). The relative weakness of a national debate on agri-environmental problems stems from the perspective which sees the modernisation of agriculture as an unfinished process and from the assumption of an essential compatibility between agriculture and the environment. Rural environmental problems are thus not only overshadowed by the crisis of agriculture but are believed to stem from it. The desertification of the countryside is seen to benefit neither agriculture nor the environment, particularly where it leads to the abandonment of traditional practices of soil, water, pasture or woodland management.

The EU agri-environmental measures (under Regulation 2078/92) were first responsible for raising the national debate on the impact of agricultural change on the environment (Patrício and Lima, 1996; Billaud *et al.*, 1997). The agri-environmental measures were 'translated' into domestic policy as a support programme for traditional family farmers based on the underlying assumption that agriculture is the backbone of the countryside. The channeling of agri-environmental concerns towards national agricultural conditions has thus been done through the preservation of existing forms of production, namely supporting extensification programmes. In this way agri-environmental questions have enlarged the range of issues posed by the profound social and economic changes that rural Portugal is experiencing.

The interlinkage between the agricultural crisis and rural depopulation has meant that agricultural policy has subsumed other programmes and issues, such as programmes of social support and regional diversity, shaped in part by the new discourse on rurality (Reis and Lima, 1998). Agri-environmental issues are located within this larger framework that is evolving towards an inclusive rural development policy, and through it they have begun to relate to regional and local conditions.

INSTITUTIONAL STRUCTURES FOR AGRICULTURE AND THE ENVIRONMENT

The institutional actors

Agri-environment policy inevitably involves a range of institutional actors. In all countries there is a well established Ministry of Agriculture but for most of these ministries agri-environment represents a new policy departure requiring them to relate to other parts of the state as well as to organised interests outside of the traditional agricultural lobby. The environment side of the equation is quite differently organised in different countries and it is evident that the strength, politically as well as in terms of institutional capacities, of environmental ministries and agencies varies considerably. The dynamics of the agri-environment policy segment very much reflect the comparative strengths and interrelationships between these two separate policy communities (these institutional structures, of course, are not independent of the way problems are framed).

The sectoral versus territorial orientations of agricultural ministries

Agri-environment policy involves a movement from sectoral policy making to spatial and territorial policy making, and the interest and adaptability of different bureaucratic structures in this new policy departure is related to their past experience in these regards. For example, some Ministries of Agriculture have much greater traditions of dealing with territorial issues (rural planning, land consolidation, soil and water conservation programmes, rural development, land drainage, agri-tourism, etc.) as well as the more normal sectoral concerns. Such bureaucratic structures have a greater institutional interest, as well as a wider policy orientation, to be able to absorb agri-environment policy demands. In doing so, they will tend to subsume them to an existing agrarian or rural discourse which portrays farming as the key component of the rural economy and environment. This is particularly noticeable, for example, in the Portuguese and Irish cases where the Ministries of Agriculture have some traditional territorial responsibilities and have absorbed agri-environment measures into their existing structures dealing with farming and rural development. In Italy too, the strong decentralisation of agricultural policy gives it a pronounced regional orientation.

At the other extreme would be the Danish Ministry of Agriculture. This traditionally has been a sectoral ministry looking after the interests of an export-oriented agriculture and food industry, rather than the countryside. Because agricultural extension services have been organised under the auspices of the farming unions and agricultural co-operatives in Denmark, the Ministry does not even have the sort of large field-level bureaucracy that some other Ministries of Agriculture traditionally have. In consequence, it has been quite willing to cede control over the regional schemes of agri-environment policy to local government - this involves not only a decentralisation of authority but the effective transference of responsibility from the agricultural policy community to the planning/local government community. This is part of a broader understanding in Danish society of agriculture as an economic sector like any other which must take responsibility for its environmental consequences.

Environmental interests in government

Arguably, the devising of effective agri-environment policies needs the involvement of organised environmental interests, particularly where these policies are intended to deliver positive environmental benefits. Unfortunately, it is evident in a number of countries that environmental ministries and agencies are politically too weak to have a significant involvement in relation to agricultural ministries. An exception is Denmark, where a very strong Ministry of Environment and Energy has played a leading role. At the other end of the spectrum is Ireland whose Ministry of Environment is mainly responsible for local government infrastructure rather than environmental protection. In between (but closer to Ireland) are the experiences of Italy and Portugal with recently created Ministries that still have to establish their authority and carve out a clear domain within the bureaucratic hierarchy. For example, in Italy environmental issues relating to agriculture may be the concern of as many as five ministries which makes co-ordinated intervention difficult and favours the much more unified perspective of the agricultural interests.

The comments made above about the relationship between sectoral and territorial perspectives for agricultural ministries also apply to environmental ministries and agencies. In some countries there is a long tradition of protected areas legislation and policy. This has tended to involve specialised national agen-

Table 15.6 Internationally recognised protected areas (% of total area, 1990)

Country	IUCN category I-III [a]	IUCN category IV [b]	IUCN category V [c]
Denmark	0.35	1.83	7.63
Ireland	0.32	0.06	-
Italy	0.42	0.92	2.97
Portugal	1.11	0.79	3.0

[a] Outstanding ecosystems, features and/or species of national scientific importance, largely or wholly protected from human exploitation or disturbance. [b] Sites or habitats essential to the continued well-being of resident or migratory species of national or global significance. Protection of nature must be the primary purpose but this does not exclude the harvesting of renewable resources. [c] Valued (semi-natural and cultural) landscapes. The IUCN definition of Category V is 'areas of land, with coast and sea as appropriate, where the interaction of people and nature over time has produced an area of distinct character with significant aesthetic, cultural and/or ecological value. Safeguarding the integrity of this traditional interaction is vital to the protection, maintenance and evolution of such an area' (IUCN, 1990).

cies, often interacting with local and regional government, but also necessitating the development of constructive relationships with farming groups. Table 15.6 gives details of the geographical extent of protected areas between the four countries, according to a common IUCN classification. While IUCN's Categories I-III cover wild areas not subject to human exploitation, Category V includes valued semi-natural and cultural landscapes where the development of policy has necessitated interaction with the agricultural sector. It will be seen that only in Denmark has there been significant experience of pursuing this type of area protection on cultivated and privately occupied land. In Ireland, in contrast, constructive interaction between protected area policy and the agricultural sector is notable by its absence.

The environmental lobby

Related to these factors, also, are the relative strengths of farming and environmental lobbies as well as their histories of interaction. The environmental lobby is only weakly developed in Portugal (Table 15.7). In Ireland and Italy there has been a well organised politics around green issues but this has not given much attention to rural environmental concerns and it has tended to be a highly politicised form of oppositional politics and demonstrations which has not been well geared up to the sort of bureaucratic lobbying, environmental monitoring and practical conservation work which would help to practically progress rural environmental programmes. Denmark again stands out in terms of having a highly organised and efficient environmental lobby: it is the only country of the four in which voluntary environmental groups can act as effective partners with state agencies in the development of agri-environment policies and programmes. In the other countries (with the exception of some of the northern regions of Italy), environmental groups tend to be marginalised and disparaged by the farming lobby.

Table 15.7   Popular action for the environment: results of a survey of national publics

Percentage of respondents who have:	Denmark	Ireland	Italy	Portugal
Been a member of an association for the protection of the environment	17	7	6	4
Financially supported an association for the protection of the environment	26	16	5	4
Taken part in a local environmental initiative	10	16	9	7
Demonstrated against a project that could harm the environment	5	7	9	6
Bought an environmentally friendly product, even if more expensive	61	43	33	23

Source: European Commission (1995).

Internalisation of agri-environmental issues by agricultural bureaucracies

These different variabilities in the strength and experience of environmental administrations and lobbies have a wide range of consequences. Where environmental ministries or agencies are strong and long established, it was typically they who took the lead in pioneering agri-environment measures in the 1980s, often under domestic legislation and often in the face of some scepticism or even opposition from the agricultural establishment. The Europeanisation of this policy field then swept it into mainstream agricultural policy, with the implementation responsibility going to agricultural ministries, leaving environmental organisations to play a watchdog role.

Where environmental ministries or agencies or groups are weak or non-existent, there are limited counter pressures to the complete internalisation of agri-environment policies within the agricultural bureaucracy (the most effective counter pressures may indeed come from the European Commission). There may be very limited public debate about the nature, purpose or even existence of such policies. It may therefore be very difficult to know whether the environmental benefits are being maximised in the design, implementation and monitoring of measures. It may also be the case that Ministries of Agriculture and rural development agencies do not recognise the possibilities, the potential benefits or the policy synergies that could arise from countryside management approaches precisely because they have no experience of the means to pursue them or the results that they bring. This may have the effect of narrowing the response to European programmes.

The political geography of implementing agri-environment measures

These different factors shape the geography of implementation of agri-environment measures under Regulation 2078/92. The emphasis in the separate national programmes reflects the different ways in which rural environmental problems have been framed. Thus Denmark focuses on organic farming and measures to improve environmental quality, whereas Portugal has a strong emphasis on the maintenance of low intensity farming systems (see Chapter 12).

The institutional structures for agriculture and environment also determine the nature of agri-environment programmes and the broader objectives they fulfil. Thus, in Portugal and Ireland, strong central agricultural bureaucracies have set thresholds of beneficial participation for agri-environmental supports to make them attractive to poorer farming cohorts and have seen these supports as complementary to other measures to bolster incomes in depressed rural regions.

The decentralised structure in Italy, in contrast, has meant that the better organised regional administrations in the north and centre have benefited more from agri-environment aids than the depressed south. Somewhat different regional priorities have also been pursued - with the south emphasising organic farming and the north and centre giving more emphasis to measures to reduce input use. As well as seeking to maximise the opportunities of a new income transfer mechanism, the implementation of the Regulation in Italy has thus reflected the national preoccupation with the possible contamination of food and water supplies by agricultural chemicals. Support for low intensity farming in high natural value areas is only a feature of alpine areas in the north but not of the marginal Apennine areas which suffer from major problems of land abandonment and overgrazing.

The distribution of European financing of agri-environment measures between the four countries is highly skewed (see Chapter 12). Ireland, Italy and Portugal are middle-ranking beneficiaries (below Germany, France and Austria), but when the level of EAGGF expenditure under Regulation 2078/92 is adjusted in relation to the utilised agricultural area (UAA) Ireland and Portugal move to the top of the league. Denmark, in contrast, is right at the bottom, receiving less than a tenth of the ECUs per ha of UAA of Ireland. Whereas, 15-20% of agricultural land is covered by agri-environment measures in Ireland, Italy and Portugal, in Denmark it is about 3%. These comparisons reveal the implicit orientation of Regulation 2078/92 towards low intensity farming systems and income support objectives. While the Regulation has thus been successful in spreading the notion of agri-environment policy from such early pioneers as Denmark to countries such as Portugal and Ireland, it has offered little assistance in tackling the problems of intensive agriculture that beset the Danes.

PROPERTY RELATIONS SURROUNDING RURAL LANDOWNERSHIP AND THE PUBLIC INTEREST

Public interest in farm production methods

Agri-environment policies imply a new set of relationships between the state, civil society and the owners and occupiers of rural land. Economists talk about the delivery of environmental goods and services as if these were like any other agricultural commodity. But they are not. There is no specific market for them although there may be a public demand, usually expressed through collective action of some sort (mainly voluntary organisations), and that demand has usually to be expressed through state regulations or grants. The environmental relations of agriculture, moreover, depend less on what is produced and more on how it is produced. This implies a public interest in production methods.

The way these issues are addressed differs markedly between countries particularly in relation to the expectations of the rights and responsibilities attached to rural landownership and the modalities for expressing the public interest in rural land. In societies where agrarian values are strong, farmers are traditionally portrayed as the guardians of the land, and in each of the countries studied this had been an element in national ideologies. It remains particularly strong in Ireland and to a lesser extent in Portugal and Italy, and in these countries it is not seen as problematic to be paying farmers income supports under the agri-environment scheme simply to continue in traditional farming. In contrast, the once strong agrarian ideology has been erased in Denmark largely through the experience of agricultural pollution.

Incentives versus controls

Conversely, it is quite evident that some countries and national systems are prepared to pursue more interventionist approaches than others and to interfere much more directly with the exercise of farmers' rights. At one extreme there is the Danish case where stringent regulations have been imposed upon farmers. One example of extensive encroachment on farmers' rights is the way in which the whole of the country has been designated a Nitrate Vulnerable Zone which entails a prohibition of cultivation alongside all watercourses. Such an intervention would be unthinkable in some of the other countries. In part this reflects the framing of the Danish agri-environment problem in terms of pollution - a problem which invites a Draconian response. But it is also clear than the Danish land use planning tradition involves a long history of intervention in landowners' property rights in the wider public interest. In interview a number of groups and officials referred to the end-game of agri-environmental policies in Denmark as being to integrate the agricultural industry fully into the land use planning system and its local controls. The clearest contrast to this would be the Irish example where farming groups have been successfully able to challenge protected area designations as being an unacceptable intrusion on farmers' property rights. In Italy too, the farming unions have been able to prevent the few mandatory policies on agri-environmental issues so far enacted from encroaching on the property rights of rural landowners.

To a certain extent, therefore, where farmers are seen to be producing disamenities or negative externalities, then controls may be invited that interfere with property rights, and where farmers are seen to be generating positive externalities, then payments may be justified. But the matter is not as simple as this and there are clearly cultural differences in the extent to which farmers' property rights are regarded as a limiting factor. Thus, while the Danes use land use planning restrictions to secure certain landscape and conservation benefits, the Irish clearly feel justified in using public grants to assist farmers to prevent water pollution. The swing issue would seem to be over that of habitat protection: on the one hand, the destruction of an important semi-natural habitat could be construed as a loss to the public interest but, on the other hand, the requirement to preserve that habitat may be represented as a restraint on commercial farming. The resolution of this conundrum is culturally very specific.

The relative strength of farmers' property rights

Why is it that some societies are more respectful of farmers' property rights than others? In part this relates to the relative degrees to which societies are agrarian or urban. Amongst the societies studied, the Danes are the most post-agrarian (and post-industrial), and have an active citizenry willing to challenge farmers' actions. In addition, farmers' property rights are more deeply embedded in certain national cultures than others. In particular, where peasant-oriented land reform movements were a feature of modern state formation then rural property rights tend to be deeply embedded in national ideologies (the prime example would be Ireland, with the lesser example of Portugal, and other examples being Finland and France).

CONCLUSIONS: DIFFERENT NATIONAL STRATEGIES TOWARDS THE DEVELOPMENT OF EU POLICIES

Countries such as Denmark which are prepared to interfere in farmers' property rights are likely to see environmental policy and regulation as more important means of resolving agri-environment problems than agricultural policy. Other countries are much more likely to seek to internalise agri-environment policy within agricultural policy. These different national strategies lead to different orientations towards European policy regimes.

On the one hand, Denmark has been particularly proactive in the development of EU environmental policy. For example, the EC Nitrates Directive (91/676/EEC) was prompted largely by Danish efforts to curb nitrate leaching. Because the Directive resembled the existing national rules restricting the application of animal manure on areas prone to leaching, its implementation in Denmark has had only limited (additional) consequences for the livestock sector. More generally, the Danish perception is that their own environmental rules are more stringent than EU ones and that they are more conscientious in applying and following both domestic and EU rules than are other countries. While Denmark has thus been a force to strengthen the environmental regulation of agriculture, it has also been critical of the effectiveness of EU agri-environmental measures in helping to achieve the objectives of environmental policy, not least the Nitrates Directive.

The other three countries, however, while enthusiastic converts to EU agri-environment policy, have all procrastinated over the implementation of the Nitrates Directive. More generally, the European environmental agenda is seen in southern Europe as being driven by northern European concerns. The response of officials in southern capitals is either the principled one of 'we fit in with these policies as good Europeans' or the cynical one of 'we respond to these policies because of European funding'. European regulations, though, do tend to command some respect (the Nitrates Directive notwithstanding) because they are regarded as tougher and more certain than national regulations in these countries (La Spina and Sciortino, 1993).

Italy, Portugal and Ireland have certainly not been pace-setters in the development of EU environmental policy. Indeed, they have sought to avoid regulatory approaches that might be seen to impinge on farmers' property rights. This has meant that they have had to backtrack over the implementation of EU environmental measures that would affect agriculture, including not just the anti-pollution oriented Nitrates Directive but also the conservation-oriented Birds Directive (79/409/EEC) and Habitats Directive (92/43/EEC).

Unlike Denmark, these countries look increasingly to agri-environment measures to resolve conflicts between agriculture and the environment. Thus, in the negotiations surrounding the Agenda 2000 reforms, they have sought to increase the resources available for the so-called Accompanying Measures (which includes agri-environment measures). Denmark, in contrast, has been a leading advocate of imposing environmental cross-compliance conditions on commodity payments under the CAP which could be a means of obliging both farmers and Member States to conform not only to mandatory environmental standards but also to additional obligations.

Agri-environment policy, as it has developed within the CAP, employs a particular means to pursue its very diverse ends, i.e. payments to farmers and their voluntary involvement in agri-environment schemes. Implicitly or explicitly, therefore, it promotes a particular model for resolving the tensions between agriculture and the environment and of the property rights that should regulate the matter. As we have seen, in some EU countries peasant-derived private property rights remain strongly entrenched. Under such conditions, it would take a strong state with considerable public legitimacy to intervene in the distribution of property rights, to claw back or create public rights. But the EU is a weak, supranational state, deficient in such legitimacy. The CAP, moreover, is a policy regime in which payment to farmers is the norm. It was perhaps inevitable, therefore, that agri-environment policy within the CAP would develop on the basis of voluntary compensatory schemes for farmers.

This approach clearly suits those countries that see the CAP as essentially a means of income transfer to rural areas, whose main rural problems are seen to come from the prospect or the reality of agricultural decline and rural abandonment and who tend to follow rather than lead the development of EU environmental policy. It suits much less those countries that see the CAP fundamentally as a market or trade discipline (or constraint), whose main rural problems are seen to come from agricultural intensification and who pursue a progressive environmental policy. For a country like Denmark, the model of property rights embodied in EU agri-environment policy also challenges the much more public-interest oriented approach it has tended to adopt nationally, including for example the maintenance of low intensity grazing systems through land use zoning restrictions on farmers' property rights.

These different national orientations are likely to be played out through the stances Member States adopt towards the development of the two separate EU policy regimes - environment policy and the CAP - with different states according greater priority or precedence to one or to the other in the resolution of agri-environment problems. However, beyond this sort of policy jostling, which is not uncommon in EU politics, there is something more fundamental at stake - namely, competing conceptions of the role of farmers and agriculture in contemporary society.

Acknowledgement

The study was commissioned by the British wildlife and countryside agencies.

REFERENCES

Allen, R. and Jones, T. (1990) *Guests of the Nation; The People of Ireland versus the Multi-nationals.* Earthscan, London.

Andersen, M.S. and Hansen, M.W. (1991) *Vandmiljøplanen: Fra Forhandlung til symbol.* Copenhagen.

Andersen, M.S. and Liefferink, D. (eds) (1997) *European Environmental Policy: the Pioneers.* Manchester University Press, Manchester.

Bagnasco, A. (1977) *Tre Italia. La Problematica Territoriale Dello Sviluppo Italiano.* Il Mulino, Bologna.

Baldock, D., Beaufoy, G. and Clark, J. (1994) *The Nature of Farming*. Institute for European Environmental Policy, London.

Baptista, F.O. (1995) Agriculture, rural society and the land question in Portugal. *Sociologia Ruralis* 35(3-4), 309-321.

Barberis, C. and Siesto, V. (eds) (1993) *Agricoltura e Strati Sociali*. Franco Angeli, Milan.

Bignall, E. and McCracken, D. (1996) Low-intensity farming systems in the conservation of the countryside. *Journal of Applied Ecology* 33, 413-424.

Billaud, J.-P., Bruckmeier, K., Patrício, T. and Pinton, F. (1997) Social construction of the rural environment. Europe and discourses in France, Germany and Portugal. In: De Haan, H., Kasimis, B. and Redclift, M. (eds) *Sustainable Rural Development*. Ashgate Publishing Limited, Aldershot, pp. 9-35.

Bordin, A., Cesaro, L., Gatto, P. and Merlo, M. (1998) Italy. In: Brouwer, F. and Lowe, P. (eds) *CAP and the Rural Environment in Transition: A Panorama of National Perspectives*. Wageningen Pers, Wageningen, pp. 241-266.

Brouwer, F.M. (1995) Pesticides in the European Union. Paper presented at the *Workshop on Pesticides*, Wageningen, 24-27 August 1995.

Caldas, J.C. (1998) Portugal. In: Brouwer, F. and Lowe, P.(eds) *CAP and the Rural Environment in Transition: a Panorama of National Perspectives*. Wageningen Pers, Wageningen, pp. 285-302.

Danish Ministry of Environment (1995) *Nature and Environmental Policy Memorandum*. Copenhagen.

Daugbjerg, C. (1998) *Policy Networks under Pressure. Pollution Control, Policy Reform and the Power of Farmers*. Ashgate, Aldershot.

De Benedictis, M. and De Filippis, F. (1998) L'intervento pubblico in agricoltura tra vecchio e nuovo paradigma: il caso dell'Unione Europea. *la Questione Agraria* 71.

European Commission (1995) *Europeans and the Environment*. European Commission, Brussels.

European Commission (1998) *The Agricultural Situation in the European Union - 1997 Report*. CEC, Luxembourg.

Fabiani, G. (1986) *L'agricoltura Italiana tra Sviluppo e Crisi (1945-1985)*. Il Mulino, Bologna.

Flynn, B. (1998) Ireland. In: Brouwer, F. and Lowe, P. (eds) *CAP and the Rural Environment in Transition: a Panorama of National Perspectives*. Wageningen Pers, Wageningen, pp. 83-98.

Hickie, D. (1996) *Evaluation of Environmental Designations in Ireland*. Heritage Council, Dublin.

Hickie, D., Turner, R., Mellon, C. and Coveney, J. (1995) *Ireland's Forested Future*. Discussion paper prepared for the Royal Society for the Protection of Birds. An Taisce and the Irish Wildbird Conservancy.

IUCN (1990) *United Nations List of National Parks and Protected Areas*. IUCN, Cambridge.

La Spina, A. and Sciortino, G. (1993) Common agenda, southern rules: European integration and environmental change in the Mediterranean states. In: Liefferink, J.D., Lowe, P.D. and Mol, A.J.P. (eds) *European Integration and Environmental Policy*. Belhaven, London.

Lima, A.V. de (1991) Velhos e Novos Agricultores em Portugal. *Análise Social* 26(111), 335-358.

Linddal, M. (1998) Denmark. In: Brouwer, F. and Lowe, P. (eds) *CAP and the Rural Environment in Transition: a Panorama of National Perspectives*. Wageningen Pers, Wageningen, pp. 185-197.

Mantino, F. (ed.) (1995) *Impresa agraria e dintorni*. Studi e Ricerche, Istituto Nazionale di Economia Agraria, Rome.

McCullagh, C. (1991) A tie that blinds - family and ideology in Ireland. *Economic and Social Review* 22l, 199-211.

McLoughlin, M. (1998) REPS - The first four years. In: *Protecting Ireland's National Heritage, REPS Conference Proceedings.* Johnstown Castle, Wexford, 3 December, 1998.

Merlo, M. and Manente, M. (1994) *Consequences of Common Agricultural Policy Reform for Rural Development and the Environment.* European Economy, Reports and Studies, no. 5.

Ministério da Agricultura (1993) *Dois Contributos para um Livro Branco sobre a Agricultura e o Meio Rural.* Secretaria Geral do Ministério da Agricultura, Lisbon.

Murphy, J. (1996) REPS not targeting wildlife. *Wings* - Journal of the Irish Wildbird Conservancy - Birdwatch Ireland. Autumn: 21.

OECD (1997) *Agri-environmental Indicators Database.* Organisation for Economic Cooperation and Development, Paris.

Oldrup, H. and Just, F. (1999) The implementation of sustainable agriculture in Denmark. In: David, C., Bernard, C. and Just, F. (eds) *Making Agriculture Sustainable.* University of Southern Denmark, Esbjerg.

Patrício, T. and Lima, A.V. de (1996) *Sociological Enquiry into the Conditions Required for the Success of the Supporting Environmental Measures Within the Reform of the Common Agricultural Policy - National Report: Portugal.* CIES/ISCTE, EU Programme for Research and Technical Development in the Area of Environment, Area III, SEER-II Project, No. EV5V-CT94-0372, Lisbon.

Portela, J. (1994) Agriculture is primarily what? In: Symes, D. and Jansen, A. (eds) *Agricultural Restructuring and Rural Change in Europe.* Agricultural University, Wageningen, pp. 32-48.

Povellato, A. (1993) Rural development and small farms. Information needs for public intervention. *Annals of Warsaw Agricultural University.* Agricultural Economics and Rural Sociology 30.

Povellato, A. (1996) The response of the Member States: Italy. In: Whitby, M. (ed.) *The European Environment and CAP Reform. Policies and Prospects for Conservation.* CAB International, Wallingford, pp. 131-154.

Reis, M. and Lima, A.V. de (1998) Desenvolvimento, Território e Ambiente. In: *Portugal, que Modernidade?* Celta, Oeiras.

Saraceno, E. (1994) Alternative readings of spatial differentiation: the rural versus the local approach in Italy. *European Review of Agriculture Economics* 21(3/4).

Sheehy, S.J., Skeffington, M. and Bleasdale, A.J. (1991) Influence of agricultural practices on plant communities in Connemara. In: Feehan, J. (ed.) *Environment and Development in Ireland.* Proceedings of UCD Conference, December 1991.

Sillani, S. (1987) Land consolidation and environment. In: Merlo, M., Stellin, G., Harou, P. and Whitby, M. (eds) *Multipurpose Agriculture and Forestry.* Proceedings of the 11th Seminar of the EAAE, 28 April-3 May 1986, Wissenschaftsverlag Vauk, Kiel, pp. 299-306.

Stanners, D. and Bourdeau, P. (eds) (1995) *Europe's Environment: The Dobríš Assessment.* European Environment Agency, Copenhagen.

Varley, T. (1996). Farmers against Nationalists: The rise and fall of Clann na Talmhan in Galway. In: Moran, G. and Gillespie, R. (eds) *Galway: History and Society.* Geography Publications, Dublin, Chapter 18.

Market Remuneration for Goods and Services Provided by Agriculture and Forestry

16

Maurizio Merlo, Erica Milocco and Paola Virgilietti

INTRODUCTION

In addition to the traditional market commodities, of food and fibres (F&F), agriculture and forestry provide, or rather produce, a range of environmental and recreational goods and services (ERGS) like pleasant landscapes, country lanes, footpaths and other amenities, habitats for various kinds of flora and fauna, and space for various sporting and recreational activities. These ERGS are generally perceived by society as public goods: people cannot be excluded and rivalry is not greatly felt.

The degree of excludability can, however, vary according to the property rights for each particular ERGS (which sometimes differ between countries),[1] the local customs and the transaction costs to enforce exclusion. The degree of rivalry is equally variable according to each particular ERGS - e.g. a landscape visible from a public road does not generally have a high level of rivalry while the access to a nature reserve can create congestion and rivalry.

With reference to the diagram of public and private goods (Figure 16.1) the ERGS supplied by agriculture and forestry are mainly located in the low excludability and low/medium rivalry area, while food, fibres and timber are private goods. Therefore forests (as a whole) and to a more limited extent farmland, can be considered an 'impure public good', where there is a joint production of both public and private benefits.

The public-good nature of ERGS can also be seen in terms of positive externalities associated with forestry. The notion of externalities raises the concept of market failure, whenever for various reasons, as is often the case with farming and forestry, positive or negative externalities cannot be internalised through incentives or taxes as advocated by Pigou (1920). Internalisation of externalities opens

---

[1] Hunting rights and related access belong to the landowners in the Anglo-Saxon and German countries, and are strictly private in nature. In contrast, in Latin countries, these rights are public and assigned by the state to individuals who pay for a permission to use the rights.

up a new perspective for ERGS, that of joint production with traditional market commodities (F&F). Complementarity/indifference/substitution represent the possible relationships between ERGS and F&F, which can be depicted in a conventional Production Possibility Curve. The lack of price for ERGS carries the status of unintended by-product with production equilibria tending towards market goods like F&F, i.e. the vertical axis.

The three criteria employed so far (public goods, externalities and joint unintended by-products) underline the non-market essence of ERGS and therefore potential market failures because of free riding, if not the tragedy of the commons.

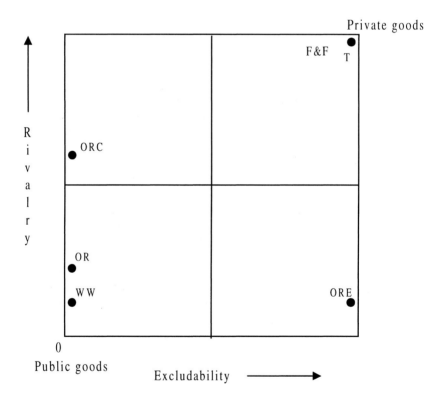

Figure 16.1   Public and private goods according to excludability and rivalry criteria

T: timber (complete excludability and rivalry)

F&F: food and fibres (complete excludability and rivalry)

WW: wildlife watching (no excludability and low rivalry)

OR: outdoor recreation with open access (no excludability and low rivalry)

ORE: outdoor recreation subject to ticket payment (excludability) without congestion (no rivalry)

ORC: outdoor recreation with open access subject to congestion (rivalry but no excludability)

Production is likely to be lower than demand, implying a loss of welfare. The obvious answer, reflected by many existing policies across Europe, has been in the field of mandatory policy, that is obliging landowners to adopt measures aimed at guaranteeing a certain level of ERGS; for instance, practices aimed at preventing landslides and avalanches in mountain areas are often applied in Alpine regions. More recently, following Pigou (1920), internalisation through financial aids based on the 'state pays' approach (OECD, 1996) has been utilised in various policies, in particular the accompanying measures, which were introduced in 1992 under the Common Agricultural Policy (CAP), e.g. Regulation 2078/92 on agri-environmental measures and Regulation 2080/92 on afforestation of farmland. This orientation seems to be confirmed by further reform, in particular Agenda 2000, where existing agri-environmental measures reaffirmed while all compensatory payments could be coupled to the provisions of ERGS. Unfortunately, this decision will be left to a large extent to Member States, and this could undermine the meaning stated by Agenda 2000. What remains to be investigated and/or extensively applied, are Coasian approaches (1960) based on the 'beneficiary pays' approach. The hypothesis is certainly appealing because it would be based on the market and, even better, would be without public cost. The prospect of creating a market for ERGS, that would transform public goods and services into private-market goods and services, lies behind the notion of integrated rural development which is a feature of the Agenda 2000 reforms of the CAP. The transformation could be helped by the implementation of the new Rural Development Regulation.

The scope of the chapter is to outline cases where ERGS have been transformed into market goods, or means have been found to remunerate the producers of ERGS in the market through complementary goods and services.

The chapter first presents a characterisation and classification of various ERGS found in Italy. The situation before transformation is outlined, while development strategies are also described including legal changes as well as the attachment of complementary goods through marketing techniques. Some considerations on possible strategies to achieve market remuneration of ERGS are put forward. The conclusions underline some possible policies and measures to achieve the transformation of ERGS into marketable environmental and recreational product (ER-Products).

ERGS AND ER-PRODUCTS CHARACTERISATION

Location

Research has been undertaken in four European Union (EU) countries, though the present chapter reports only on results from the Italian case studies. In order to gather the necessary information on ERGS, a product-oriented questionnaire

```
01:Mushroom picking permits in Borgotaro (Parma, Emilia Romagna)
02:Consortium of the Borgotaro Mushroom in Borgotaro (Parma, Emilia Romagna)
03:Golf course in the Cansiglio plateau (Treviso, Veneto)
04:Car park in Val Visdende (Belluno, Veneto)
05:Cross-country and alpine skiing in the Cansiglio plateau (Treviso, Veneto)
06:Forest access on mount Nevegal (Belluno, Veneto)
07:Ecological and ethnographic museums in the Cansiglio plateau (Treviso, Veneto)
08:Mushroom picking permits in the Cansiglio plateau (Treviso, Veneto)
09:Cross-country skiing in the Gares Valley (Belluno, Veneto)
10:Guided tours in the Foreste Casentinesti National Park (Grosseto, Tuscany)
11:Stelvio National Park-Camping site in the Rabbi Valley (Trento, Trentino Alto-
   Adige)
12:Car park in San Cassiano (Bolzano, Trentino Alto-Adige)
13:Certification of timber production sustainability in the Fiemme Valley (Trento, Tren-
   tino)
14:Cross-country skiing on mount Nevegal (Belluno, Veneto)
15:Cross-country skiing in Falcade (Belluno, Veneto)
16:Car park and picnic area around the Barcis Lake (Pordenone, Friuli Venezia-Giulia)
17:Forest access through the Cimoliana Valley (Pordenone, Friuli Venezia-Giulia)
18:Riding school in Val Canzoi (Belluno, Veneto)
19:Mushroom picking permits in the Fiemme Valley (Trento, Trentino Alto-Adige)
20:Guided visits in the Stelvio National Park (Trento, Trentino Alto-Adige)
21:Resonance Timber Label in the Canale Valley (Udine, Friuli Venezia-Giulia)
22:Specialisation courses 'Università nel Bosco' at the Gola del Furlo (Persaro,
   Marche)
23:Forest access to Bosco Fontana (Mantova, Lombardia)
24:Picnic areas in Val Canzoi (Belluno, Veneto)
25:Skiing concessions in the Cortina Valley (Belluno, Veneto)
26:Renting of tourist infrastructures in the Cortina Valley (Belluno, Veneto)
27:Montella Chestnut DOC FGO (Avellino, Campania)
28:Access to the Maremma Regional Park (Grosseto, Toscana)
29:Sponsorship by WWF and SNAM to the Abruzzo National Park (L'Aquila, Abruzzo)
```

Figure 16.2   List of the Italian ER-Products documented and their geographical
location

was jointly developed. Some 29 ERGS case studies were investigated in Italy, as reported in Figure 16.2. They were concentrated in northern Italy, but cases were also included that illustrated important experiences of central/southern Italy.

Categories

The ER-Products supplied are classified in Table 16.1 into three broad categories: recreational, environmental and traditional products. The majority of products fall into the recreational category. The majority of the so-called recreational products are given by organised or 'structured' activities, taking advantage of a rural environment, which could have been carried out elsewhere, for example car parks and various sports. Table 16.2 distinguishes between 'non-structured' and 'structured' goods and services. Actually, the only pure environmental good, for which it has

Table 16.1 Typology of ER-Products

Category	Type of product	Number of case studies documented
Recreational	Car parks	2
	Cross-country skiing	4
	Picnic areas	2
	Mushroom picking permits	3
	Riding schools and other sporting activities (horse-tourism, golf clubs)	2
	Renting of tourist infrastructures	3
	Forest access	4
	Guided visits and museums	2
Environmental	Sponsorship	2
	Environmental courses	1
Traditional	Timber quality labels	2
	Other traditional products linked with the environment	2

Table 16.2 Classification of ER-Products according to 'structures'

Goods and/or services	Conditions for market creation	Payment vehicles [a]
1. Environmental goods and services (non-structured)		
• Picking mushrooms, truffles and other non-timber forest products	Property rights definition and assignment	Access card, tickets, etc.
• Hunting and fishing	Property rights definition and assignment [b]	Access cards, consumption invoices, etc.
• Access (trekking, skiing, birdwatching, climbing)	Property rights definition and assignment [b]	Access cards subscriptions, concessions, etc.
• Water regime and treatment	Property rights definition and assignment (management agreement)	Consortia, subscriptions, etc.
• Landscape images	Property rights definition and assignment	Advertising property rights

Table 16.2   Continued

Goods and/or services	Conditions for market creation	Payment vehicles [a]
2. Additional/complementary services with respect to environmental goods and services (structured)		
• Picnic sites, service structures, etc.	Licences and permits [b]	Access cards, tickets, subscriptions, etc.
• Footpaths for various purposes, etc.	Licences and permits [b]	Access cards, tickets, subscriptions, etc.
• Car parks, caravan sites	Licences and permits	Tickets
• Tourism and educational centres, laboratories and museums	Licences and permits	Tickets, subscriptions, etc.
• Sports centres and recreation facilities	Licences and permits	Access cards, tickets, subscriptions, etc.
• Hire of sporting equipment, riding facilities, stables	Licences	Access cards, tickets, subscriptions, etc.
• Farmhouse, accommodation, meals and beds	Licences and permits	Consumption invoices
3. Traditional quality products linked to the area and production techniques		
• Traditional quality products linked to environmental quality	Property rights definition/ assignment, labels and trade marks [b]	Consumption invoices
• Traditional quality products linked to environmentally friendly techniques	Property rights definition/ assignment labels and trade marks	Consumption invoices

Source: Merlo and Ruol (1996).
[a] Intermediary figures may exist between the landowner and the consumer (various levels of managers, tenants and contractors) responsible for marketing environmental products and services. [b] Cases in which collective agreements between supply and demand organised in associations or consortia may facilitate the creation of a market.

been possible to create a market was given by 'sponsorship' of specific important environments - for example the Parco Nazionale d'Abruzzo with its well known brown bear. However, it could be argued that what is sold is an environmental image, structured in the advertising market.

Conceptually similar is the case of traditional forest products where, thanks to *appellation d'origine* and certification schemes, the consumer is paying for the environment through a traditional product. The 'structure' is given in this case by the certification scheme.

Ownership and management

The land where ERGS has found market remuneration is mainly publicly owned (state government or regional and local authorities), including 18 cases out of 29. The other cases fall under common-property ownership or the modern version given by consortia, associations and trusts between private and sometimes public subjects. Only two of the cases covered private land (Table 16.3). This is not surprising as far as Italy is concerned. The creation of a market for ERGS requires large spaces that are not compatible with the small fragmented private holdings, which are very common in Italy. Only public land, or various forms of land owning associations, can provide the appropriate size for producing and marketing ER-Products. Economies of scale are particularly felt in producing ER-Products, even more than in traditional agricultural and forest products.

The hypothesis that the production of ER-Products can find more effective application in 'common land' of adequate size, with a legal status midway between state and private, must not be neglected. Pareto (1896) showed that forestry and grazing in mountain areas were more easily managed by common property regimes. The 'clubs theory' of Buchanan (1955) supports this line of thought. After all, the services provided by tennis courts or swimming pools are not so different from certain ERGS provided by agriculture and forestry; Glück (1995) uses the term 'toll goods', implying some sort of remuneration through excludability.

Table 16.4 gives a picture of the rather diversified management status of ER-Products. Cases of collective public management predominate. However, the role of private contractors, sometimes organised as associations or co-operatives, must

Table 16.3 Legal status of the land where ER-Products are produced

Legal status	Number of ER-Products
Public:	
Federal [a]	8
State [a]	5
Local authority	5
Common property	7
Consortia and associations	2
Private	2

[a] National and Regional Parks are included in this category; however, they are not landowners, but institutional bodies that manage the whole designated area through a specific management plan which must be complied with by all the landowners.

Table 16.4   Management status of ER-Products

Type of management	Number of ER-Products
Private landowner	1
Contractor	4
Consortium	4
Co-operative	1
Association	2
Common properties	5
Public bodies	4
Park authority	3
Other	5
Total	29

not be overlooked. They have often been the driving force behind the transformation, management and marketing of ER-Products.

THE STATE OF ERGS BEFORE TRANSFORMATION

In general, before the commercial transformation had taken place, there was a countryside (farmland or forests and rural environments) that could be used almost for free by visitors with little competition (rivalry). As shown by Tables 16.5 and 16.6, excludability was not much in evidence (only in 5 cases out of 29) while rivalry, although perceived in half the cases, remained rather limited - only in 2 cases out of 29 was it felt to be a relevant factor. It is clear from the results that a large part of the countryside where ERGS occurred had the status of public-collective goods. These were usually positive externalities connected to agriculture and forestry and the rural environment.

Pure public goods are characterised by 'non-rivalry of consumption' and 'indivisibility of benefits' (Samuelson, 1955). Excludability, meanwhile, is a boundary

Table 16.5   Excludability before transformation

Excludability	Number of case studies [a]
No	23
Very low	-
Low	2
Relevant	1
Very relevant	2
Total	28

[a] Not applicable for one case.

to free use, makes the good divisible and therefore, at least potentially, marketable.

The concept of rivalry overlaps that of congestion. In fact a good is rival when the consumption by one individual detracts all the possibilities of consumption of the same unit of the good to other consumers. Congestion can be considered, in general terms, as 'the situation in which one individual's consumption reduces the quality of service available to others' (Cornes and Sandler, 1986). In this chapter the reference is rather to Randall's (1987) definition of congestible goods: 'any good that can be enjoyed by many individuals but is subject to a capacity constraint and for which the fixed cost of provision far exceeds the marginal cost of adding additional users until the capacity constraint is approached'. From this definition, most club goods can be congestible. It is clear that congestion, diminishing the possibility of consumption to other people, causes an increase in rivalry, but rivalry can exist independently from congestion, if the good is excludable and divisible. Rivalry, for club goods and services, can exist at two levels: an internal one (among people using the good and/or services, where congestion can be a problem) and an external one (among potential users, depending on intrinsic good characteristics). For example, considering mushrooms picking permits, there can be rivalry to get the permission, if the number of permissions is limited and demand is greater than supply (external rivalry), and, once the permits have been issued, rivalry can exist among mushrooms pickers, due to congestion (internal rivalry). As far as possible, the two components (internal and external) are evaluated separately and added up in the final assignment of rivalry level. Of course this classification has mostly an explanatory and descriptive meaning.

Table 16.6 Rivalry before transformation

Rivalry	Number of case studies [a]
No	14
Very low	3
Low	9
Relevant	2
Very relevant	-
Total	28

[a] Not applicable for one case.

CAUSES AND CONDITIONS FOR THE COMMERCIAL TRANSFORMATION OF ERGS

The development of ERGS into ER-Products can imply the creation of a new product or the upgrade or revalorisation of an existing product. In general, a 'new' product implies a shift in the control over the resource such as a modification of property rights. This is the case of mushroom picking permits, or the creation of new facilities in a certain area, particularly through co-operation between several landowners. The introduction of new labels and trade marks or sponsorship also constitutes creation of a new product. An upgrade, in contrast, is more related to the reorganisation of the marketing of an existing product, with new forms of payment, advertising, certification, etc. The survey showed that ER-Products are new in 53% of the case studies, while in the remaining cases there was an upgrade of already existing products.

Product development is evenly distributed over the past decades. This applies particularly to recreational products where the emergence of the demand was not so recent, dating back to the 1960s and 1970s. The upgrade of existing products, including the use of *appellation d'origine*, seems to have concentrated in the 1990s, and this is undoubtedly due to EU Regulations (2081/92 and 2082/92) that have extended to various products the legislation once applied only to cheeses and wines.

The impetus for ER-Product development is rather diverse, as shown by Table 16.7, which refers only to the upgrade of existing products. The increasing cost of the provision of the existing products, or other economic pressures such as decreasing demand and prices or unfair competition, have often been the driving force for development (50%). In several cases, however, it has been the knowledge of new opportunities that have pushed towards changes, such as legal innovation, re-assignment of property rights, market niche exploration, etc.

Table 16.7   Causes for upgrading existing ER-Products

Causes of development	No. of ER-Products [a]
Increasing costs	5
Decreasing demand/price/unfair competition	4
Others:	
• Creation of National Parks with good organisational and marketing potential	2
• Increasing demand	1
• Introduction of a new law on mushroom picking	1
• Re-assignment of property rights over alpine huts	1
• Demand niches on environmentally friendly products	1
Total	15 [a]

[a] Multiple answers possible.

Table 16.8 again confirms that the success factors behind ER-Product development (both new and upgrades) are given by 'opportunity exploitation'. Considering the crucial issue of competition between ER-Products and from similar ERGS available locally, the survey shows that in general it is not possible to get free access to a similar product in the area. Where free substitution possibilities do exist, the new development has been possible thanks to additional advantages/benefits made available to the consumers, for example:

- a mix of services, all linked together and well organised;
- the high quality of the new service provided;
- the closeness of the new service to tourist destinations.

Table 16.8   Success factors in the development of ER-Products

Development success factor [a]	No. of ER-Products [a]
• The quality of the product or service supplied (professionality of guides, good cross-country track, etc.)	7
• Lack of competition (no other car parks, no other camping sites, etc.)	4
• Proximity to tourist destinations	4
• Popularity of mushroom picking in Italy	4
• Structural effects, i.e. large set of activities available in the area, well organised and linked together	2
• Beauty of the landscape and of the area in general	2
• Presence of local demand	2
• Marketing activity of National Parks with respect to guided visits	2
• Characteristics of the service that satisfies certain categories of users (very easy cross-country skiing track for families and beginners)	1
• The object of sponsorship (the brown bear for the WWF)	1
Total	29 [a]

[a] Multiple answers were possible; however, in several cases no answer was given.

As shown by Table 16.9 the idea for the product development, in a substantial number of cases (19), was self-generated by the agriculture-forestry owner or manager. There are, however, other cases where imitation and external information have played an important role. The same situation applies to the provision of the know-how, which has developed internally to the local managing enterprise in the majority of cases. Total or partial use of external know-how is, however, also common.

In general the personnel involved with the new initiatives, as was the case with the know-how, were already available. This reduced any difficulty in organising the initiative. Large facilities (e.g. buildings and sporting facilities) for the

Table 16.9 Provenance of the idea for the development of ER-Products

Sources of development idea	Number of ER-Products [a]
Landowner/forest manager	19
Imitation	6
Mass media	3
Specialised press	3
Customers	1
Others including: University, Provincial or Regional regulation on mushroom picking permits, Forest Administration and Municipality	13
Total	45 [a]

[a] Multiple answers possible.

product's development were needed by more than half of the case studies (17 out of 29). In the majority of cases these were not however easily available and some kind of investment had to be made, but, in general, the investment was not extensive.

TRANSFORMATION TECHNIQUES FROM ERGS TO ER-PRODUCTS

Excludability is a necessary condition for the transformation. In principle, it can be brought about by modifying the legal status of the ERGS, and therefore the property rights that surround them. Such changes, even if possible, may be quite difficult to implement, because they generally require changes in legislation. People, moreover, might reject these changes if consolidated traditional rights are affected. Alternatively, a '*de facto*' excludability can be established in a softer way by using marketing techniques. In this case ERGS consumption is linked to additional (complementary) services. In effect, ERGS are coated in additional goods, that are remunerated by the market, and existing rights are not changed. Of course, excludability can be more easily achieved when a high complementarity between ERGS and ER-Products exists.

The case studies show that before transformation many ERGS were open to the public for free (Table 16.5). The situation, however, was in general rather unsatisfactory because of pressure on natural resources, free riding, poor management, and lack of complementary services. Some sort of legal changes therefore have been introduced. This enabled support for substantial management of the ERGS, and their transformation into marketable ER-Products. In 19 cases product development did require some kind of legal change (Table 16.10).

Going through the individual cases, however, it shows rather clearly that no dramatic changes of property rights have taken place. In general, it has been a matter of adjustments/clarifications mainly linked to permissions by Local Authorities for selling private products and services attached to the quality of the environment. Once again the point is that what is really sold in the market is a product complementary to the ERGS; the latter's legal status has not changed, and could not have been, given consolidated property rights.

Table 16.10 Legal changes that had occurred in the transformation of ERGS into ER-Products

Legal changes [a]	Number of ER-Products			
	Traditional	Recreational	Environmental	Total
No	1	8	1	10
Yes	3	16	-	19
Total	4	24	1	29

[a] The following legal changes were reported:
- access allowed with payment of a ticket authorised by local authorities;
- cross-country skiing - modification of the Town and Country Planning in order to be able to use private land for the track. The track layout is indicated on the plan and fencing off of private property is forbidden. No compensation is provided to private owners;
- right to label the product acquired through certification, origin label, or other trade mark;
- regional or provincial law in the case of mushroom picking permits that rules that the access to the woods for mushroom picking is possible only for permit holders;
- creation of Regional and National Parks.

As shown in Table 16.11, the transformation has, in any case, remarkably increased excludability. In 17 cases out of 26, measures have been taken to prevent free riding and therefore to enforce excludability or, better, to regulate access to ER-Products. Rather interestingly, transaction costs to prevent free riding have been considered very relevant only in four cases. It appears that 'normal' measures are in general sufficient. This fact can be related to the nature of the ER-Products provided, and paid for by the consumers, that are usually market products with which they are already familiar (car parks, various facilities, etc.).

Before the transformation, rivalry was an element in 14 cases out of 28 but in only two of these did it reach a relevant level. Afterwards the situation has changed rather deeply and in all cases a certain level of rivalry now prevails (Table 16.12). In the few examples that faced a major congestion problem, the transformation was aimed at reducing it, by regulating the use of the resource (internal rivalry). In any case, the increase in excludability changes the economic 'status' of the good, shifting towards private goods, and therefore causes an increase in (external) rivalry, that generally exceeds the opposite effect of congestion reduction. Moreover, in certain cases, transformation has increased internal rivalry because access to the ER-Products has been facilitated thanks to new structures. In conclusion, an increase in rivalry was almost always registered. However, only in five cases have the ER-Products achieved the status of a private good (mainly traditional products linked to environment quality). The status of club goods is by far the most frequent, due to the intrinsic characteristics of the goods and services offered (car parks, picnic sites, cross-country skiing).

Table 16.11 Excludability before and after transformation (number of case studies) [a]

	Before trans- formation	After transformation					
		No	Very low	Low	Relevant	Very relevant	Total
No	23	-	2	-	-	21	23
Very low	-	-	-	-	-	-	-
Low	2	-	-	-	-	2	2
Relevant	1	-	-	-	-	1	1
Very relevant	2	-	-	-	-	2	2
Total	28	-	2	-	-	26	28

[a] Not applicable for one case.

Table 16.12 Rivalry before and after transformation (number of case studies) [a]

	Before trans- formation	After transformation					
		No	Very low	Low	Relevant	Very relevant	Total
No	14	-	4	2	5	3	14
Very low	3	-	-	2	1	-	3
Low	9	-	-	1	8	-	9
Relevant	2	-	-	-	-	2	2
Very relevant	-	-	-	-	-	-	-
Total	28	-	4	5	14	5	28

[a] Not applicable for one case.

'Complementarity' behind transformation of ERGS into ER-Products: the main philosophy for product development

According to the survey, legal changes in the status of the ERGS represent the first preliminary step in the transformation process. The core of the transformation

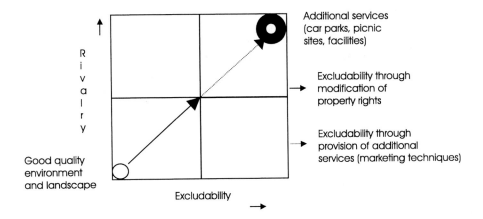

Figure 16.3 Transformation paths from ERGS to ER-Products

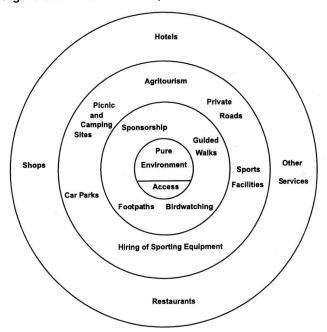

Figure 16.4 Complementarity between ERGS and ER-Products
Source: Elaborated from Mantau (1995)

process is made up by the application of marketing techniques, based on the complementarity between ERGS and ER-Products. In fact, in most cases, the presence of additional goods and services was necessary to create a market for the ERGS. Figure 16.3 exemplifies the transformation process: the ERGS, initially a public

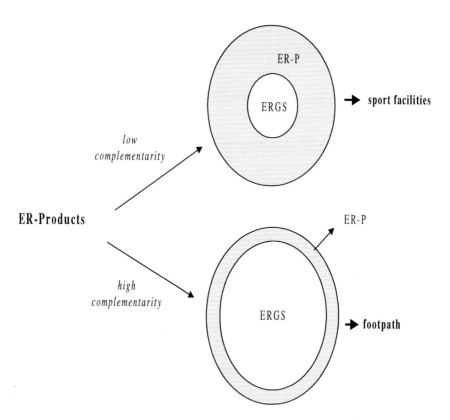

Figure 16.5   Complementarity between ERGS and ER-Products

good, is coated with traditional market goods such as a car park, tourist and sporting facilities, even traditional farm and forest products, whose quality and image is linked to the environment and landscape, where they are produced.

Thus the payment for ERGS is generally quite indirect - through the complementary/additional ER-Products. Figure 16.4 outlines the situation where particularly attractive or interesting environments and landscapes are paid for through different kinds of goods and services. Access in itself and on its own does not in general justify any form of payment. Only facilities like those shown in the first ring allow payment to take place to a certain extent. The second ring considers even more 'structured' services with higher added value, and even more the third one. The facilities included in the first ring present a higher complementarity with the pure environment and landscape. Complementarity decreases in moving to the second and third rings.

At the limit, when complementary is very high the ERGS and the ER-Products overlap almost completely, except for the payment that the latter requires (Figure 16.5). A case in point would be access to a nature reserve paid by the visitor. Before transformation access was free, but difficult because of the lack of footpaths. After transformation the access must be paid for at least to cover the cost of providing the footpath. In other cases the ERGS are different from the ER-Products paid for by the consumer, such as specific sporting facilities, camping sites, holiday cottages, etc. However, the ER-Products draw upon the qualities of the attached ERGS (the nice landscape or the interesting environment). In other words people pay to use certain facilities or to have a good with certain characteristics.

Possible failures of transformation strategies

Amongst the 29 case studies surveyed, only one clear failure can be counted. A common property in a tourist area decided to introduce a fee to a previously unorganised parking site that had been used for free. This resulted in the former customers complaining very vociferously and in the municipality removing the permission to ask for a parking fee.

Managers of ER-Products reported knowledge of nine failures due to various causes, in particular lack of financial resources. Property rights problems were indicated in at least four cases. One of these involved the obstruction of the one major road leading into an Alpine valley with a barrier in order to prevent free access and to enable the collection of an entrance fee. Since the road was public, as soon as people started to complain, the municipality ordered the landowner (a common property) to remove the barrier. Subsequently, thanks to a different transformation strategy, where payment is asked for parking, the case has turned out to be a success.

A different type of failure needs to be pointed out regarding mushroom picking permits. In 1992 a State Act required each region to produce a regulation on mushroom picking. The regions adapted the State Act to their particular situation, but with one common characteristic: public bodies were entitled to gather the revenues derived from issuing mushroom picking permits, thus bypassing private forest owners who were not remunerated. There have been only two cases of private owners avoiding this situation due to their status and size of their property, both of them being common properties: Magnifica Comunità della Val di Fiemme and Comunalie Parmensi were able to prove their legal and technical capabilities in managing ER-Products.

TRANSFORMATION STRATEGIES TO DEVELOP ER-PRODUCTS

Transformation paths to achieve market remuneration for ERGS can be very different, as shown by Figure 16.6 which summarises the development of ER-Products. The paths may include modifications in the legal status of ERGS (thick line) or the application of marketing techniques (dotted line) or a combination of the two.

Depending on the situations before and after transformation (initial and final levels of excludability and rivalry) and on the importance of the two mechanisms implied (legal changes and marketing techniques) some common paths can be identified:

a. *Partial excludability where payments are left to visitors* (2 cases): this is the case, for instance, when the payment for the use of a picnic site is voluntary. Some sort of excludability is in place, but nobody is really forced to pay for it. Also if a service is offered by two different organisations, one for free and the other for a price, the source of any potential excludability lies in the superiority of the service offered by the more organised service.

b. *Excludability due to legal changes alone* (5 cases): such as those regarding mushroom picking where permits and tickets have been introduced. Rivalry can be high, but cases with low levels of rivalry are not uncommon. This for example is the case with mushroom picking outside the main season, when there are very few pickers. The transformation path is characterised by rather strong legal innovations and a lack of marketing.

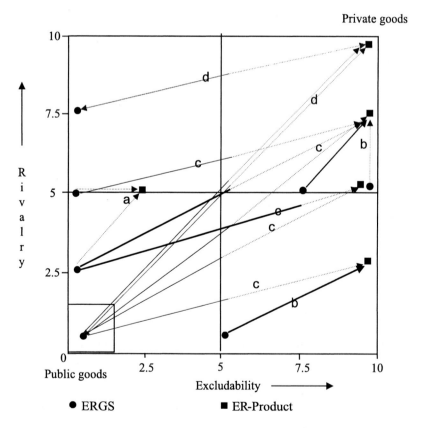

Figure 16.6   Development paths of ERGS into ER-Products

c. *Excludability due to legal changes coupled with marketing techniques* (16 cases): determined by very moderate legal changes together with effective marketing actions. Even if the levels of rivalry are quite high, after transformation a better use of the resource is recorded. The majority of the Italian cases fall under this category which corresponds to transformation into club goods, including products like car parks, picnic sites and sporting facilities. In the Italian situation, these represent the most popular and accepted (by public opinion and consumers) cases, with correspondingly good returns for the management body.

d. *Excludability due to marketing techniques of traditional products and sponsorships* (almost two-thirds of Italian case studies): the environmental quality becomes a market good through different devices which range from sponsorships to the sale of certified traditional product. Excludability in these cases can be identified with legal factors, such as the right to use the environmental image, or a certification scheme.

In the Italian situation the main shortcoming of any development strategy is given by the fact that those producing ERGS are not always those managing the ER-Products. Compensation amongst economic actors at area level must therefore be considered, otherwise landowners providing the basic environment are not stimulated to maintain it. They feel they are the object of 'free riding' both by consumers and by those managing the ER-Products.

CONCLUSIONS

The following issues seem to be important to achieve market remuneration of ERGS through their commercial transformation into ER-Products:

- The type of product is particularly important; in general those linked to recreation can more easily find market remuneration, while the environment alone is not so easily transformed into a marketable good.
- Large areas of consolidated ownership are conditional to ER-Product development. It is therefore not surprising that product development takes place mainly on public land or various forms of common property.
- Management of ER-Products is however mostly provided by various forms of associations or by contractors able to achieve some minimum amount of land required. Public and private owners are less active: the latter lack adequate size, the former entrepreneurship. The critical land mass is the surface for producing ERGS/ER-Products. A park for deer, for example, may require at least 30,000 ha.
- Before development excludability is in general very low, while rivalry varies from low to medium levels; commercial transformation increases mainly excludability, so ERGS and ER-Products complementary to them are mainly provided as 'club goods'.

- Increasing management costs, and other economic pressures, represent possible reasons for explaining ER-Product development. Opportunity exploitation is, however, the driving force behind development.
- Legal changes are generally needed for ER-Product development; yet property rights are not very much affected. It mainly concerns permissions granted by local authorities.
- Excludability is certainly achieved after transformation, the concept, however, is applied mainly to the ER-Products that are complementary to ERGS.
- Measures aimed at enforcing exclusion and preventing free riding are also taken, though in general they do not entail high costs.
- The main strategy behind development is given by 'complementarity' between ERGS and ER-Products, what is sometimes referred to as the 'structures' necessary for recreation and access to the environment.
- Failures are not rare and can be due to property rights violation as well as a lack of demand and/or poor management.
- Development strategies sometimes follow the rationale to achieve ERGS remuneration through traditional products whose image is influenced by the environment where they are produced - another form of complementarity.
- The main shortcoming of any development strategy is given by the fact that those producing ERGS are not always those managing the ER-Products (hotels, sporting facilities, etc.). Compensation amongst economic actors at an area level must therefore be considered.

Acknowledgement

The present paper is part of an EU-financed research programme on 'Niche Markets for Recreational and Environmental Services-RES' (FAIR – CT95-0743) undertaken in collaboration with the University of Hamburg (co-ordinator Prof. Udo Mantau), the University of Wien (Prof. Walter Sekot) and IBN-DLO (Dr Harrie Hekhuis and Dr Kees Van Vliet).

REFERENCES

Buchanan, J.M. (1955) An economic theory of clubs. *Economica* 32, 1-14.
Coase, R. (1960) The problem of social cost. *Journal of Law and Economics* 3, 144-147.
Cornes, R. and Sandler, T. (1986) The theory of externalities, public goods, and clubs, 1st edition. Cambridge University Press, United States of America, 303pp.
Glück, P. (1995) Forest policy means for non-timber production in Austria. In: Salberg, B. and Pelli, P. (eds) *Forest Policy Analyses - Methodological and Empirical Aspects.* EFI Proceedings No. 2, pp. 119-129.
Mantau, U. (1995) Forest policy means to support forest outputs. In: Salberg, B. and Pelli, P. (eds) *Forest Policy Analyses - Methodological and Empirical Aspects.* EFI Proceedings No. 2, pp. 131-145.
Merlo, M. (1995) Common property forest management in northern Italy: a historical and socio-economic profile. *Unasylva* 180(46), 58-63.

Merlo, M. and Ruol, G. (1996) Foreste e conservazione della natura: gli sviluppi della politica forestale italiana. *Monti e Boschi* 2, 11-14.

OECD (1996) *Amenities for Rural Development*. OECD, Paris.

Pareto, L. (1896) (Italian edition, 1949) *Corso di Economia Politica*. Einaudi, Turin.

Pigou, A.C. (1920) *Economia del benessere*. UTET, Turin.

Randall, A. (1987) *Resource Economics - an Economic Approach to Natural Resource and Environmental Policy*. 2nd edition. John Wiley & Sons, United States of America.

Samuelson, P.A. (1955) A diagrammatic exposition of a theory of public expenditure. *Review of Economics and Statistics* 36, 387-389.

Part IV

Outlook

Economic and Environmental Impacts of Agenda 2000

17

Werner Kleinhanss

INTRODUCTION

Within Agenda 2000 the Commission of the European Union (EU) made proposals for a further reform of EU policies, including agricultural market and price policy, agri-environmental policy and structural policy (European Commission, 1997; Commission of the European Union, 1998). Within the market and price policy proposals the following measures might be of importance with regard to environmental objectives:

- the reduction of intervention prices might induce lower intensities;
- the decoupling of compensation payment will affect the distribution of crops;
- stricter environmentally oriented rules for extensification premiums in the beef sector will be used;
- Member States will be allowed to restrict compensation payments related to environmental indicators (principle of cross-compliance).

Further, agri-environmental programmes will become more important.

The objective of this chapter is to assess the economic and environmental impacts of Agenda 2000. The quantitative assessment is based on the Commission's proposals of March 1998. In the event, the European Council of March 1999 made significant amendments to these proposals. The implications of the final decision compared to the proposals will be discussed in the conclusions. Agri-environmental measures of 2078/92 are excluded due to lack of information about the future specification of the numerous programmes (Dabbert *et al.*, 1998). The assessment is based on a sample of farms representing the structural characteristics of farming in Germany. General conclusions on the pros and cons of the different policy measures with regard to agricultural and other policy objectives will be drawn.

The policy measures in Agenda 2000 are rather complex, inducing adaptations at farm level depending on type of production, size and location. For this reason a 'Linear Programming Model' is used allowing a detailed specification of the different policy measures and the modelling of adaptation strategies by changes of intensity, scale and allocation of production.

MODEL, SCENARIOS AND DATA

For the farm-level assessment an optimisation model for representative farms has been used (Isermeyer *et al.,* 1996; Kleinhanss, 1996). It has been developed with the aim of technology and policy assessment taking into account the structural characteristics of farming in Germany. The model includes the most important activities of crop and livestock production. As far as farm-specific data is available, input and output coefficients are specified on the basis of farm accounting data, otherwise normative or statistical data are used. Variable inputs of crop production are determined on the basis of normative demand functions and they are adjusted to the farm accounts. Decision variables with regard to policy instruments of Common Agricultural Policy (CAP) (e.g. the small producer schemes, different levels of beef premia depending on livestock density) are specified as mixed-integer variables. Adaptation possibilities via changes in the intensities of crop production and modified fattening periods for beef production are included. Input-output coefficients are externally modified with regard to the scenario conditions, e.g. the time-scale, technical progress, etc. The objective function is the maximisation of gross margins; therefore the model predicts short-term adaptations of farms. It allows the assessment of changing economic conditions by comparison with the base situation.

Farm sample and data

For the underlying assessments a sample of 973 farms, randomly selected from the 45,000 farms included in the farm accounting system of the Land-Data, has been used. Farms are selected with regard to a projection of farm structure for the year 2005 by farm types, size-class and regions (level of the Länder). The matrices for the individual farms are automatically processed with the help of a matrix generator and solved with an integrated Mixed Integer Problem (MIP) solver. Results are aggregated depending on relevant criteria, e.g. farm type, size, region, etc. Because of missing weighting factors, the aggregation of the results at sector level is not yet possible.

Environmental indicators

The underlying approach gives only some rough indications on the state of the environment and the environmental impacts of policy measures, because only rather weak environmental indicators can be derived from farm accounting data (Nieberg, 1996) and additional information is not available because farm data are anonymous. With regard to water pollution problems the livestock densities, the level of fertiliser use and pesticides input can be used. Beyond these criteria mineral balances for nitrogen are calculated as a simple gross balance between total fertiliser input (mineral and organic fertilisers without losses) and the crops' uptake. Referring to Shannon,[1] an index of diversity is introduced to characterise the complexity of cropping patterns. This index (Hs) is expressed as follows:

$$Hs = -1 \times \sum_{i=1}^{n} \left(\frac{b_i}{a} \times \ln \frac{b_i}{a} \right)$$

a = total area of all cropping activities, $i = 1, ..., n$;
b_i = area of activity i;
n = number of activities included.

From this equation, Hs_{max} can be derived assuming an equal distribution of all activities.[2] $E = Hs/Hs_{max}$ indicates the level of diversity relative to the maximum level. Overall, 13 different activities are included after aggregation of crops with similar characteristics, e.g. winter wheat and winter barley, or corn and silage maize. Referring just to Agenda 2000, where most changes occur in the arable sector, this index should be restricted to arable land. In that case, though, comparisons between cropping and cattle farms would not be possible. For this reason the index is based on total land use, where grassland is split up into two activities - roughage fodder production and fallow. Due to the high proportion of grassland on cattle farms the diversity is, by definition, lower for these farms. For this reason the use of the indicator will focus on comparisons between policy scenarios.

Description of scenarios

For the assessment, six policy scenarios are defined, two of them referring to the base situation (FAA *et al.*, 1997a). All scenarios are determined for the year 2005. The policy measures of the post-1992 CAP, prevailing in 1995/96, are taken as the base situation. These CAP measures are characterised by product-specific compensation payments in the arable sector (cereals, oilseeds, protein crops and set-aside), as well as for beef and suckler cows. Supply control in the arable sector is mainly realised with obligatory set-aside. Depending on the future development of world market prices two alternative base scenarios are defined:

- An increase of world market prices up to the actual level of EU-intervention prices for cereals (Base 1); under this condition cereals can be exported on the world market without restrictions,[3] and a 5% obligatory set-aside is assumed to be necessary to receive compensation payments.
- World market prices below the actual EU-intervention prices of cereals; to fulfil export restrictions for cereals a 27% obligatory set-aside is assumed (Base 2).

---

[1] Schaefer, 1993 cited by Meudt (1996).
[2] Schubert, 1991 cited by Meudt (1996).
[3] GATT restrictions for meat and poultry products are implicitly included in the market model (GAPsi; see also FAA *et al.*, 1997a) which was used for the definition of price conditions.

With reference to these price conditions two scenarios for Agenda 2000 are assessed:

- Agenda 1: High world market prices for cereals; to avoid overcompensation, area based payments in the arable sector are almost the same as those for cereals in Base 1.
- Agenda 2: Low world market prices for cereals; reduction of cereal prices equal to the change of intervention prices, and higher area-based payments in the arable sector.

The following conditions are valid for both scenarios of Agenda 2000:

- decoupled area payments for cereals, oilseeds and set-aside on the basis of reference yields for cereals;
- eligibility of forage maize for area payments;
- no set-aside obligation, but voluntary set-aside between 10 and 33% of the base areas;
- reduction of prices for beef meat by 30% and for milk by 15%;
- a 1.3% higher milk quota;
- increase of headage payments for male cattle and suckler cows;
- the introduction of premiums for dairy cows related to farm-specific milk yields (virtual cows).

Additional premiums for extensive livestock will only be paid for grazing beef and suckler cows and in accordance with density restrictions based on real livestock units. With regard to farm structural conditions in East Germany the. following measures are of importance: the general introduction of ceilings in the beef sector (up to 90 fattening bulls eligible for beef premiums), and degressive payments above 100,000 ECU per farm.

The options of partially decoupled payments in the beef and dairy sector are assessed under conditions of Agenda 1, i.e. the transformation of beef and cow premiums within the national 'envelope' into premiums for grassland (Agenda 1 GP). Premium restrictions on the basis of cross-compliance are not included in the model calculations.

The price conditions of the scenarios are determined by a partial equilibrium model for EU-15 (GAPsi) and a regionally differentiated model for the German agricultural sector (RAUMIS) (Frenz and Manegold, 1995; Isermeyer *et al.*, 1996; FAA *et al.*, 1997b). Based on these models and the above-named 'representative farm model', Agenda 2000 has been assessed in a rather broad context - the sectoral, regional and farm level (Kleinhanss *et al.*, 1998). Here, only the results derived from the farm-level assessments are discussed.

Impacts of policy instruments should be measured with regard to the base scenarios; therefore only those scenarios have to be compared for which the same conditions (i.e. the development of world market prices for cereals) are valid. For didactic reasons we put all the graphs describing the main results of the different scenarios together. For this reason it is necessary to relate relative changes to scenario of Base 1. Thus, it is also possible to explain the possibilities and limits of supply control by obligatory set-aside under conditions of Base 2.

IMPACT OF POLICY CHANGES

Environmental problems resulting from agricultural production are mainly influenced by the intensity, distribution and concentration of crop and livestock production; in Germany they are mainly location specific. With the underlying farm sample it is only possible to identify environmental impacts of policy changes depending on various structural characteristics, i.e. type of production, intensity levels, etc. In the following, farms are aggregated by farm type and size classes. Farms larger than 500 ha are located in the eastern part while the smaller ones are mainly located in the western part.

Table 17.1   Land use, environmental and economic indicators for different farm types

Farm type	All farms	Cropping	Dairy and beef	Pig	Mixed
Number of farms	973	348	462	97	66
Average farm size (ha)	87.1	142.3	60.8	40.4	48.5
Structure of land use (share of total UAA in %)					
Cereals	59.4	73.2	34.1	70.9	54.3
Oilseeds	7.0	6.6	7.0	11.4	9.1
Forage maize	4.1	1.2	9.5	1.2	6.7
Set-aside	4.3	4.5	3.5	7.3	5.5
Thereof non-food	1.7	2.2	1.0	1.2	0.8
Livestock production in 100 kg					
Beef meat	60.9	29.9	95.4	13.3	52.2
Pig meat	48.6	37.6	10.8	219.8	120.4
Milk	1,054.8	316.4	1,880.1	28.2	679.7
Fertiliser and pesticides (per total UAA)					
Nitrogen fertiliser (kg)	129.3	129.0	128.7	137.1	129.2
Nitrogen manure (kg)	40.6	10.5	86.2	71.5	69.8
Crop protection (DM)	142.9	166.6	98.6	163.0	141.7
Environmental indicators					
Nitrogen surplus kg ha^{-1}	54.8	25.5	96.9	97.4	86.0
Diversity index % [a]	54.9	53.4	56.4	54.9	63.0
Economic results (in 1,000 DM)					
Gross margin	149.8	202.3	122.7	116.1	111.3
Transfer payments	47.3	83.9	27.4	24.7	26.7

[a] % of max. diversity.
Source: BEMO.

Structural characteristics, the level of production, environmental indicators and income are given in Table 17.1, for the different farm types, as well as for the whole sample. Further information on size classes is given in Figures 17.1 to 17.3. With regard to environmental indicators the following tendencies can be identified:

- Due to rather extensive crop production, the usage of fertiliser and pesticides is significantly lower on large farms; at the same time nitrogen supply from animal manure is also lower. Therefore there is a downward tendency of nitrogen surpluses by farm size. Cropping farms show low nitrogen surpluses of less than 50 kg ha^{-1} while for pig and mixed farms it is higher than 100 kg ha^{-1}.
- Land use is not at all diversified: cereals are the dominant crop on cropping and pig farms. The diversity index varies between 50 and 60% for cropping and pig farms, and between 40 and almost 70% for dairy and beef farms. With some exceptions the diversity index increases with farm size, indicating that small farms are more specialised in land use than large farms. The relatively low level of this index indicates a considerable shortfall from maximum diversity.

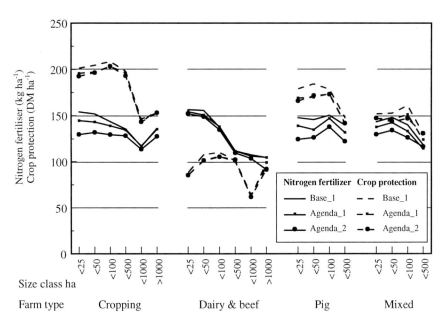

Figure 17.1 Intensity of nitrogen fertiliser and pesticides use in different scenarios (related to UAA, excluding set-aside)
Source: Kleinhanss *et al.* (1998).

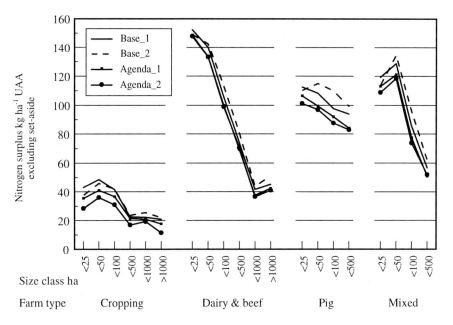

Figure 17.2 Impact of different policy scenarios on nitrogen surpluses (related to UAA, excluding set-aside)
Source: Kleinhanss *et al.* (1998).

Maintaining the existing CAP

The post-1992 CAP is mainly characterised by reductions of intervention prices for the most important arable crops, partially decoupled compensation payments, supply control by obligatory set-aside and restrictions of compensation payments by eligible base areas or livestock categories. Supply control instruments have to be determined with regard to price conditions and the influence of technical progress and structural developments on supply and demand. Under conditions of favourable world market prices, cereals can be exported without export restitutions; therefore the set-aside share can be low - for the underlying scenario a minimum level of 5% of base areas is assumed (Base 1).

Assuming less favourable world market prices for cereals (world market prices lower than EU-intervention prices), subsidised cereal exports are restricted to 22 million tonnes. To reach this level a minimum set-aside share of 27% of the base area has to be introduced (Kleinhanss *et al.*, 1998) in the year 2005 (Base 2). Allocation of production, intensity and farm income will be affected; there are also environmental implications. The main consequences for crop production are:

- Large farms cannot react otherwise than to set-aside arable land at least to the minimum required to be eligible for compensatory payments. Under unfavourable natural conditions voluntary set-aside will also be realised. About

15% of set-aside in the western part and 30% in the eastern part are used for non-food production. Cereal and food-oilseed areas will be reduced as well.

- The larger the minimum set-aside share is, the more small farms use the small producer scheme with no set-aside obligation, but equal premiums for cereals, oilseeds and protein crops. Therefore oilseed production will almost be abandoned and the resulting area will be used for cereals. Consequently crop rotation will become less diversified.

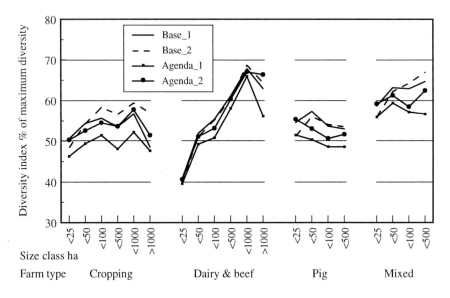

Figure 17.3 Impact of different policy scenarios on diversity of land use
Source: Kleinhanss *et al.* (1998).

Although there is a reduction of agro-chemical inputs depending on the set-aside share, intensity will not at all be reduced on the remaining areas (Table 17.2). As manure spreading on set-aside areas is not allowed (except where the land is used for non-food production), the supply of organic nitrogen on cropping areas increases. Therefore the following environmental implications can be expected:

- Nitrogen surpluses increase on medium and large cattle and beef, pig and mixed farms due to the reduction of manure spreading areas. The opposite occurs with small farms due to the reduction of set-aside as farmers enter the small-producer scheme.
- Diversity of land use increases on large farms, while it decreases on small farms due to the partial substitution of cereals for oilseeds.

Thus the environmental impacts would be mixed: on the one hand there is a tendency towards more diversified land use; on the other hand pollution problems from agro-chemicals might increase.

IMPACTS OF AGENDA 2000

Changing economic conditions under Agenda 2000 affect the arable as well as the beef and dairy sectors. Changes in the intensity and volume of production, as well as in incomes, will be induced.

Impacts on production

The principal conclusion for the arable crops sector is that the economic competitiveness of all production lines (and set-aside) favoured by the high compensatory payments under the former market regime will be reduced with the introduction of standardised area-based payments. Considering the whole sample of farms and referring to Base 1 the following changes due to Agenda 1 are predicted (Table 17.2):

- Without the set-aside obligation, the set-aside area will be reduced by about 90%. Voluntary set-aside will only be realised on rather large farms with low yields; this is especially true for dairy and beef farms in the size class 500 to 1,000 ha.
- Oilseed production will become less competitive due to the drastic reduction of premiums by about 450 DM ha^{-1}. On the other hand, premiums boundaries related to arable land - being introduced in the New Länder - will be abandoned. Due to these restrictions, non-food oilseed production covers about 30% of set-aside areas in the base situation. Under Agenda 2000 non-food oilseeds will be substituted by food oilseeds. Further, the price relationship between oilseeds and cereals is important: with relative prices of 1.8 to 1 in Agenda 1 the oilseed area decreases by 20%, while with relative prices of 2 to 1 in Agenda 2 oilseed production even increases. The conclusions are that non-food oilseed production will be abandoned and food oilseed production will be reduced significantly in Western Germany, but will rather increase in the Eastern part.
- The area of fodder maize will be reduced by about 9% on average due to changes in the beef sector.

Under conditions of decoupled and standardised premiums, cereal production will become more competitive. The area will be extended by about 10% on average, varying between 0 and 35% between the farm groups. In some farms the share of cereals in arable crop rotations increases to 90%, such that negative impacts on yields and pesticide use might be induced.

The economics of milk production will not change significantly, although there are income losses due to undercompensating cow premiums. With the exception of a few farms with low milk yields, milk production will be extended through the additional quota both for young farmers and for mountain regions.

Beef production is negatively affected. Price reductions are not compensated for veals and heifers, therefore these activities will become less competitive. If changes in market prices are equal to the reduction of intervention prices, the beef premiums do not completely compensate income losses. In addition to that, the economic conditions will become less favourable due to the following farm-specific factors:

- Additional premiums for extensive beef and suckler cow production will only be paid for up to 1.4 grazing Livestock Units per ha, calculated on the basis of actual livestock and without taking into account arable forage crops.
- Upper limits for grazing livestock densities and of 90-head for beef premiums will become an economic barrier, because headage premiums will become much more favourable than area payments for fodder maize, which in the base situation are claimed for fattening bulls produced beyond the above restrictions. Although these restrictions may be justified with regard to

Table 17.2   Impacts of policy scenarios on production and environmental indicators - average of all farms

	Relative changes compared to Base 1 (%)			
	Base 2	Agenda 1	Agenda 1 GP [a]	Agenda 2
Land use				
Set-aside	327	-91	-90	-75
Cereals	-19.3	9.3	10.0	3.0
Food-oilseeds	-37.8	-21.0	-19.7	9.4
Forage maize	-2.5	-9.1	-22.5	1.7
Beef production	0.7	-9.2	-19.6	-5.7
Environmental indicators				
(related to UAA excluding set-aside)				
Nitrogen fertiliser	1.3	-1.5	-0.3	-5.9
Nitrogen surplus	8.8	-7.0	-8.6	-12.0
Crop protection	0.0	-1.5	-1.3	-2.4
Environmental indicators				
(related to total UAA)				
Nitrogen fertiliser	-11.8	0.8	1.9	-4.4
Nitrogen surplus	-5.2	-4.8	-6.6	-10.6
Crop protection	-12.9	0.8	0.9	-0.8
Diversity index	4.0	-7.1	-8.3	0.1

[a] Agenda 1 GP takes up the option within the national 'envelope' of transforming beef and cow premiums into grassland premiums.
Source: BEMO.

environmental objectives, the efficiency and structural development of beef production will be negatively affected.

• Degressive payments by farm will reduce beef premiums.

The last-mentioned factor will mainly affect large farms in Eastern Germany; beef production will be reduced by 15-30% on these farms. Small and medium-sized farms even increase beef production by 3-5%. On average beef production will be reduced by 9%.

Decoupling of beef premiums will affect beef production even more. For the underlying scenario (Agenda 1 GP) it is assumed that the EU-component of beef premiums will be paid as headage premiums, while the national component will be transformed into grassland premiums. Intensive beef fattening based on maize silage will be affected most (with a reduction of headage premiums and no compensation payments for fodder maize). As a result beef production will be reduced by 20% on average with intensive beef fattening based on maize silage and concentrated feed being reduced, and partly replaced by more extensive beef production systems (grazing animals, grass silage, low use of concentrated feed).

Environmental impacts

Changes in environmental impacts are induced by the above-mentioned changes in crop and livestock production. Referring to the use of agro-chemicals on cropping areas (excluding set-aside), the amount for crop protection decreases by 2% on average. This is mainly induced by the substitution of cereals for rape-seed (see Figure 17.1 and Table 17.2). As this substitution mainly takes place on farms of less than 500 ha, the change in intensity is greater than for the larger farms. Almost the same response holds for mineral nitrogen fertiliser (Agenda 1). Under conditions of Agenda 2, fertiliser intensity is reduced further due to significantly lower cereal prices.

Concerning nitrogen surpluses and diversity, contrary reactions can be expected. Nitrogen surpluses will be reduced by 7% under the Agenda 1 scenario or by 12% under Agenda 2, compared with Base 1. The most important changes are realised in small and medium-sized cropping farms and in pig farms. This can be explained by the increase of manure spreading areas due to reductions in set-aside. On dairy and beef farms as well as on mixed farms nitrogen surpluses are reduced by 5 to 10 kg ha^{-1}; there are no significant differences in the effects between size classes and cereal price levels under the Agenda scenarios.

Beside the adaptations in the arable sector, the reduction in beef production has an influence too. The conclusions are that nitrogen surpluses will be reduced under the conditions set by the Agenda 2000 proposals, but the changes are more important on cropping farms, which are characterised by a relatively low level of nitrogen surpluses.

The consequences for diversity of land use depend on the different price levels of the Agenda scenarios. Under conditions of favourable world market prices for cereals and the resulting partial substitution of oilseeds by cereals, the diversity index drops by about 5 points for cropping farms and between 2 and 7 points for other farm groups. If such a substitution does not take place, which is the case for scenario Agenda 2, the diversity index will not change at all.

Concerning the overall environmental impacts of the Agenda 2000 proposals it can be concluded that positive effects could be anticipated with regard to the use of agro-chemicals and related pollution problems. On the other hand, decoupled premiums favour cereal production and will therefore tend to reduce the diversity of land use.

Economic impacts

Changing economic conditions always induce economic impacts. For example, under the post-1992 CAP a high share of set-aside in Base 2 induces significant income losses: large farms are affected most because they cannot react other than to set-aside land, whereas small farms can apply for the small-producer scheme without set-aside. Like with the Agenda 2000 proposals will induce economic impacts, influenced by the level of market prices, transfer payments, farm adaptations and the changing role of control measures restricting farm adaptations (Figure 17.4).

Referring to Agenda 1, distinctions can be made between farm size and farm types. Pig farms could improve their gross margins by about 1%. Mixed farms will have income losses of 3 to 4% of gross margins. In cropping farms of less than 500 ha, income reductions are in the range of 2 to 3% of gross margins, while large farms located in Eastern Germany will experience drastic income

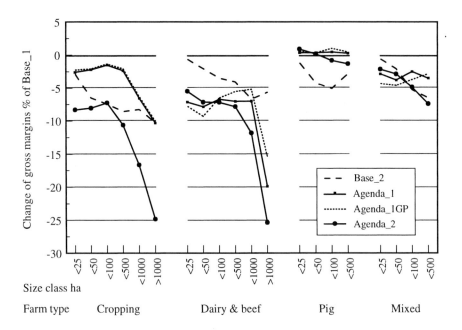

Figure 17.4 Income effects of policy scenarios
Source: Kleinhanss *et al.* (1998).

losses due to degressive payments. Income losses are generally more important in dairy and beef farms due to undercompensating premiums in the beef and dairy sector. Farms larger than 1,000 ha are affected further by degressive payments and the upper limit of 90-head for beef premiums inducing a drastic reduction of beef production.

Under conditions of Agenda 2 additional income losses can be expected, because output reductions due to 20% lower cereal prices are only half compensated. The level of transfer payments will increase and therefore a larger share of farms will be affected by degressive payments. Cropping as well as dairy and beef farms less than 500 ha will have income losses of about 7 to 10% of gross margins, and on larger farms gross margins will decrease by 25%. A differential impact such as this is likely to provoke the consideration of the partition of large farms into smaller ones, although the propriety of that particular adaptation strategy and the transaction costs involved remain unclear.

Under the conditions proposed by Agenda 2000 the level of transfer payments will significantly increase, especially in the beef and dairy sectors. Difficulties of control and monitoring will arise over headage premiums. These could partly be solved by decoupled payments through transforming headage premiums into premiums for grassland. Another option in Agenda 2000 not included in the model is the restriction of transfer payments on the basis of environmental conditions (cross-compliance).

CONCLUSIONS AND RECOMMENDATIONS

This chapter presents a first assessment of the economic and environmental effects of the market and price policy measures proposed in Agenda 2000. The assessment is based on a farm sample presenting structural characteristics of farming in Germany. The principles of the arable crop regimes are the reduction of intervention prices for cereals, decoupled area-based payments and the end of obligatory set-aside. Set-aside and oilseed production will be reduced in favour of cereals. In the beef and dairy sector significant price reductions are partly compensated by headage premiums. Economics of beef production will be influenced by the level of headage premiums, ceilings for the number of eligible animals and reduction of the premiums paid due to budget constraints or the application of a degressive principle. In particular, those farms affected by ceilings in the cattle sector or degressive payments will have high income losses. Concerning environmental effects, contradictory tendencies are identified. On the one hand, nitrogen surpluses will be reduced to a certain degree; on the other hand, crop production will become less diversified.

The final decisions on Agenda 2000 by the European Council did not go as far as what the European Commission had proposed and what has been assessed above. The changes will mainly have economic implications. Income losses will be lower due to lower price decreases - e.g. for cereals, higher headage premiums - and the stepwise and delayed implementation of policy reforms for most products. The position of large farms has largely improved: degressive payments are not introduced and Member States are allowed to introduce other ceilings for beef premiums than the 90-head beef restriction. The relative income effects in East Germany will therefore be similar to those in the rest of the country.

Compared to the original Agenda 2000 proposal, no major changes are foreseen in terms of the projected impact on the environment. The only exception might be the extensification of beef production. This would be induced by favourable premiums for extensive livestock holdings. Therefore, livestock densities might be reduced, especially in large-scale beef and dairy farms. However, a partial decoupling of beef headage premiums by transforming the national premiums into grassland premiums will have no major economic impacts, due to the reduction in the budget available within the national envelope. Therefore, the beneficial impacts of this measure, described in Table 17.2, might not be realised.

REFERENCES

Commission of the European Union (1998) Agenda 2000, Legislation proposals, http://europa.eu.int/rapid/start/cgif/..... 18 March.

Dabbert, S., Kilian, B. and Umstätter, H. (1998) *Umweltwirkungen der Agenda 2000 im Agrarbereich.* AGRA-EUROPE 19/98 vom 11.05.1998, Sonderbeilage 1-4.

European Commission (1997) *Agenda 2000. Eine stärkere und erweiterte Union,* Band I, Kom (97) 2000, Brussels, July 1997.

FAA, FAL-BW and FAL-MF (1997a) *Modellrechnungen zu Auswirkungen der Agenda 2000 in der deutschen Landwirtschaft.* Arbeitsbericht 7/97, Institute of Farm Economics, Federal Agricultural Research Centre, Braunschweig.

FAA, FAL-BW, FAL-MF and IAP (1997b) Modellrechnungen zur Weiterentwicklung des Systems der Preisausgleichszahlungen. In: *Arbeitsmaterial der Forschungsgesellschaft für Agrarpolitik und Agrarsoziologie* e.V. 2, Bonn.

Frenz, K. and Manegold, D. (1995) Auswirkungen von GAP-Reform und GATT-Auflagen auf Erzeugung und Verbrauch von Getreide, Ölsaaten und Hülsenfrüchten in der EU-Modellrechnung. In: Frenz, K., Manegold, D. and Uhlmann, F. (eds) *EU-Märkte für Getreide und Ölsaaten.* Reihe A: Angewandte Wissenschaft 439, Münster, pp. 185-344.

Isermeyer, F., Hemme, T., Hinrichs, P., Kleinhanss, W., Heitman, D., Schefski, A., Fasterdig, F. and Manegold, D. (1996) *Software Use in the FAL 'Model Family'.* Institute of Agricultural Policy. 'International Workshop on Software Use in the Agricultural Sector Modelling', 27-28 June, Bonn.

Kleinhanss, W. (1996) Auswirkungen unterschiedlicher produktgebundener bzw. produktneutraler Transferzahlungen im Rahmen der EU-Agrarmarktregelungen. *Landbauforschung Völkenrode* 4, 198-211.

Kleinhanss, W., Osterburg, B., Manegold, D., Seifert, K., Cypris, Ch., Hemme, T., Jacobs, A., Kreins, P. and Offerman, F. (1998) *Auswirkungen der Agenda 2000 auf die deutsche Landwirtschaft - Eine modellgestützte Folgenabschätzung auf Markt-, Sektor-, Regions- und Betriebsebene.* Institut für Betriebswirtschaft der Bundesforschungsanstalt für Landwirtschaft, Arbeitsbericht 2/98, Braunschweig.

Meudt, M. (1996) Abbildung der potentiellen Naturraum- und Biotopausstattung. In: Heinrichsmeyer, W., Isermeyer, F., Neander, E. and Manegold, D. (eds) *Entwicklung des gesamtdeutschen Agrarsektormodells RAUMIS 96.* Endbericht zum Kooperationsprojekt. Forschungsbericht für BMELF (94 HS 021), unpublished, Bonn/Braunschweig.

Nieberg, H. (1996) Möglichkeiten und Grenzen der Verwendung von Agrar- und Umweltindikatoren. In: Heinrichsmeyer, W., Isermeyer, F., Neander, E. and Manegold,

D. (eds) *Entwicklung des gesamtdeutschen Agrarsektormodells RAUMIS 96. Endbericht zum Kooperationsprojekt. Forschungsbericht für BMELF* (94 HS 021), pp. 146-164, Bonn/Braunschweig, unpublished.

Agenda 2000: A Wasted Opportunity?

18

Philip Lowe and Floor Brouwer

BACKGROUND TO THE AGENDA 2000 REFORMS OF THE CAP

Between 1997 and 1999, the European Union (EU) engaged in another round of reform of the Common Agricultural Policy (CAP), following on the MacSharry reforms of 1992. This was part of Agenda 2000 which sought to establish the overall budget for the EU for the period 2000-2006. Additional changes to the CAP were seen to be necessary to prepare for the imminent enlargement of the EU and for the reopening of world trade negotiations on agriculture. There was also concern within the European Commission to respond to the broadening public demands on agriculture and the countryside. The Commission set out the following objectives for a newly reformed CAP: increased competitiveness; high standards of food safety and quality; ensuring a fair standard of living for the agricultural community; the fuller integration of environmental goals; the creation of alternative job and income opportunities in rural areas; and simplification of EU legislation and administration (CEC, 1997).

PROGRESS TOWARDS AN INTEGRATED RURAL POLICY

Any overall assessment of Agenda 2000 must consider to what extent it takes us towards a more legitimate and sustainable model of support for rural areas, that addresses contemporary socio-economic needs and consumer and environmental concerns. An alternative, integrated rural policy, would be characterised by the following arrangements:

- markets would largely determine the income that farmers receive from growing crops and raising livestock (with a basic level of support retained for emergency or unusual conditions);

- farmers would receive sufficient support for the environmental management functions of agriculture;
- rural development would be given greater promotion, to assist in the economic adjustment of rural areas and to help improve rural incomes and employment.

Such a model was put forward by the group of experts, chaired by the agricultural economist Alan Buckwell, who were charged by the Agricultural Directorate of the European Commission to outline the principles that might guide the transition of the CAP towards the integration of environmental and rural development objectives (European Economy, 1997). The Buckwell Group mapped out a series of stepwise transitions whereby the CAP could be transformed into an integrated rural policy over the medium term (see Figure 18.1). This would involve the progressive liberalisation of the various Common Market Organisations (CMOs) entailing the payment of time-limited compensation to producers affected by price cuts. The MacSharry reform initiated this process, converting some of the indirect costs of supporting protected and managed markets into direct subsidies to farmers. However, the Buckwell Group saw those changes as only the first step in the transformation of the CAP. Subsequent steps should involve not only the dismantling of the panoply of supply controls and the decoupling of compensation payments from production, but also steady reductions in these payments and the switch of public resources to support both the environmental management functions of agriculture and the socio-economic development of rural areas.

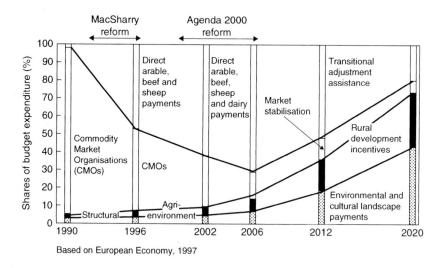

Based on European Economy, 1997

Figure 18.1   From CAP to integrated rural policy

AGENDA 2000: THE COMMISSION'S PROPOSALS

The original proposals for Agenda 2000 put forward by the Commission in July 1997 were influenced, to some extent, by this kind of thinking (CEC, 1997). Mostly, though, they were concerned with changes to the CMOs, particularly the dairy, arable and beef regimes. The Commission's intention was largely to continue the reforms initiated in 1992 by further reducing price support towards world prices, moving away from certain supply controls (e.g. by setting the obligatory rate of set-aside at zero but retaining dairy quotas) and increasing direct compensation payments for farmers. However, stress was also laid on the role of farmers in maintaining the countryside. It was suggested that the agri-environment Regulation be reinforced and better funded to give it a prominent role in supporting sustainable development in rural areas and meeting society's environmental demands. The possibility was also raised of transforming Less Favoured Areas (LFAs) policy into a basic instrument to maintain and promote low output farming systems. The most radical proposal was to combine these two measures - the agri-environment regulation and LFA policy - with rural development measures, to create a new instrument under the CAP to support integrated rural development across the EU. This horizontal set of measures, it was proposed, would be financed under the Guarantee Section of the EAGGF (FEOGA) but would be implemented in a decentralised way at the appropriate level, at the initiative of Member States. The proposal therefore took forward the idea trailed by the Commission at the 1996 Cork Conference of a flexible and programmed approach to the promotion of a sustainable rural policy within the CAP that would be responsive to the diversity of rural needs and environmental circumstances across the EU. In keeping with this new approach to rural policy inside the CAP and throughout the EU, Agenda 2000 proposed that established regional supports for poorer rural regions should be concentrated onto a smaller area.

In key respects, therefore, the Commission's Agenda 2000 did set out the basic parameters that could guide the transition from the CAP to Integrated Rural Policy. However, the Commission did not bite the bullet of proposing that compensation payments to farmers should be time limited. Undoubtedly, the negative reaction of several Member States to the Cork Conference had signalled strong political resistance to any move that could be seen as taking money away from farmers (Lowe *et al.*, 1996). The Commission was therefore constrained in the proposals it could feasibly bring forward. As a result, though, the Agenda 2000 proposals did not establish the critical resourcing linkage that the Buckwell Group envisaged whereby a steady reduction in production subsidies could be used to fund the build-up of alternative rural policy supports. Indeed, in resource terms, the Commission was obliged to pursue its nascent rural policy in parallel with, but effectively detached from, the reform of the CAP commodity regimes. The medium-term resourcing of the integrated rural development instrument was thus left unclear which meant that references to it becoming 'the second pillar of the CAP' (CEC, 1998, para. 2.4) seemed wishful thinking. With the prospect of limited additional funding for rural supports (including agri-environmental measures) but a major expansion of compensation payments to farmers - with no indication that these payments would not be indefinite - the Commission had to concede the extension of the option of placing environmental conditions on live-stock payments to arable and set-aside payments too.

AGENDA 2000: THE OUTCOME

From 1997 to 1999 there were intense negotiations over the Agenda 2000 proposals between the European Commission and the Member States. As well as the divide between liberalising and protectionist positions and the pursuit of disparate national agricultural interests, CAP reform also got caught up in a growing debate over the EU's future budget. Prospective new demands on the budget arising from EU enlargement were met by calls to restrict the growth of the budget and to rebalance the financial burden between Member States. With the CAP accounting for about half of the EU's budget, it could not be insulated from this debate. Indeed, one of the main pressures to reform the CAP came from recognising that it would be extremely expensive to extend the existing CAP to the new entrant countries. The Commission's own proposals to reform the CAP involved an increase in expenditure to compensate farmers for reduced support but to stabilise spending from 2004 onwards. In order to reach agreement, as the negotiations with Agriculture Ministers proceeded, the Commission had to make various concessions to its original proposals, thereby increasing the cost further. This aroused growing impatience from Finance Ministers and heads of government who eventually ruled that there should be no overall increase in the agricultural budget.

The negotiations over CAP reform were concluded in March 1999 by heads of government meeting in Berlin. The Summit made a number of changes to an agreement reached by the Agriculture Council a few weeks earlier. These were mainly to reduce the projected expenditure on the CAP and included restrictive limits on the amount of money that could be spent on rural development and the environment. The main CAP budget was limited to ECU 40.5 billion per annum. However, heads of government disagreed as to how this should be achieved. One group, including France and the UK, were in favour of a gradual reduction in the level of direct payments which farmers were to receive, with some of the savings transferred to the rural development budget (which includes agri-environment). This approach, known as degressivity, was opposed by Germany and other Member States who preferred to cut expenditure by postponing reform of the dairy regime, scaling back price cuts for cereals and retaining set-aside. This group prevailed in the final conclusions since it proved difficult to agree on a way of making degressivity work acceptable to all Member States. The Berlin agreement on Agenda 2000 thus significantly watered down the reform package and inevitably there was some loss of the sense of direction and coherence of the Commission's own proposals (Tangermann, 1999).

AGENDA 2000: THE ENVIRONMENTAL PERSPECTIVE

Political agreement on the Agenda 2000 reforms was thus achieved in March 1999. The changes made to the CAP are significantly less ambitious than what the Commission had earlier proposed and they are to be revisited in 2002. The key changes continue in the direction set by the MacSharry reforms of 1992, of price cuts linked to increases in direct compensatory payments. The commodity sectors covered are arable crops, beef and wine (dairy reform was put off until 2005). There are also certain important procedural changes that could potentially affect all supported sectors, including the attachment of environmental conditions to farm payments and the possibility to modulate (i.e. to cream off a proportion of) the CAP subsidies to individual farmers in order to fund countryside management schemes. Finally, measures to promote agricultural and rural development and the management of the countryside are consolidated and made subject to a new decentralised programming procedure under the new Rural Development Regulation (1257/99).

Agenda 2000 proved to be only a very partial reform of the CAP which arguably failed to meet its main economic and political objectives or to respond adequately to the broadening public demands on agriculture and the countryside. It certainly failed to live up to the expectations it had aroused. In the words of one NGO:

> This reformed CAP will continue to reward intensive production with the largest share of the budget being given over to farmers who engage in agricultural practices which are damaging Europe's rural environment. Little has been achieved to encourage a shift from production towards more environmentally sustainable ways of farming. New resources have not been made available to enable farmers to find new sources of income (WWF, 1999, p.6).

Because the reform does not reduce the level of subsidisation of the agricultural sector and does not decouple compensation payments from production, it is likely to be subject to fierce challenge in the coming WTO round when the EU will face strong pressures to further liberalise policy and especially to reduce area aids. In addition, the heavy levels of agricultural subsidy would be expensive to introduce into the countries that are seeking EU membership. The European Commission insists that it would be inappropriate to pay compensatory payments to farmers in the new Member States for cuts in price support which have occurred in the EU. However, it will be hard to sustain differential levels of agricultural subsidy in a single market, once the EU has been enlarged. For these reasons, it is likely that further changes to the CAP will be needed within the next few years.

From an environmental and countryside perspective, the Agenda 2000 results are disappointing because they:

- prolong many production subsides which continue to encourage intensification or have other damaging consequences;
- fail to fulfil the promise to establish rural policy as a 'second pillar' to the CAP;
- postpone reform of the EU dairy regime whose intensive production contributes to nitrate problems and farm waste pollution;

- leave unreformed the regimes for many Mediterranean products and for sheep meat.

There are, nevertheless, some positive features which should allow environmental considerations to be taken forward, notably:

- the new framework of the Rural Development Regulation and its opportunities to promote the integrated and decentralised planning of agri-environment, agricultural and rural development measures, though within severe financial constraints;
- opportunities to apply environmental conditions where direct commodity payments are made;
- a new approach to LFAs where payments will be made on an area, rather than a headage, basis and be better targeted at environmental objectives;
- national discretion in the application of a proportion of direct payments (the national envelope) to the beef sector (and eventually the dairy sector) which potentially could support extensive grazing for environmental purposes;
- national discretion to modulate the total CAP subsidies for individual farmers in order to increase expenditure on agri-environment, LFAs, afforestation and early retirement schemes.

A common characteristic of all these features is the significant degree of national discretion involved in their implementation. We consider the first two of these features separately in following sections, but turn next to consider the remaining features within a discussion of the resources for environmental supports. Then the scope is considered for achieving an integrated system of supports specifically focused on environmental objectives for LFAs. For agriculture elsewhere, much greater emphasis will need to be given to promoting environmental standards through a variety of mechanisms, including the application of environmental conditions in relation to commodity payments. Finally, the Rural Development Regulation offers the opportunity to promote the decentralised planning of countryside management and development.

RESOURCES FOR ENVIRONMENTAL SUPPORTS

Agenda 2000 commits few additional resources directly for environmental supports. Indeed, agri-environment expenditure as a component of the Rural Development Regulation is subject, in principle, to a freeze on spending until 2006 (see Figure 18.2). This contrasts with the situation following the 1992 reforms of the CAP when expenditure on agri-environment and other accompanying measures was allowed to rise year on year in response to the take-up of relevant schemes by Member States. CAP expenditure on the rural development and accompanying measures is expected to reach ECU 4.38 billion in 2000, from the Guarantee Section of the EAGGF (FEOGA), around 10.5% of the total budget. By the year 2002, the proportion of the budget will have actually

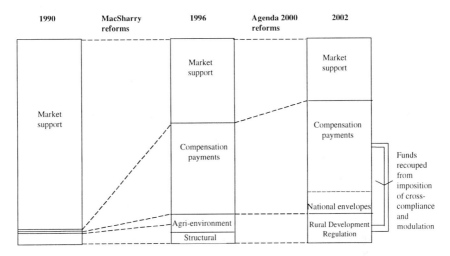

Figure 18.2   The changing architecture of the CAP, 1990-2002

fallen to 9.9%. Even by 2006, rural development and the accompanying measures will still account for no more than 10.5% of the CAP budget. Even though both the Rural Development Regulation and the commodity supports are within the EAGGF (FEOGA) Guarantee Section, decisions taken at the Berlin summit effectively segregate the expenditures on them, which will prevent any transfer of funding out of commodity support, even if savings are made in the latter.

The prospects for additional resources for environmental ends now depends on the possibilities to reorient certain production-related payments: namely, through the re-use of monies recouped from the imposition of modulation and cross-compliance conditions on commodity payments; and through the 'greening' of Hill Livestock Compensatory Allowances (HLCAs) (i.e. the additional payments to support livestock farming in LFAs). The national envelopes for beef payments could also in principle be used to support environmentally beneficial farming systems.

Under the so-called horizontal regulation (1259/99), which applies to all the commodity regimes, Member States are required to define appropriate environmental conditions to attach to commodity payments to farmers, as well as proportionate penalties - through forfeiture of payments - for farmers who infringe these conditions (see below). Member States are also authorised to modulate direct payments per farm in relation to either employment on the farm, farm profitability criteria, or the total amount of state aids received. The funds accrued from the withholding of payments under either measure will remain available to the particular Member State as an additional support for certain measures under the Rural Development Regulation, namely agri-environmental measures, LFA, early retirement and afforestation. However, the resources that may become available from penalising farmers for transgressing environmental cross-compliance conditions are likely to be neither significant nor reliable. Modulation, though, could yield significant resources by reducing supports to large farms (typically arable ones). Member States are entitled to modulate up to 20% of compensation payments to farmers.

There are also the changes to HLCA payments. Compensation allowances for farmers in LFAs will in future be conditional on the use of sustainable farming practices. For livestock, they will be paid per ha, rather than in the form of head-age payments. Furthermore, there is a new provision whereby compensation allowances can be paid to farmers who are subject to restrictions arising from Community environmental legislation. These changes amount to a significant re-orientation of LFA policy towards environmental objectives.

A final element of resource that could be inflected towards environmental objectives is the 'national envelope' within the beef regime. A small proportion of beef direct payments can be distributed by Member States at their discretion, within certain EU rules, to give greater flexibility in addressing regional dispari-ties and to encourage extensive production. Support from these envelopes can be either in the form of headage payments or acre payments, based on the area of pasture on a farm. Area payments directed towards more extensive producers on permanent or semi-permanent pasture would be the best option environmentally in most circumstances.

AN INTEGRATED SUPPORT SYSTEM FOR LFAs

These changes, if properly orchestrated, could provide an integrated support sys-tem at least for LFAs that would be environmentally sustainable. However, even within the LFAs, because HLCAs are only a fraction of the subsidies that hill farmers receive, any effective greening of overall support policies would need to be complemented by the use of cross-compliance and the national envelope. An integrated support system for the LFAs under Agenda 2000 could therefore look like Figure 18.3.

The HLCAs would need to be redesigned as area payments and redirected to support the kind of farming systems that could deliver high biodiversity and land-scape goods. The funds specifically available for this are very limited and could neither give sufficient support to the farming nor cover the costs of positive envi-ronmental management. This is why careful co-ordination is needed with the application of commodity payments on the one hand, and with agri-environment and conservation management payments on the other. An additional possibility is opened up with the new LFA regulation which allows for payments to compen-sate for environmental restrictions. This could mean elements of LFA support being inserted into the top layer to fund some management of protected sites.

The critical requirement will be for careful co-ordination between the layers to ensure that the rules and resources applied complement rather than duplicate one another. In this regard, the Rural Development Regulation, as the main in-strument for programming the middle layer in correlation with both national resources and CAP commodity payments, assumes strategic significance (see be-low).

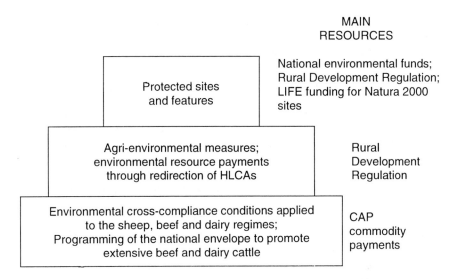

MAIN
RESOURCES

| Protected sites and features | National environmental funds; Rural Development Regulation; LIFE funding for Natura 2000 sites |

| Agri-environmental measures; environmental resource payments through redirection of HLCAs | Rural Development Regulation |

| Environmental cross-compliance conditions applied to the sheep, beef and dairy regimes; Programming of the national envelope to promote extensive beef and dairy cattle | CAP commodity payments |

Figure 18.3   An integrated support system for LFAs

A different model would need to be applied, under Agenda 2000, to the environmental protection of lowland farming in general and arable agriculture in particular - one of special sites and targeted agri-environment schemes resting on a base of environmental standards and legislation. For these areas, therefore, the opportunities to apply environmental conditions in relation to commodity payments may offer a crucial set of basic safeguards.

ENVIRONMENTAL STANDARDS AND CROSS-COMPLIANCE

Where farmers receive direct support, Member States are under a general obligation to take 'the environmental measures they consider to be appropriate in view of the situation of the agricultural land use or the production concerned and which reflect the potential environmental effects'. Member States have considerable latitude in deciding what is appropriate and may choose from one of the following measures:

- support for agri-environmental schemes;
- general mandatory environmental requirements;
- specific environmental requirements as a condition for direct payments (what is commonly referred to as cross-compliance).

Member States may also decide on proportionate penalties to apply for environmental infringements involving, where appropriate, the reduction or even the cancellation of direct payments to errant farmers.

Member States already had the option, under the 1992 MacSharry reforms, of applying environmental cross-compliance to livestock payments, but most had not done anything about it. Under Agenda 2000, action by Member States is no longer optional - there is a formal obligation on them to specify appropriate environmental measures. Moreover, they must do so in relation to all sectors (not just livestock) that benefit from direct payments to farmers. Potentially, this undoubtedly represents a major extension, across sectors and countries, of the principle of attaching environment conditions to farm payments. Indeed, where there is the will, there is now the scope to take a comprehensive approach to specifying environmental requirements for supported agricultural sectors. In principle, this should involve identifying the variety of ways in which supported farm production can damage the environment and the combination of regulations, cross-compliance and incentives to prevent it.

Member States, though, are left with considerable discretion over how to proceed. In some respects, this is quite appropriate as the environmental relations of agricultural production vary considerably by farming system and region. However, within the discretion ceded to them, it would seem quite possible for Member States to adopt, if they choose, a rather formularistic and cursory stance to their responsibilities. The way is certainly open for divergent approaches in terms of both the degree and the nature of the action taken. The very limited response to the cross-compliance option under the MacSharry reforms does not indicate enthusiasm for the measure within Member States. Without a stronger, more co-ordinated and more accountable common framework for action, some states are likely to succumb to lobbying aimed at blocking any effective steps towards environmental protection that might adversely affect the competitiveness of domestic producers. States already differ in the significance they accord agrienvironmental problems and in their relative preference for controls or incentives in dealing with them (see Chapter 15, this volume). This reflects in part differences in the objective nature and severity of the problems they face. But it also reflects the relative strengths of agricultural and environmental interests which vary markedly between states. That makes problematic the achievement of an EU-wide reference level for good agricultural practice and may well result in variable environmental standards and different public responses to agrienvironmental problems between Member States.

THE RURAL DEVELOPMENT REGULATION

The Regulation lays the basis for a Community rural development policy which aspires to be a 'second pillar' (to that of commodity management) within the CAP. It draws together a number of existing regulations and agricultural support measures (Figure 18.4). Rural development, thus conceived, embraces both farm and non-farm developments as well as agri-environment measures and forestry, and has the following strategic objectives:

- supporting a viable and sustainable agriculture and forestry sector at the heart of the rural economy;

The new regulation is based on and supersedes a number of existing regulations and support measures:

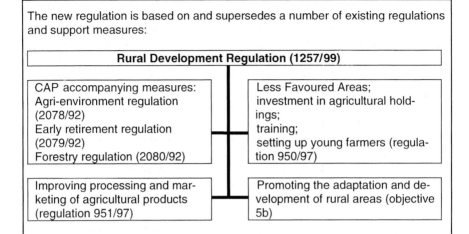

Under the new Regulation the aims of rural development policy are as follows (Article 2):

- the improvement of structures in agricultural holdings for the processing and marketing of agricultural products;
- the conversion and reorientation of agricultural production potential, the introduction of new technologies and the improvement of product quality;
- the encouragement of non-food production;
- sustainable forest development;
- the diversification of activities with the aim of complementary or alternative activities;
- the maintenance and reinforcement of a viable social fabric in rural areas;
- the development of economic activities and the maintenance and creation of employment with the aim of ensuring a better exploitation of existing inherent potential;
- the improvement of working and living conditions;
- the maintenance and promotion of low-input farming systems;
- the preservation and promotion of a high nature value and a sustainable agriculture respecting environmental requirements;
- the removal of inequalities and the promotion of equal opportunities for men and women, in particular by supporting projects initiated and implemented by women.

Figure 18.4   The new Rural Development Regulation

- developing the territorial, economic and social conditions necessary for maintaining the rural population on the basis of a sustainable approach;
- maintaining and improving the environment, the countryside and the natural heritage of rural areas.

The new Regulation allows considerable discretion to Member States to programme the different elements together holistically and in ways responsive to the diversity of rural conditions and circumstances. They must 'ensure the necessary balance' between the measures used from the Regulation (Article 43). Agri-environment measures are the only compulsory element which will have to be implemented throughout the territories of all Member States.

Member States are required to submit integrated Rural Development Planning Documents to the Commission by January 2000 to cover spending for the period 2000-2006. The Regulation specifies that 'Rural development plans shall be drawn up at the geographical level deemed to be the most appropriate' (Article 41). This leaves it entirely to the Member States to decide on the appropriate internal level for implementation. Thus, programming could be at the national level if states chose to do so, although this would not be in the spirit of the decentralised approach the Commission is promoting. The guidance on the implementation of the Regulation says 'emphasis must be on participation and a 'bottom up' approach'.

The Rural Development Regulation has thus a number of novel, and potentially significant aspects. First of all, rural development policy is established as a central feature of the CAP, co-financed by the Community from the CAP Guarantee Fund. Secondly, it is to be a horizontal policy, covering all rural areas. Thirdly, an integrated legal framework is set up for farm and rural development and agri-environmental measures, with the implementation subject to decentralised multi-annual programming. Fourthly, support is potentially available for

---

As far as the promotion of adaptation and development of rural areas is concerned, the new Regulation will provide support for the following measures:
- land improvement;
- reparcelling;
- setting up of farm relief and farm management services;
- marketing of quality agricultural products;
- basic services for the rural economy and population;
- renovation and development of villages, protection and conservation of the rural heritage;
- diversification of agricultural activities and activities close to agriculture, to provide multiple activities or alternative incomes;
- agricultural water resources management;
- development and improvement of infrastructure connected with the development of agriculture;
- encouragement for tourist and craft activities;
- protection of the environment in connection with agriculture, forestry and landscape conservation as well as with the improvement of animal welfare;
- restoring agricultural production potential damaged by natural disasters and introducing appropriate prevention instruments;
- financial engineering.

Figure 18.5 Article 33 'Measures for the adaptation and development of rural areas'

non-farmers and non-farming activity from the CAP (under Article 33 of the Regulation - see Figure 18.5).

The new Regulation has been enthusiastically welcomed by environmental and rural organisations. In the words of one NGO:

> The new Rural Development Regulation is a major opportunity to rebalance support to rural areas. It is a blueprint for an innovative and more integrated approach to rural development. The new regulation, which must work in harmony with regional and other rural policies, has the potential to deliver real economic, social and environmental benefits (Royal Society for the Protection of Birds, 1999, p. 6).

However, initially, there are few additional resources to implement the Regulation (see Table 18.1). The Berlin agreement does not allow for any real growth in the budget for the Regulation between 2000 and 2006, which will make it difficult for Member States to implement it effectively. The hope, in the longer term, must be that monies saved from agricultural support could be made available for integrated rural development. There is no assurance, though, that this will happen. Member States, however, do have the discretion to modulate commodity payments in order to expand the resources available under the Regulation. Whether

Table 18.1 Indicative financial allocations [a] under the Rural Development Regulation for the period 2000-2006 from the EAGGF Guarantee Fund [b]

Member State	Allocation (annual average in millions of ECU at 1999 prices)	% total allocation
Austria	423	9.7
Belgium	50	1.2
Denmark	46	1.1
Finland	290	6.7
France	760	17.5
Germany	700	16.1
Greece	131	3.0
Ireland	315	7.3
Italy	595	13.7
Luxembourg	12	0.3
Netherlands	55	1.3
Portugal	200	4.6
Spain	459	10.6
Sweden	149	3.4
UK	154	3.5
Total	4,339	100

[a] These may be adjusted by the Commission within the first 3 years and within the limits of the overall resources available. [b] Total resources will be higher because of the need for Member State part-funding.

or not they choose to do so will be a significant test of their interest in the progressive reform of the CAP.

In the short and medium term, the most significant implications of the Regulation concern potential changes in procedure that could lay the basis for new institutional structures for rural development programming and support, around which over time the larger CAP could be transformed.

REFERENCES

CEC (1997) *Agenda 2000: For a Stronger and Wider Union.* Bull. E.U. Supplement 5/97. Commission of the European Communities, Brussels.

CEC (1998) *Proposals for Council Regulations (EC) Concerning the Reform of the Common Agricultural Policy.* COM(1998) 158 final. Commission of the European Communities, Brussels.

Directorate-General of Agriculture, European Commission (1999) *CAP Reform - A Policy for the Future.* Brussels: DGVI.

European Economy (1997) *Towards a Common Agricultural and Rural Policy for Europe.* Directorate-General for Economic and Financial Affairs, Reports and Studies, Vol. 5. Brussels.

Lowe, P., Rutherford, A. and Baldock, D. (1996) Implications of the Cork Declaration. *Ecos,* 17(3/4), 42-45.

Royal Society for the Protection of Birds (1999) *Agenda 2000 Common Agricultural Policy Reforms: Implications for the Environment.* RSPB, Sandy.

Tangermann, S. (1999) Agenda 2000: tactics, diversion and frustration. *Agra Europe* (28 May 1999) pp. A1-A4.

WWF (1999) *Agenda 2000 - A Way Forward.* WWF, Brussels.

Index